Detection of Pathogens in Water Using Micro and Nano-Technology

Detection of Pathogens in Water Using Micro and Nano-Technology

Giampaolo Zuccheri and Nikolaos Asproulis

Publishing
London • New York

Published by

IWA Publishing
Alliance House
12 Caxton Street
London SW1H 0QS, UK
Telephone: +44 (0)20 7654 5500
Fax: +44 (0)20 7654 5555
Email: publications@iwap.co.uk
Web: www.iwapublishing.com

First published 2012

British Library Cataloguing in Publication Data
A CIP catalogue record for this book is available from the British Library

ISBN: 9781780401089 (Paperback)
ISBN: 9781780401096 (eBook)

Contents

Preface ... xiii
Contributors ... xvii

Chapter 1
Overview of European regulation and standards on microbiological water analysis ... 1
Sabine Müller and Jonathan Loeffler

1.1 Introduction ... 1
1.2 European Regulation on Microbiological Analysis of Drinking Water ... 1
1.3 European Regulation on Microbiological Analysis of Recreational Water ... 2
1.3.1 Case of non-treated recreational water (seas, rivers, coastal water...) ... 2
1.3.2 Case of treated recreational water (swimming & spa pools etc.) ... 3
1.4 European and International Standards for Microbiological Water Analysis ... 3
Abbreviations ... 6
References ... 6

Chapter 2
Risk analysis of bio-terroristic attacks on drinking water systems ... 7
Christian Mittermayr

2.1 Introduction ... 7
2.2 Definitions ... 8
2.3 Risk Analysis for Terrorism ... 9
2.3.1 Critical infrastructure and key asset inventory ... 10
2.3.2 Criticality assessment ... 10
2.3.3 Threat assessment ... 10

2.3.4 Vulnerability assessment 12
2.3.5 Risk evaluation 12
2.4 Risk Analysis of Bio-Terroristic Attacks on Drinking Water Systems 12
2.4.1 Threat assessment 12
2.4.2 Criticality assessment 21
2.4.3 Vulnerability assessment 21
2.5 Risk Estimation 22
2.5.1 Hierarchical risk estimation 22
2.5.2 Quantitative microbial risk assessment 25
2.5.3 Data sources 26
References 27

Chapter 3
Sample collection procedures for Online Contaminant Monitoring System ... 33
Miloslava Prokšová, Marianna Cíchová and Lívia Tóthová

3.1 Introduction 33
3.2 Microbial Monitoring of Drinking Water 34
3.3 Sampling Plan 36
3.3.1 What should the sampling plan include? 37
3.4 New Security Approaches for Drinking Water 40
3.4.1 Water sources 41
3.4.2 Raw water transport 41
3.4.3 Treatment plants 42
3.4.4 Service reservoirs and distribution 42
3.5 New Approach of Online Contamination Monitoring Device (OCMD) 43
3.5.1 Location of OCMD in water system 44
References 45

Chapter 4
A device to extract highly diluted specimens out of large volumes of water for analysis in lab-on-a-chip detection systems 47
Christoph Zeis

4.1 The Need for a Macro-to-Micro-Fluidic Interface 47
4.1.1 The DINAMICS EU research project 47
4.1.2 Device needed for generating input for analytics 47
4.2 Principles of Separation 48
4.2.1 Survey of concentration methods 48
4.3 The Needle in the Haystack 52
4.4 Technical Description of the DINAMICS Concentration Apparatus 53
4.4.1 First draft of the dead-end filtration system 54
4.4.2 Tangential (cross-flow) filtration system 56
4.4.3 Steps towards a continuous working device 56
4.5 Conclusions 59
References 59

Chapter 5
Sustainable DNA/RNA release methods for in-line waterborne pathogen screening devices . . . 61
Hunor Sántha

Definitions and glossary . . . 61
5.1 Introduction . . . 61
5.2 Survey on Pathogen-Lysis/Cell-Disruption Methods . . . 63
5.2.1 Examples for devices utilizing electric field . . . 65
5.2.2 Examples for devices harnessing mechanical impact . . . 67
5.2.3 Chemical impact based methods (i.e. no cell-wall digestive enzymes added) . . . 72
5.3 Considerations for Water Samples and In-line (Quasi Continuous) Operation . . . 72
5.4 The Realised Case in The DINAMICS Project . . . 76
References . . . 79

Chapter 6
The microsystem based core of the DINAMICS water testing system: Design considerations and realization of the chip units . . . 81
Theo T. Veenstra

6.1 Microfluidics in DINAMICS . . . 81
6.2 Layout of the Fluidic Parts of DINAMICS . . . 81
6.3 Microfluidic Components in DINAMICS . . . 83
6.3.1 The valve component . . . 83
6.3.2 The mixer component . . . 84
6.3.4 The DNA isolation component . . . 86
6.3.5 The PCR reaction component . . . 88
6.3.6 The hybridization chamber component . . . 90
6.3.7 Fabrication – Technology platform . . . 90
6.4 Current Status . . . 92
6.5 Conclusion . . . 93
References . . . 93

Chapter 7
Electrochemical biosensor strategies for pathogen detection in water security . . . 95
Daniele Gazzola, Simone Bonetti and Giampaolo Zuccheri

7.1 Introduction . . . 95
7.2 Biosensor Strategies for Pathogen Detection in Water Security . . . 95
7.2.1 Conventional laboratory analysis . . . 96
7.2.2 Biosensor analysis . . . 97
7.2.3 Conclusions . . . 101
7.3 Common Electrochemical Detection Systems for DNA Biosensors . . . 101

7.3.1 Amperometry 102
7.3.2 Voltammetry 104
7.4 The On-Chip Simplified Electrochemical Technique Developed in the DINAMICS Project 106
7.4.1 On-chip voltammetry 106
7.4.2 Electrochemical reporter 107
7.4.3 Measurement principle 107
7.4.4 Interpretation of the measurements 108
7.4.5 Results 109
7.4.6 Conclusions 111
References 111

Chapter 8
***Biochemical and Nanotechnological Strategies for Signal Enhancement in the Detection of Nucleic Acids with Biosensors* 113**
Alessandra Vinelli, Manuele Onofri and Giampaolo Zuccheri

8.1 Introduction 113
8.1.1 Rationale of the classification 114
8.2 Enhancement Methods Based on Enzymatic (or Catalyzed) Reactions 115
8.2.1 Peroxidase to enhance the signal of nucleic acids detection 116
8.2.2 One to several instances of alkaline phosphatase for the electrochemical detection of nucleic acids 116
8.2.3 Terminal transferase to grow DNA at the recognition site 118
8.2.4 Signal enhancement of fluorescence through the use of a nickase 118
8.2.5 RNase H as a target recycling operator for RNA-based sensors 119
8.2.6 Nucleic acid sequence-based amplification (NASBA) 120
8.2.7 Strand displacement amplification 121
8.2.8 Loop mediated isothermal amplification (LAMP) 121
8.2.9 Metal nanoparticles as active enhancer labels 122
8.3 Enhancement Methods Based on Nanoparticles or Nanostructures 122
8.3.1 Methods that employ liposomes 122
8.3.2 Fluorescent nanoparticle labels 123
8.3.3 Colorimetric assays with metal nanoparticles 123
8.3.4 Signal amplification by conjugate breakdown: the bio-barcode assay 124
8.3.5 Quantum dots instead of organic dyes 124
8.4 Method Based on DNA Nanostructures 124
8.4.1 Cascade signal amplification by the combined use of rolling circle amplification 125
8.4.2 Branched DNA signal amplification 126
8.4.3 Aptamers and DNAzymes 127
8.4.4 Hybridization chain reaction and surface-initiated DNA polymerization 127
8.5 Conclusions and Perspectives 130
References 130

Chapter 9
Computational modelling of aqueous environments in micro and nanochannels 135
D. Mantzalis, K. Karantonis, N. Asproulis, L. Könözsy and D. Drikakis

9.1 Introduction 135
9.2 Effects of Physical Characteristics 137
9.2.1 Surface roughness 137
9.2.2 Surface stiffness 137
9.2.3 Wetting – surface energy – contact angle 138
9.2.4 Shear rate – pressure 138
9.3 Computational Approaches 139
9.3.1 Atomistic modelling 139
9.3.2 Continuum modelling 141
9.4 Liquid Flow in Confined Geometries 150
9.4.1 Flow behaviour in nanochannels 150
9.5 Molecular Modelling of Water 157
References 158

Chapter 10
Computational recipes of transport phenomena in micro and nanofluidics 163
N. Asproulis

10.1 Introduction 163
10.2 Modelling Approaches 164
10.2.1 Modelling multiple scales 164
10.2.2 Brownian motion 165
10.2.4 Continuum scale diffusion 167
10.3 Meta-Modelling for Macromolecules 169
10.4 Hybrid Continuum-Molecular Models 170
References 171

Chapter 11
Multi-detection of waterborne pathogens in raw and treated water samples by using ultrafiltration concentration and DNA array technology 173
Sophie Courtois, Anne Cajon, Aurore Romey, Fanny Poyet and Claude Mabilat

11.1 Introduction 173
11.2 Improved and Simplified Method for Concentrating Viral, Bacterial, and Protozoan Pathogens 174
11.2.1 Technical challenges for a universal concentration protocol 174
11.2.2 Protocol 177
11.3 Integrated Protocol for Nucleic Acid Extraction, Amplification and Sequence Identification Through High Density Microarray 178
11.4 Results 181

11.4.1 Recovery from 30 L-initial volume to final concentrate ... 181
11.4.2 Impact of BSA blocking and elution agents on waterborne pathogen recovery using two-step ultrafiltration protocol from 30 L-drinking water ... 182
11.4.3 Back volume calculation and comparison with other available detection methods ... 184
11.4.4 Multi-detection of waterborne pathogens by DNA Chip hybridization ... 184
11.5 Conclusions ... 186
References ... 187

Chapter 12
Detection and enumeration of waterborne mycobacteria ... 191
Joseph O. Falkinham III

12.1 Ecology of Waterborne Mycobacteria ... 191
12.1.1 Mycobacterial diseases ... 191
12.1.2 Mycobacterial habitats ... 192
12.1.3 Transmission of mycobacteria ... 192
12.2 Physiological Ecology of Waterborne Mycobacteria ... 192
12.2.1 The lipid-rich mycobacterial envelope ... 193
12.2.2 Consequences of the slow growth of mycobacteria ... 193
12.2.3 Viable but unculturable mycobacteria ... 194
12.3 Risk Analysis and Source-Tracking Environmental Mycobacteria ... 194
12.3.1 *Mycobacterium avium* and the candidate contaminant list ... 194
12.3.2 Risk analysis for mycobacteria ... 195
12.3.3 Source-tracking and DNA fingerprinting ... 195
12.4 Sampling and Sample Treatment Strategies for Mycobacterial Detection and Enumeration ... 195
12.4.1 Sampling strategies ... 195
12.4.2 Sample treatment ... 196
12.4.3 Sample concentration methods ... 196
12.5 Mycobacterial Detection or Enumeration ... 197
12.5.1 Detection or enumeration ... 197
12.5.2 Culture, PCR, or qPCR ... 197
12.5.3 Culture of mycobacteria ... 197
12.5.4 PCR-detection and qPCR enumeration of mycobacteria ... 198
References ... 198

Chapter 13
New molecular technologies for the rapid detection of Legionella in water ... 203
E. Soria, M. A. Yáñez, R. Múrtula and V. Catalán

13.1 Introduction ... 203
13.2 Immunodetection and *Legionella* Fast Detection ... 204

13.3 *Legionella* Detection using Microfluidics ... 205
13.3.1 Microarray platforms using antibodies ... 205
13.3.2 Microarray platforms using DNA ... 207
13.4 Future Research Directions ... 211
References ... 212

Chapter 14
Detection of virus in the water environment ... 213
Johan Nordgren and Lennart Svensson
14.1 Introduction ... 213
14.1.1 The waterborne viruses ... 213
14.1.2 Transmission of virus in the water environment ... 215
14.1.3 Virus in wastewater treatment plants ... 218
14.1.4 Monitoring of virus in sewage water – an epidemiological tool ... 218
14.2 Concentration of Virus from Water Samples ... 219
14.2.1 Concentration based on ionic charge (electrostatic adsorption/elution) ... 219
14.2.2 Concentration based on particle size separation (ultrafiltration) ... 222
14.2.3 Other concentration techniques ... 223
14.3 Detection and Quantification Methods ... 225
14.3.1 Cell culture assays ... 225
14.3.2 Molecular assays (PCR and real-time PCR) ... 227
14.3.3 ICC-PCR and detection of viral mRNA ... 228
14.4 Perspectives ... 228
References ... 229

Chapter 15
Design of PCR primers for the detection of waterborne bacteria ... 237
Julien Gardès and Richard Christen
15.1 Introduction ... 237
15.2 The Target Genes ... 237
15.2.1 rRNA genes ... 238
15.2.2 Housekeeping genes ... 238
15.2.3 Pathogenicity genes ... 238
15.2.4 Deep sequencing ... 238
15.3 Design of PCR Primers ... 239
15.3.1 The features of PCR primers ... 239
15.3.2 The softwares for designing PCR primers ... 240
15.4 DNA-Based Detection Technologies ... 243
15.4.1 Specific detection ... 243
15.4.2 Global detection: the sequencing ... 250
15.5 Conclusions ... 251
References ... 252

Chapter 16
Fluid structure and boundary slippage in nanoscale liquid films 255
Nikolai V. Priezjev

16.1 Abstract 255
16.2 Introduction 255
16.3 Molecular Dynamics Simulation Model 257
16.4 Results 260
16.4.1 Fluid density, velocity, and temperature profiles 260
16.4.2 Shear viscosity and slip length 262
16.4.3 Friction coefficient versus slip velocity 264
16.4.4 Friction coefficient and induced fluid structure 267
16.5 Conclusions 272
References 273

Chapter 17
Understanding slip at the nanoscale in fluid flows using atomistic simulations 277
T. E. Karakasidis and A. Liakopoulos

17.1 Introduction – Definition of Slip 277
17.1.1 Continuum theory and slip 277
17.1.2 Incorporating velocity slip in continuum models 278
17.2 Importance of Slip 278
17.3 Experimental Measurement of Slip 280
17.4 Atomistic Simulations 280
17.4.1 Methodological issues 281
17.4.2 Property calculations 284
17.4.3 Transport properties 284
17.4.4 Slip velocity/length calculation 285
17.5 Atomistic Simulations Results About Slip 286
17.5.1 Wall roughness effects 288
17.5.2 Effect of periodic wall patterns 289
17.5.3 Effect of nanostripes 292
17.6 Conclusions 294
References 294

Preface

The microbiological safety of drinking-water is one of the key environmental determinant of health. Assurance of drinking-water quality has been a pillar of primary prevention for more than 150 years and continues to be the foundation for the prevention and control of waterborne diseases. Roughly 1 every 8 people in the world still lack access to safe drinking water, according to the World Health Organization and UNICEF. Meanwhile, water use has increased by more than twice the rate of the world population growth during the past century.

The population of the industrialized world trusts the quality of the drinking water that distribution systems provide, but it is now becoming a fact that the microbiological safety on drinking water can no longer be taken for granted. Waterborne diseases are one of the major world-wide threats to public health, despite significant advances in water and wastewater treatment technology (World Health Organization, 2003). Waterborne diseases are estimated to be responsible for 4% of all deaths and 5.7% of the total disease burden worldwide. The occurrences of natural waterborne disease outbreaks as result of a failure in the conventional water treatment barriers were documented. Outbreaks have occurred in developed countries such as the United States (Cryptosporidium, Milwaukee, 1993) and Canada (*Escherichia coli* O157:H7, Walkerton, Ontario, 2000) and also in the United Kingdom and Europe. In Europe, in 2007, 17 waterborne outbreaks were reported by 8 countries (European Food Safety Authority, 2009), probably under-reporting the true number. These involved 10912 cases, with 232 hospitalisations. The main biological risks involved were Campylobacter, Norovirus, Giardia and Cryptosporidium.

Erratic and extreme precipitation events can overwhelm water treatment facilities and lead to Cryptosporidium outbreaks due to oocysts infiltrating drinking-water reservoirs from springs and lakes and persisting in the water distribution system for a long time despite vigorous and repetitive flushing of the system. A study from England and Wales found that 20% of waterborne outbreaks in the past century were associated with a sustained period of low rainfall, compared with 10% associated with heavy rainfall. Droughts or extended dry spells can reduce the volume of river flow possibly increasing the concentration of effluent pathogens posing a problem for the clearance capacity of treatment plants.

In Europe, flooding has rarely been associated with an increased risk of waterborne disease outbreaks, but a few exceptions exist in the UK, Finland, the Czech Republic, and Sweden. An outbreak of

Cryptosporidium hominis in November 2010 in northern Sweden (in Östersund) is held responsible for about 12700 cases: in samples from 174 cases, Cryptosporidium was confirmed; the water supply tested positive for Cryptosporidium, both in raw and potable water, very likely due to sewage water being released to the lake serving as reservoir for drinking water. The recommendation to boil drinking water was lifted only in February 2011 after an upgrade of the water treatment plant (ECDC, 2011).

Deliberate sabotage of large municipal water supplies is possible while difficult, especially due to the large amount of biological agents needed. Nevertheless, criminal acts perpetrated on smaller water supplies have been recorded and taking into account the complexity of water systems many possible access points for deliberate contamination acts can be identified. Breakdown in water supply safety may lead to large scale contamination and potentially to detectable disease outbreaks. Other breakdowns and low-level, potentially repeated contaminations may lead to significant sporadic diseases, but it is unlikely for these to be associated with the drinking water source by public health surveillance.

The consequences of the use of water of non-potable quality may be severe on food processing facilities and public health and they will depend not only on the direct use of the water but also on the subsequent processing of potentially contaminated materials. Water of a quality that may be tolerated occasionally in drinking water supply may be unacceptable for some uses in the food industry. Inefficient management of the water quality may result in a significant financial impact on food production, for example, through product recalls. Currently, there is no single method to collect, process, and analyze a water sample for all pathogenic microorganisms of interest. In fact, water is currently monitored through infrequent batch measurements procedures, which unfortunately, can miss transient, but problematic water safety events. Some of the difficulties in developing a universal method include the physical differences between the major pathogen groups (viruses, bacteria, protozoa), efficiently concentrating large volume water samples to detect low target concentrations of certain pathogen groups, removing co-concentrated inhibitors from the sample, and standardizing a culture-independent endpoint detection method. Integrating the disparate technologies into a single, universal, simple method and detection system would represent a significant advance in public health and microbiological water quality analysis. Recent advances in sample collection, on-line sample processing and purification, and DNA microarray technologies may form the basis of a universal method to detect known and emerging waterborne pathogens. Thus, specific detection methods are still required in order to trace the origin of etiological agents, identify lapses in water treatment, and identify new quality control processes and procedures.

By the time the results of microbiological assays on the water supply are available, thousands of people may have consumed the water and become sick. As also conveyed by the Technical Task Force of the WHO/UNICEF (Villié-Morgon, France 16–18 November 2010) new tools should be used for rapid assessments. The need of a fast response to some of the threats related to drinking water pushes towards the development and validation of molecular-based and biosensor-based methods which are expected to speed up significantly the analysis with respect to state-of-the-art cell culturing. Still, technological challenges must be overcome. Microtechnology and nanotechnology can come in help: the speed and performance of microelectronics, the versatility of microfluidics, the self-assembly of nanostructures for the making and functioning of biosensing surfaces.

In this context, the DINAMICS EU FP6 Collaborative Research Project put together the efforts of university laboratories, research institutes and private companies to deliver technology for point-of-need automated networkable microbiological analysis systems. The end-result was the development of technology for a prototype of a fully automatic system that could collect a large volume of water from the supply (also directly from the tap), concentrate the nano- and micro-particulate comprising the pathogens, lyse the cells, extract the nucleic acids and expose them to the sensing surface of an electronic (multi-pathogen) biosensor. The software managing the entire analytical process could then send a message exploiting the mobile telephone network, in the case of positive detection of pathogens.

This book reports on some of the results and technologies of DINAMICS, through a number of chapters written by the project participants. Sabine Müller and Jonathan Loeffler (Steinbeis Europa Zentrum) give an overview of the European regulations for drinking water. Christian Mittermayr (Lambda, GmbH) describes the intricacy of microbiological risk assessment, the mathematical modelling that can turn analytical results in the input for decision making. Miloslava Prokšová and collaborators (at the Slovak Water Research Institute) tell about the sound procedures for water sampling. Christoph Zeis (Provenion Engineering) describes an automatic system for concentrating pathogens out of large volumes of water. Hunor Santha and co-workers (University of Technology and Economy of Budapest) illustrate their cell-lysis device. Theo Veenstra (LioniX, BV) zooms in on microfluidics devices designed to do wonders such as DNA extraction, on-chip PCR, mixers and valves, hybridization chambers with electrodes. Daniele Gazzola and co-workers (University of Bologna) review the characteristics of electrochemical biosensors and report on the type of such biosensor developed within DINAMICS. Alessandra Vinelli and collaborators (University of Bologna) reviews some of the modern nucleic acids technologies that can be used to enhance the signal coming from the recognition of pathogenic nucleic acids. Dimitris Mantzalis and collaborators provide an overview of water modelling approaches in both continuum and molecular framework. Nikolaos Asproulis discusses the various numerical techniques employed for simulating transport phenomena within micro- and nano-fluidic devices.

To make the picture more complete, researchers who did not participate in DINAMICS were invited to contribute their views and their results. Many enriched this book. Sophie Courtois (Suez Environment) describes the process from water concentration to microarrays developed within the HealthyWater EU Project. Joseph Faulkinham (Virginia Polytechnic Institute and State University) tell about quantitating Micobacteria. Vicente Catalan and co-workers (LabAqua) tells about new methods for the detection of Legionella. Johan Nordgren and co-workers (Linköping University) tell about how to detect viruses in water. Richard Christen and co-workers (Université de Nice) tells about how to design PCR primers to detect waterborne bacteria. Nikolai Priezjev provides fruitful insights on the boundary slippage in nanoscale liquid films. Theodoros Karakasidis and Antonios Liakopoulos discuss about the slip phenomena noticed within micro and nano-fluidic devices along with the contributing factors.

Of course, the technology and knowledge described in this book alone are not enough to revolutionize the microbiological safety testing of drinking water. Still, the process has started and we would not be surprised if micro- and nanotechnology will soon lead the molecular detection of pathogens into a mature technology-driven field. In a similar way as chemical analysis takes full advantage of automation nowadays, our cities and homes might be protected in the future against waterborne infections. We trust we and the authors of this book contributed to this goal.

Giampaolo Zuccheri
(Bologna, Italy)
Nikolaos Asproulis
(Cranfield, UK)

Contributors

Chapter 1

Sabine Müller
Jonathan Loeffler
Steinbeis-Europa-Zentrum der Steinbeis Innovation gGmbH
Haus der Wirtschaft
Erbprinzenstrasse 4-12
D-76133 Karlsruhe
Germany

Chapter 2

Christian Mittermayr
Lambda GmbH
Gewerbepark 2
Rainbach
Austria, A-4261

Chapter 3

Miloslava Prokšová
Marianna Cíchová
Lívia Tóthová
Water Research Institute
Slovak National Water Reference Laboratory
Arm.gen.L.Svobodu 5
812 49 Bratislava – Slovak Republic

Chapter 4

Christoph Zeis
Provenion GmbH
Spannleitenberg 1
85614 Kirchseeon
Germany

Chapter 5

Hunor Sántha
Budapest University of Technology and Economics
Department of Electronics Technology
H-1521, Hungary
P.O. box: 91

Chapter 6

Theo T. Veenstra
LioniX BV
PO Box 456
7500 AL Enschede
The Netherlands

Chapter 7

Daniele Gazzola
Università degli Studi di Bologna
Centro Interdipartimentale di Ricerca
Industriale Scienze della Vita e Tecnologie per la Salute
Via Tolara di Sopra
50 – Ozzano Emilia
Bologna
Italy 40064

Simone Bonetti
Università degli Studi di Bologna
Dipartimento di Biochimica "G. Moruzzi"
Via Irnerio
48 – Bologna
Italy 40126

Giampaolo Zuccheri
Università degli Studi di Bologna
Dipartimento di Biochimica "G. Moruzzi"

(currently: Università degli Studi di Bologna,
Department of Pharmacy and Biotechnology)
Via Irnerio
48 – Bologna
Italy 40126

Chapter 8

Alessandra Vinelli
Università degli Studi di Bologna
Dipartimento di Biochimica "G. Moruzzi"
Via Irnerio
48 – Bologna
Italy 40126

Manuele Onofri
Università degli Studi di Bologna
Centro Interdipartimentale di Ricerca Industriale Scienze della
Vita e Tecnologie per la Salute
Via Tolara di Sopra
50 – Ozzano Emilia
Bologna
Italy 40064

Giampaolo Zuccheri
Università degli Studi di Bologna
Dipartimento di Biochimica "G. Moruzzi"
(currently: Università degli Studi di Bologna
Department of Pharmacy and Biotechnology)
Via Irnerio
48 – Bologna
Italy 40126

Chapter 9

Dimitrios Mantzalis
Konstantinos Karantonis
Nicolaos Asproulis
László Könözsy
Dimitris Drikakis
Dept. of Fluid Mechanics and Computational Sciences
Cranfield University
Bld. 83
Cranfield University
Cranfield
Beds MK43 0A

Chapter 10

Nicolaos Asproulis
Dept. of Fluid Mechanics and Computational Sciences
Cranfield University
Bld. 83, Cranfield University
Cranfield, Beds MK43 0A

Chapter 11

Sophie Courtois
Anne Cajon
Aurore Romey
Fanny Poyet
Claude Mabilat
SUEZ ENVIRONNEMENT
38, Rue du President Wilson
78230 Le Pecq – France

Chapter 12

Joseph O. Falkinham III
Department of Biological Sciences
Virginia Polytechnic Institute and State University
Derring Hall
Blacksburg, VA, 24061-0406, U.S.A.

Chapter 13

Elena Soria
M. Adela Yáñez
Raquel Múrtula
Vicente Catalán
Labaqua
C/Dracma, 16-18
Polígono Industrial Las Atalayas
03114 Alicante, Spain

Chapter 14

Johan Nordgren
Div. Molecular Virology
Linkoping University
581 85 Linköping, Sweden

Lennart Svensson
Div. Molecular Virology
Linkoping University
580 85 Linköping, Sweden

Chapter 15

Julien Gardès
Institute of Signaling
Developmental Biology and Cancer Centre de Biochimie
Faculté des Sciences
Université de Nice, France
6109 Nice cedex 2, France

Richard Christen
Institute of Signaling, Developmental Biology and Cancer Centre de Biochimie
Faculté des Sciences
Université de Nice, France
06108 Nice cedex 2, France

Chapter 16

Nikolai V. Priezjev
Dept. of Mechanical Engineering, Michigan State University
2465 Engineering Building
East Lansing, MI 48824-1226

Chapter 17

Theodoros E. Karakasidis
Athanassios Liakopoulos
Department of Civil Engineering
School of Engineering
University of Thessaly
Pedion Areos, 38334 Volos, Greece

Chapter 1

Overview of European regulation and standards on microbiological water analysis

Sabine Müller and Jonathan Loeffler

1.1 INTRODUCTION

Currently, there are three European Directives laying down rules for members states in the field of microbiological water analysis: the Drinking Water Directive (DWD) 98/83/EC on the quality of water intended for human consumption, the Bathing Water Directive 2006/7/EC on the management of bathing water quality and the Directive 2008/105/EC on environmental quality standards in the field of water policy. All are directly referring to specific European/International quality standards (EN ISO standards) concerning sampling, detection and validation methods and systems, as well as to specific World Health Organization (WHO) guidelines.

1.2 EUROPEAN REGULATION ON MICROBIOLOGICAL ANALYSIS OF DRINKING WATER

The Council Directive 98/83/EC, also called Drinking Water Directive is to date the main legislative document considering (microbiological) quality assessment of water intended for human consumption.

Published in 1998, the Directive sets the general obligation that drinking water as well as all water used in food production undertaking must be wholesome and clean in order to ensure protection of consumer's health. For this purpose, the document sets several quality standards (microbiological, chemical and organoleptic parameters), which are to a large extent based on WHO guidelines (World Health Organization, 2008). Although more than 25 chemical parameters are listed in the directive, there are only 3 microbiological parameters set for drinking water analysis. These are corresponding to values for *Escherechia coli* (*E. coli*), Enterococci and *Pseudomonas aeruginosa* (*P. aeruginosa*), all set to 0 bacterium/100 mL water sample (0/250 mL water sample in case of bottle-water). An additional general microbiological parameter concerns the counting of bacteriologic colonies at 22°C and 37°C, sets respectively to 100 bacteria/mL and 20 bacteria/mL. Beside that there are around 20 so-called indicator parameters which values need to be fixed only for monitoring purposes or if any failure to meet the precedent parametric values occurs. Such parameters are for example organoleptic or radioactivity parameters, conductivity, turbidity, as well as numbers of *Clostridium perfringens* (*C. perfringens*) (including spores) and coliform bacteria (values set to 0/100 mL or 250 mL).

Specifications for the analytical methods that have to be used in respect to these parameters are given in the directive and are, with exception of *C. perfringens*, directly referring to CEN/ISO standard methods. Hence, detection and enumeration of *E. coli* and Coliform bacteria, intestinal enterococci and P. aeruginosa have to be performed via specific membrane filtration techniques specified respectively in EN ISO 9308-1, EN ISO 7899-2 and EN ISO 16266. Bacteriological colony count has to be realized by inoculation in a nutrient agar culture medium as described in EN ISO 6222.

With respect to the DWD, Member States have to regularly monitor drinking water quality. The European Council recommends establishment of specific monitoring programmes to check that requirements of the directive are fulfilled. Two kinds of monitoring have to be foreseen: check monitoring and audits monitoring (reported to the Commission). Sampling specifications (sample points, frequency, and volume) are provided by the directive. These are depending on volume of distributed water and purpose of water distribution. The microbiological parameters listed before are all concerned by the monitoring requirements.

While translating the Drinking Water Directive into their own national legislation, the Member States can include additional requirements and parameters that are relevant within their territory (they have to set value for additional parameters and specifications for new analytical methods) as well as adopt more stringent or supplement standards if it's required. But Member States are not allowed to set lower standards as the level of protection of human health should be the same within the whole European Union. In specific cases like in case of contamination suspicion, additional parameters can also be provisionally included in the monitoring programmes. New analytical methods developed by member states should in any case ensure reliability, reproducibility and comparability of results.

Last important issue of the directive concerns the obligation for each member states to provide to consumers adequate and up-to-date information on their drinking water quality. Complementing the regular information to consumers, drinking water quality has to be reported to the European Commission every three years.

A revision of the Drinking Water Directive by the European Commission is under preparation since 2009. Based on an impact assessment study for the revision performed in 2008 by an external consultant and recommendations from stakeholders involved in the preparation process for the revision, the revised Directive should be published in 2012. The European Commission proposes to modify the list of parameters and specifications and also to introduce a risk-based approach in the directive. Other changes should concern the conditions of application of the directive and obligation of reporting to the EC on the basis of the size of the water supplies.

In order to include in the coming European regulations specifications in line with the most recent technological and scientific developments, the EC specifically launched in the last years (and especially in the FP7 Framework Programme) calls for project proposals directly related to development of new standard methods for microbiological water analysis. Hence more specific and sensible techniques will probably be introduced in the future as standards in this field.

1.3 EUROPEAN REGULATION ON MICROBIOLOGICAL ANALYSIS OF RECREATIONAL WATER

1.3.1 Case of non-treated recreational water (seas, rivers, coastal water …)

To ensure Public health safety in case of recreational water-related activities, the European parliament and council adopted the *Directive 2006/7/EC* regulating the management of bathing water quality. The directive concerns only surface water where a large number of people are expected to bathe but does not apply for treated water surface like swimming pools, spa pools and confined waters subject to treatment or used for therapeutic purposes.

Like for the drinking water directive, the bathing water directive sets microbiological parameters to be measured in order to qualify waters quality and classify the waters in respect to the results; two micro-organisms are concerned by this directive: Intestinal Enterococci and *E. coli*, and standard methods for their analysis are proposed: Intestinal Enterococci should be analysed in reference to ISO 7899-1 or ISO 7899-2 methods and *E. coli* should be analysed in reference to ISO 9308-3 or ISO 9308-1 methods. In both cases, the standards refer on the one hand to a miniaturized method for surface and waste water analysis and on the other hand to a membrane filtration method. Other methods are allowed if demonstration is done by competent authorities that produce results equivalent to standard methods. Moreover, new (better) methods are recommended to be developed in the light of scientific and technical progress. Besides that, strict sampling requirements (number of sampling per bathing season, sampling locations) are specified in the document. As previously mentioned, public must be informed in case of abnormal water pollution, risky for human health.

The Directive 2006/7/EC is complementing the Directive 2000/60/EC further amended by the *Directive 2008/105/EC*, establishing a framework for Community action in the field of water policy and also indicating recommendations for some biological elements which should be analysed.

Other recommendations concerning monitoring and quality assessment of such kind of bathing water are available in guidelines published by the WHO (World Health Organization, 2003).

1.3.2 Case of treated recreational water (swimming & spa pools etc.)

The microbiological quality of water in swimming baths, spa pools and hydrotherapy pools is not governed directly by European legislation, but each pool manager is required to ensure the health and safety of employees and pool users. For this purpose, the *WHO* has published international *guidelines* for the safety of swimming pools and similar recreational-water environments, including standards for minimizing microbial and chemical hazards (World Health Organization, 2006). Chapter 3 of the guidelines is dealing with the main microbial hazards responsible for waterborne diseases in such environment (six types of viruses, seven different bacteria including *E. coli*, *P. aeruginosa* and *Staphylococcus aureus* (*S. aureus*), but also protozoa and fungi species). Issues dealing with sampling requirements and parametric values for each of these microorganisms for monitoring water quality are covered in Chapter 5 of the book.

1.4 EUROPEAN AND INTERNATIONAL STANDARDS FOR MICROBIOLOGICAL WATER ANALYSIS

In addition to the standards directly specified in the European Directives and previously mentioned, numerous standards are published either specifying analytical methods (generally one standard for each type of bacteria or virus or other microorganisms) or providing guidance for sampling methods and sample handling. The standards are either generalist (not restrictive to one source of water) or considering specific kinds of water systems (drinking water, bathing water, coastal water, marine water, swimming pool, etc.) and sample nature (like sludge). Most of them are published by the Technical Committee TC147 "water quality"/Subcommittee SC4 "microbiological methods" as well as from the Technical Committee TC34 "food products"/Subcommittee SC9 "microbiology". Relevant examples are given in the Table 1.1.

Table 1.1 Overview of relevant standards for microbiological water analysis.

Standard Reference	Title
Standards for microbiological water analysis	
ISO 15839:2003	Water quality – On-line sensors/analysing equipment for water – Specifications and performance tests
EN ISO 17994:2004	Water quality – Criteria for establishing equivalence between microbiological methods
ENV ISO 13843:2001	Water quality – Guidance on validation of microbiological methods
ENV ISO 13530:1998	Water quality – Guide to analytical quality control for water analysis
EN ISO 8199:2007	Water quality – General guidance on the enumeration of micro-organisms by culture
EN ISO 10705-1:2001	Water quality – Detection and enumeration of bacteriophages – Part 1: Enumeration of F-specific RNA bacteriophages
EN ISO 10705-2:2001	Water quality – Detection and enumeration of bacteriophages – Part 2: Enumeration of somatic coliphages
EN ISO 11731-2:2008	Water quality – Detection and enumeration of Legionella – Part 2: Direct membrane filtration method for waters with low bacterial counts
EN ISO 16266:2008	Water quality – Detection and enumeration of *Pseudomonas aeruginosa* – Method by membrane filtration
ISO 19250:2010	Water quality – Detection of *Salmonella* spp.
EN ISO 6222:1999	Water quality – Enumeration of culturable micro-organisms – Colony count by inoculation in a nutrient agar culture medium
EN ISO 7899-1:1998/AC 2000	Water quality – Detection and enumeration of intestinal enterococci in surface and wastewater – Part 1: Miniaturized method (Most Probable Number) by inoculation in liquid medium
EN ISO 7899-2:2000	Water quality – Detection and enumeration of intestinal enterococci – Part 2: Membrane filtration method
EN ISO 9308-3:1998/AC 2000	Water quality – Detection and enumeration of Escherichia coli and coliform bacteria in surface and wastewater – Part 3: Miniaturized method (Most Probable Number) by inoculation in liquid medium
prEN ISO 9308-1 → 2012-12	Water quality – Detection and enumeration of Escherichia coli and coliform bacteria – Part 1: Membrane filtration method
ISO 6461-2:1986	Water quality – Detection and enumeration of the spores of sulfite-reducing anaerobes (clostridia) – Part 2: Method by membrane filtration
EN ISO 19458:2006	Water quality – Sampling for microbiological analysis
EN ISO 5667-1:2006/AC:2007	Water quality – Sampling – Part 1: Guidance on the design of sampling programmes and sampling techniques
EN ISO 5667-16:1998	Water quality – Sampling – Part 16: Guidance on biotesting of samples
EN ISO 5667-19:2004	Water quality – Sampling – Part 19: Guidance on sampling in marine sediments

(*Continued*)

Table 1.1 Overview of relevant standards for microbiological water analysis (*Continued*).

Standard Reference	Title
EN ISO 5667-23:2011	Water quality – Sampling – Part 23: Guidance on passive sampling in surface waters
EN ISO 5667-3:2003/AC:2007	Water quality – Sampling – Part 3: Guidance on the preservation and handling of water samples
ISO 7704:1985	Water quality – Evaluation of membrane filters used for microbiological analyses
Specific standards for characterization & sampling of sludges	
CEN/TR 15175:2006	Characterization of sludges – Protocol for organizing and conducting inter-laboratory tests of methods for chemical and microbiological analysis of sludges
CEN/TR 15214-1:2006	Characterization of sludges – Detection and enumeration of Escherichia coli in sludges, soils, soil improvers, growing media and biowastes – Part 1: Membrane filtration method for quantification
CEN/TR 15214-2:2006	Characterization of sludges – Detection and enumeration of Escherichia coli in sludges, soils, soil improvers, growing media and biowastes – Part 2: Miniaturised method (Most Probable Number) by inoculation in liquid medium
CEN/TR 15214-3:2006	Characterization of sludges – Detection and enumeration of Escherichia coli in sludges, soils, soil improvers, growing media and biowastes – Part 3: Macromethod (Most Probable Number) in liquid medium
CEN/TR 15215-2:2006	Characterization of sludges – Detection and enumeration of Salmonella spp. in sludges, soils, soil improvers, growing media and biowastes – Part 2: Liquid enrichment method in selenite-cystine medium followed by Rapport-Vassiliadis for semi-quantitative M
EN ISO 5667-13:1997	Water quality – Sampling – Part 13: Guidance on sampling of sludges from sewage and water treatment works
EN ISO 5667-15:2009	Water quality – Sampling – Part 15: Guidance on the preservation and handling of sludge and sediment samples
Swimming pool requirements	
EN 15288-2:2008	Swimming pools – Part 2 Safety requirements for operation

Besides that, three standards (ISO 13843, ISO 17994, ISO 7704) dedicated to the guidance and definition of procedures to validate and/or compare several microbiological methods are published, as well as a specific standard (ISO 15839) describing the performance testing of on-line sensors/analysing equipment for water. This standard is applicable to most sensors/analysing equipment, but it is recognized that, for some sensors/analysing equipment, certain performance tests cannot be carried out. This International Standard defines an on-line sensor/analysing equipment for water quality measurements, defines terminology describing the performance characteristics of on-line sensors/analysing equipment and specifies the laboratory and field test procedures to be used to evaluate the performance characteristics of on-line sensors/analysing equipment.

Finally, laboratories qualified for analysis of drinking water should fulfil the requirements for an accreditation procedure preferably according to EN ISO 17025.

ABBREVIATIONS

CEN : European Committee for Standardization
CEN TC : CEN Technical Committee
DWD : Drinking Water Directive
EC : European Commission
EN/prEN: European Standard/draft European Standards
ISO : International Organization for Standardization
WHO : World Health Organization

REFERENCES

Europa.eu, http://europa.eu/legislation_summaries/index_en.htm (last accessed January 2012)

European Committee for Standardization, http://www.cen.eu/cen/Members/Pages/default.aspx (last accessed January 2012)

World Health Organization, http://www.who.int/en (last accessed January 2012)

World Health Organization (2008). Guidelines for Drinking Water Quality [electronic resource]: incorporating 1st and 2nd addenda, Vol. 1, Recommendations – 3rd ed. Geneva, ISBN 978 92 4 154761 1 (WEB version) (NLM classification: WA 675).

World Health Organization (2003). Guidelines for Safe Recreational Water Environments, Volume 1, Coastal and Fresh Waters. Geneva, ISBN 92 4 154580 1 (NLM classification: WA 820).

World Health Organization (2006). Guidelines for Safe Recreational Water Environments, Volume 2, Swimming Pools and Similar Environments. Geneva, ISBN 92 4 154680 8 (NLM classifi cation: WA 820).

Chapter 2

Risk analysis of bio-terroristic attacks on drinking water systems

Christian Mittermayr

2.1 INTRODUCTION

Risk assessment emerged as a tool in the United Stated federal government, especially in the Environmental Protection Agency, in the 1970s and 1980s. That use led to a landmark publication on risk assessment [primarily human health risk assessment) by the National Academy of Sciences (National Academy of Sciences, 1983)].

Starting from models for chemical contamination of water models, the quantitative microbial risk assessment (QMRA) evolved. The use of dose-response modelling for quantitative microbial risk assessment, and the alignment of QMRA stages with the National Research Council risk assessment paradigm took stage in the 1980s. The first report on dose-response modelling for QMRA was by Haas (1983) who evaluated the risk from waterborne bacteria and viruses. Continuously, new data become available extending the QMRA to more microorganisms and making the models more accurate and reliable. Nowadays, QMRA has become a routine tool for assessing microbial contamination health risks in food and water.

In the case of a wilful rather than accidental release of hazardous material, like in the case of a terrorist attack, the traditional approaches to risk assessment will not be sufficient anymore (Fedra, 2008). The main difference from conventional risk assessment methods is the initial event probability, which is no longer probabilistic, that is, related to concepts such as:

- The probability of human error;
- The mean time between failure of technical components;
- The occurrence of pathogens in the source water.

Terrorist attacks change the risk assessment equations (Fedra, 2008) by:

- Selection of sources for maximum "planned" impact, often symbolic choice of targets;
- Affecting event probabilities by purposeful triggering of an otherwise probabilistic event;
- The potential interference with emergency measures (e.g. synchronized attacks) designed to limit the effectiveness of emergency services as much as to maximize publicity.

Information on *a priori* probabilities for an attack is clearly in the realm of military intelligence, police and secret service rather than engineering.

This report tries to include risk assessment on terrorism into QMRA by pointing on how these models are interdependent and it also tries to point out how the QMRA models have to be made more complex by including the actual structure of the water supply system and a vulnerability analysis. This report first tries to give definitions used in risk assessment methodology (Section 2), and gives a short review on risk assessment method on terrorism (Section 3) including the steps Criticality, Threat and Vulnerability Assessment. The section Risk Analysis of Bio-terroristic Attacks on Drinking Water Systems (Section 4) will focus on the threat assessment, particularly the assessment of the biological agents as threat for the drinking water system. The final section, Risk Estimation (Section 5) describes a hierarchical process of risk estimation, that describes interdependencies between risks at different levels, QMRA and sources for the necessary data.

2.2 DEFINITIONS

Risk assessment has been developed in many different areas and therefore terminology varies amongst the various fields. The definitions given below will be used for this report even when in the cited literature different terms are used.

Risk. The ISO 31000 (2009)/ISO Guide 73:2002 definition of risk is the "effect of uncertainty on objectives". This rather abstract definition translated into a more tangible language defines risk as the chance that a hazardous event or a chosen action or activity (including the choice of inaction) will lead to a undesirable outcome (loss). In the case of public health, the undesirable outcome are harmful effects to human health or even death. Other definitions are: risk is the potential for an unwanted outcome resulting from an incident, event, or occurrence, as determined by its likelihood and the associated consequences. Risk may manifest at the strategic, operational, and tactical levels (Department of Homeland Security, 2008).

Scenario (Risk). A scenario is a hypothetical situation comprised of a hazard, an entity impacted by that hazard, and associated conditions including consequences when appropriate. A scenario that has occurred or is occurring is an incident.

Hazard. A hazard is a natural or man-made source or cause of harm or difficulty. Alternatively, it can be defined as a source of potential danger or adverse condition. An *accidental hazard* is created by negligence, error, or unintended failure. An *intentional hazard* is created by a deliberate action or a planned course of action. A *natural hazard* is created by a meteorological, environmental, or geological phenomenon or combination of phenomena (Department of Homeland Security, 2008).

Threat. A threat is a natural or man-made occurrence, individual, entity, or action that has or indicates the potential to harm life, information, operations, the environment and/or property. For the purpose of calculating risk, the threat of an intentional hazard is generally estimated as the likelihood of an attack being attempted by an adversary; for other hazards, threat is generally estimated as the likelihood that a hazard will manifest. *Threat assessment* is the process of identifying or evaluating entities, actions, or occurrences, whether natural or man-made, that have or indicate the potential to harm life, information, operations and/or property (Department of Homeland Security, 2008).

Hazard versus Threat. A hazard differs from a threat in that a threat is directed at an entity, asset, system, network, or geographic area, while a hazard is not directed. A hazard can be actual or potential.

Risk Analysis versus Risk Assessment. There is an unfortunate inconsistency in usage between two communities importantly involved in understanding the risk of terrorist events: intelligence analysts and risk analysts. In the *intelligence community*, it is customary first to gather information about an opponent's intentions and capabilities and then to use this information to present a statement of the current situation. The first step is usually called "analysis," and the second step is called an "assessment" of the situation. The *risk and decision community* reverses these definitions: the first step of gathering information is usually called "assessment," while the second step – the process of using this information and combining it in such a way that a decision maker can make better decisions – is usually called "analysis".

Risk Assessment. Risk Assessment is defined by the ISO/IEC Guide 73 as the overall process of *risk analysis and risk evaluation*. Risk assessment characterizes the nature and magnitude of risks from threats or hazards. Risk assessment is the determination of quantitative or qualitative value of risk related to a concrete situation and a recognized threat (or hazard). In the context of public health, risk assessment is the process of quantifying the probability of a harmful effect to individuals or populations from certain human activities or hazardous events.

Risk Analysis. Risk analysis is the systematic use of available information to identify hazards and to estimate the risk to individuals or populations, property or the environment. According to ISO 31000, risk analysis consists of 5 steps:

- *Risk Identification* sets out to identify an organisation's exposure to uncertainty
- *Risk Description*: The objective of risk description is to display the identified risks in a structured format.
- *Risk Estimation* can be quantitative, semi-quantitative or qualitative in terms of the probability of occurrence and the possible consequence. Risk estimation is the process used to produce a measure of the level of risk being analysed.
- *Risk Analysis methods and techniques*: A range of techniques that can be used to analyse risks.
- *Risk Profile*: One of the outputs of the risk analysis process, it gives a significance rating to each risk and it provides a tool for prioritising risk treatment efforts. It ranks each identified risk so as to give a view of the relative importance.

Risk Pathway. The risk pathway is the potential pathway from the hazard(s) of interest to the outcome(s) of interest. The elucidation and description of such pathways is essential for a risk assessment.

Risk Evaluation. When the risk analysis process has been completed, it is necessary to compare the estimated risks against risk criteria which the organisation has established (taking into account factors such as socioeconomic and environmental aspects). Risk evaluation is used to make decisions whether each specific risk should be accepted or treated (Fowle & Dearfield, 2000; Hokstad *et al.* 2009; Rosén *et al.* 2007).

Risk Management. Risk management is the systematic application of management policies, procedures and practices to the tasks of analysing, evaluating and controlling risk. Risk assessment provides "INFORMATION" on potential risks, and risk management is the "ACTION" taken based on consideration of that and other information.

2.3 RISK ANALYSIS FOR TERRORISM

While there exist numerous approaches for risk assessment of food and drinking water with respect to human health, as by Benford (2001), Dawson (2003), Dufour *et al.* (2003), Koopmans and Duizer (2003), Thoey *et al.* (2003), Larson *et al.* (2006), Dechesne and Soyeux (2007), Schroeder *et al.* (2007), ILSI (2008), Parkin (2008), Riha (2009), USDA & EPA (2011), not so many publications focus on intentional contamination of drinking water by terrorists. The last decade brought a considerable body of literature on risk analysis of terrorist attacks that form the foundation for any specific scenario. Materials can be found in IRM (2002), Tuduk (2004), FEMA (2005), Willis *et al.* (2005), Masse *et al.* (2007).

The risk analysis approach used in this report uses the following sequential steps described in Leson (2005):

(1) Critical infrastructure and key asset inventory (consider what can be threatened and what must be protected).
(2) Criticality analysis (a set value, it determines the ultimate importance of the asset).
(3) Threat analysis.
(4) Vulnerability analysis (identification of weaknesses).
(5) Risk Evaluation (as defined above), Leson (2005) uses the term "risk calculation".

Leson (2005) defines the steps 2–4 as *assessment*, while it is termed here *analysis*, according the definitions above. The steps 1–4 combine the steps *Risk Identification* and *Risk Description* in a different way according to the parameters Criticality, Threat and Vulnerability.

2.3.1 Critical infrastructure and key asset inventory

This step considers what can be threatened and what must be protected. This is usually done on a national, regional or organisational level. Generally, the assets have to be defined, but in specialized literature the assets are presumed to be known, so does for example, the EPA (2009) microbial risk analysis approach does not explicitly ask for an asset evaluation since it is silently understood that it is about the drinking water system at hand. Since this report focuses on the drinking water system, no further considerations are given.

2.3.2 Criticality assessment

Criticality assessment considers the *consequences* of the loss of or serious damage to assets. The measure of criticality, or *asset value*, determines the ultimate importance of the asset. The loss can economic or the loss of lives, but an assets value is also determined by its visibility and symbolic value.

Shock also has to be considered in the criticality assessment. Shock combines the health, psychological, and collateral national economic impacts of a successful attack on the target system, as explained in Catlin and Kautter (2007). The psychological impact of an attack will be increased if there are a large number of deaths or the target has historical, cultural, religious or other symbolic significance. Psychological impact will be increased even further if victims are members of sensitive subpopulations such as children or the elderly.

2.3.3 Threat assessment

Most civil applications are dealing with hazards and not with threats according to the definition above, as confirmed by EPA's microbial risk assessment EPA (2009) or the WHO's water safety plans (Bartram *et al.* 2009), where this step is called *hazard identification*. When considering security issues, the term threat is more appropriate since it includes the intent and directedness of a man-made hazard.

When modelling the risk, the distinction becomes even easier to understand. Accidental hazards are usually random, even when they include human error and man-made products. The frequency of occurrence is commonly described statistically by distribution functions. Threat on the other hand is estimated as the likelihood of an attack being attempted by an adversary. Intentional hazards have to be treated differently and psychological, social and political factors have to be included in the assessment. Mostly qualitative methods will have to be used.

Threat assessment of terrorist attacks has to consider the adversaries, their tactics and their choice of weapon.

Adversary

First all adversaries should be listed and then characterized by several parameters:

- *Type of adversary*: Terrorist, activist, employee, other.
- *Category of adversary*: Foreign or domestic, terrorist or criminal, insider and/or outsider of the organization.
- *History of Threats*. What has the potential threat element done in the past, how many times, and was the threat local, regional, national, or international in nature? When was the most recent incident and where, and against what target?
- *Objective of adversary*: Theft, sabotage, mass destruction (maximum casualties), socio-political statement, other.
- *Number of adversaries*: Individuals, groups or "cells" of operatives/terrorists, gangs, other.
- *Range of adversary tactics*: Stealth, force, deceit, combination, other.
- *Capabilities or Expertise of adversary*: Knowledge, motivation, skills, weapons and tools. The general level of skill and training that combines the ability to create the weapon and the technical knowledge of the systems to be attacked. Knowledge and expertise can be gained by surveillance, open source research, specialized training, or years of practice in the industry.

Terroristic tactics

Terrorists act rationally to reach their destructive goals. They follow principles of human behaviour, and can be analysed by methods from social psychology, game theory, and network analysis (Leson, 2005; Rios, 2010). Woo (2008) describes the terrorists' selection of weapons and attack modes to be dominated by accessibility to the weapon. Terrorist usually choose weapon modes and targets, against which the technical, logistical and security barriers to mission success are least.

Woo (2008) models the terrorists' target selection process with these rules:

- Terrorists may substitute one target with another, according to the relative security of the targets.
- Local security enhancement transfers threat elsewhere.
- Terrorist attacks are geographically focused, with attack likelihood decreasing logarithmically for descending target tiers.

Fedra (2008) classifies the mode of terroristic attack as such:

- Direct attacks: paramilitary, explosives, suicide attacks, water supply, food chain, biological agents
- Man-made accidents: transportation system (air, rail)
- Indirect attacks: dams, chemical installations, nuclear establishments
- Denial of service (DOS): water, energy, communication

Weapons

The *Strategic Homeland Infrastructure Risk Assessment (SHIRA)* (DHS & FBI, 2008) analysis is based on a defined set of 15 *Identified Terrorist Attack Methods* that combine both the *weapon category* (biological attack, conventional, …) and the chosen target (Population, Building, Livestock, …) and tactics (direct, indirect, …). Four of the 15 attack methods include biological agents:

- Biological Attack: Contagious Human Disease
- Biological Attack: Noncontagious Human Disease
- Biological Attack: Livestock and Crop Disease
- Food or Water Contamination

A more detailed analysis of threat identification for the case of water contamination is done below. It deals mainly with the choice of weapon, that is pathogen by the attacker.

2.3.4 Vulnerability assessment

A vulnerability assessment identifies weaknesses that may be exploited by terrorists. It evaluates the potential vulnerability of the assets against the identified threats. Vulnerability is measured by the ease of accomplishing an attack. It also considers what possibilities for interventions are already in place that might thwart an attack and may suggest options to eliminate or mitigate those weaknesses.

Among others location, accessibility and recognisability are important factors to consider when determining vulnerability:

Location: Geographic location of potential targets, entry and exit routes; location of target relative to public areas, transportation routes, or easily breached areas.

Accessibility: Accessibility is the openness of the target to the threat. It describes how accessible a target is to the adversary; how easy it is for someone to enter, operate, collect information, and evade response forces. It also determines:

- o *Detectability of the attack*
- o *The volume of a contaminant that can be injected without undue concern of detection*
- o *The amount of readily available information on the target*

Recognisability: The ease by which an attacker can identify the target without confusion with other targets

2.3.5 Risk evaluation

Almost every available risk evaluation technique addresses the following three questions to aggregate the information obtained in each of the assessment steps:

- What is the likely impact if an identified asset is lost or harmed? (Criticality)
- How likely is it that an adversary will attack? (Threat)
- What are the most likely vulnerabilities that the adversary will use to target the identified assets? (Vulnerability).

In "Failure Modes and Effects analysis" (FMEA), a safety engineering method, the three components of risk estimation are *Severity, Occurrence* and *Detection Rate*, which sometimes are more readily understood. So, risk evaluation combines *criticality, threat, and vulnerability* assessment to generate a risk profile for an asset using often the simple *risk equation*: Risk = Criticality × Threat × Vulnerability.

2.4 RISK ANALYSIS OF BIO-TERRORISTIC ATTACKS ON DRINKING WATER SYSTEMS

2.4.1 Threat assessment

Deliberate food and water contamination remains the easiest way to distribute biological agents (microorganisms or biological toxins) for the purpose of terrorism. Because biological agents are often easily accessible, can be delivered concealed, are easy to transportation, have high potency and are difficult to identify there is a risk that they are used to gain political advantages by terrorists.

Usually a small number of large drinking water utilities located primarily in urban areas provide water services to the majority of a countries population. These systems represent the greatest targets of opportunity for terrorist attacks, while the large number of small systems are less likely to be perceived as key targets by terrorists who might seek to disrupt water infrastructure systems (Stephenson, 2004; Copeland, 2010).

The *type and category of adversary* as well as the type of impact sought will have an important role on the agent-selection principles. For examples, for non-state entities, accessibility, not overall aggressiveness and stability in storage, might be the dominant criterion in their choice of agent. Therefore, the rank order in which public health authorities assess the different agent threats may not be the same as that of military authorities (WHO, 2004; DHS & FBI, 2008).

The WHO (2004) divides the criteria for the pathogen selection into two broad categories

- Acquisition, production and dissemination of the contaminants
- Biological properties of the contaminant and its interaction with the host

but they have to be considered together, since many of these criteria show interactions.

Technical constraints on use

The ability to conduct attacks using contagious agents requires the possession of a viable pathogenic bacteria or virus in sufficient quantity to cause a communicable disease. Characteristics of acquisition, production and dissemination are determined by:

- Access to Agent
- Ease of cultivation
- Storage security
- Stability during handling and storage
- Dissemination method
- Establishing field dosages
- Routes of exposure

Access to agent/sources of agent

Gaining access to biological agents never appears to have been a significant limiting factor. In fact, acquiring biological agents has usually proven to be relatively easy. In a few cases, pathogens were acquired from culture collections, usually legitimately but sometimes not, while the perpetrators usually produced toxins.

The effectiveness of biological agents depends heavily on the specific strain of the organism, and it may be difficult to acquire the more dangerous strains relying on natural sources. Despite efforts to restrict the illicit acquisition of biological agents, it is likely that terrorists and criminals will be able to obtain the agent that they want when they want it. If unable to acquire from a legitimate culture collection or a medical supply company, they can steal it from a laboratory. If unable to steal it, a group with the right expertise could culture the agent from samples obtained in nature. Many biological agents are endemic, and a skilled microbiologist would have little difficulty in culturing an agent from material taken from the environment (Carus, 2001).

Ease of cultivation

Some pathogens are easy to culture, while others are extremely difficult. Some agents would require further processing to use in an attack. Particularly, viruses are very hard to culture in big quantities, while bacteria can easily be produced in large quantities in simple containers. Skill is one of the most important parameter to grow larger quantities of an agent.

Storage security

The use of highly contagious diseases may be the easiest way to cause mass casualties, from a technical perspective. However, such agents pose risks for the group using them and could have an impact on people that the perpetrators do not want to affect (Carus, 2001).

Stability during handling and storage

Some biological agents die or lose virulence once released. The implications of these variables will differ from one organism to the next and the environmental conditions. Perpetrators have to have sufficient knowledge about this factors in order to launch an successful attack (Carus, 2001).

Dissemination methods

The WHO (2004) lists several dissemination methods for biological agents:

- Airborne (inhalation)
- Drinking water (ingestion)
- Food (ingestion)
- Arthropod vectors
- Direct

Based on his analysis of over 200 past incidents of chemical and biological terrorism, McGeorge (1986) concludes that "Dissemination devices or means have typically been very simple in design or procedure and of corresponding low efficiency." The three most common dissemination routes according to Purver (1995) were:

(1) Contaminated food or drink (43%)
(2) Contaminated consumable products (13%)
(3) Contaminated water supplies (12%)

If a simple dissemination technique is employed, relatively little technical expertise is needed. For example, limited skill is required if the perpetrators can inject agent directly into their victims or if they seek to contaminate food. Many pathogens that have had a significant impact on human life, such as *Vibrio cholerae* and *Salmonella typhi*, are water-borne. Thus, using a municipal water systems to disseminate pathogens might seem promising to terrorists. On the other hand, it is very difficult to infect a large population through deliberate contamination of water supplies (Carus, 2001).

Establishing field dosages

The ability to generate a minimum concentration of agents over a predictable area must be evaluated.

Routes of exposure

The pathogens can enter the body in several ways:

- Through the respiratory system
- Through the skin
- Through the oro-nasal mucosal tissue and the conjunctiva
- Through the digestive system

The biological agents can enter the *digestive system* via contaminated food or drinking-water, by hand–mouth contact after touching contaminated surfaces, or by swallowing of respiratory mucus after the accumulation of larger aerosol particles in the nose, throat and upper airways. Of all exposure routes, entry through the digestive system is the easiest to control, provided that the contaminated sources are known (or at least suspected). Simple hygienic measures and control of supplies of food and drinking-water can significantly reduce the risk of exposure (WHO, 2004).

Characteristics of biological agents

The chief characteristic of biological agents is their ability to multiply in a host. It is this that gives them their aggressive potential. Contraction of the disease results from the multifactorial interaction between the biological agent, the host (including the latter's immunological, nutritional and general health status) and the environment (e.g. sanitation, temperature, water quality, population density). The WHO (2004) lists the following relevant biological characteristics:

- Infective Dose/Dose Response Curve
- Lethality
- Virulence
- Incubation period (rapidity of effect)
- Contagiousness (infectivity) and mechanisms of transmission
- Stability of Storage (resist degradation during handling and storage)
- Resistance to Environment and Water Treatment (chlorine, chloramines, ...)
- Resistance to medical Treatment of Patients

Infective dose/dose–response curve

Dose–response models estimate of the probability of an infection given a specific dose of a defined pathogen. The concept of the single-hit principle, where even a single pathogen may be able to cause infection and disease, supersedes the concept of a (minimum) infectious dose that is frequently used in older literature.

Available dose–response data have been obtained mainly from studies using healthy adult volunteers. However, adequate data are lacking for vulnerable subpopulations, such as children, the elderly and the immune-compromised, who may suffer more severe disease outcomes. It is crucial to note that any model will depend very heavily on the strain, particularly the presence of the virulence factors, and on the health of the individual host (Carus, 2001).

The infectious doses are independent of victim bodyweight because the pathogen reproduces in the host, so the potency of the pathogens on a weight basis exceeds that of the most toxic chemicals; between a few and a few thousand viable organisms is all that is required to produce infection in many cases (Purver, 1995).

Lethality

Lethality reflects the ability of an agent to cause death in an infected population. The case-fatality rate is the proportion of patients clinically recognized as having a specified disease who die as a result of that illness within a specified time, for example, during outbreaks of acute disease (WHO, 2004).

Virulence

Virulence is the relative severity of the disease caused by a microorganism. Different strains of the same species may cause diseases of different severity. Some strains of *Francisella tularensis*, for example, are much more virulent than others (WHO, 2004).

Incubation period

The incubation period is the time elapsing between exposure to an infective agent and the first appearance of the signs of disease associated with the infection. This is affected by many variables, including the agent, the route of entry, the dose and specific characteristics of the host (WHO, 2004).

Contagiousness

For those infections that are contagious, a measure of their contagiousness is the number of secondary cases arising under specified conditions from exposure to a primary case. The mechanisms of transmission involved may be direct or indirect. Thus transmission may, for example, result from direct contact between an infected and an uninfected person, or it may be mediated through inanimate material that has become contaminated with the agent, such as soil, blood, bedding, clothes, surgical instruments, water, food or milk (WHO, 2004).

Stability of storage

Stability of storage determines how long under which conditions the biological agents can stored and transported (WHO, 2004).

Resistance to environment & treatment

The agents have to be resistant to environmental factors such as water temperature, lack of nutrients in drinking water, surface forces, residual concentration of disinfectants and so on. In the case the contamination takes place before the water treatment plant, the microorganisms have to be resistant to for example, chlorination or other treatment methods. On the other hand, treatment plants can be overwhelmed by large spikes in pathogen concentration and render the treatment ineffective (Clark, 1980; Purver, 1995).

Resistance to medical treatment of patients

Terrorists also might consider how easy it is to treat an infection as soon as it is detected and identified. The shock might be even bigger when the public learns that there is no efficient treatment available to those that have been exposed. Most bacterial infections can be easily treated by antibiotics unless terrorists chose or genetically-engineered pathogens that are resistant to many antibiotics. Very often there is no medication for viruses and only very few to eukaryotic parasites, or only with significant side-effects.

Selection of pathogens

In the present study, the potential contaminants for the water supply system have been compiled by applying a two-step procedure proposed by the WHO (2004) amended with extra criteria. The procedure first tries to compile a list with the broadest possible coverage. In most instances, the lists do not consider the dissemination method or prioritize airborne dissemination. In the second step, the list is narrowed down to those agents that are of concern for the water supply system.

The list of harmful pathogens with the broadest possible coverage information is compiled from Field (2004) and Schmid *et al.* (2008) using data

- From *broad treaty definitions* of biological weapons
- From *agent lists* that have been negotiated to facilitate treaty implementation, or proposed therefore
- From historical data about agents that
  - Have been weaponised or stockpiled in recent times
  - Are known to have been used as weapons
  - Have been used in non-state entities/biocrimes
- From *epidemiological data on natural occurring drinking-water and food contaminations*

Several publications have assessed the potential of pathogens for use in contaminating the water distribution system (US Army, 1998; Burrows & Renner, 1999; Hickmann, 1999; Khan *et al.* 2001; Field, 2004; Shea & Gottron, 2004; Ottaviani *et al.* 2005; Gleick, 2006; Nuzzo, 2006; EPA, 2007; Winston & Leventhal, 2008; Clark, 2010; Burrows & Birkmire, 2011). Adding epidemiological data on natural drinking-water contaminants, a final list of potential biological contaminants for the drinking-water system has been compiled (see Table 2.1). A similar effort was conducted by the Austrian Agency For Health And Food Safety (AGES) (Schmid *et al.* 2008) which found a total of 180 species considered as potential bio-terroristic agents. 55 pathogens were included in the short list but no special consideration was given to pathogens that could potentially be used for contaminating the drinking water system.

International treaties

The broadest catchment of agents of concern, and therefore the starting point for the selection process, is to be found in the treaties that outlaw the possession of biological and chemical weapons. The BioWeapons Covention (BWC), which is a legal instrument, contains lists which have been developed for inclusion in the BWC Protocol. The purpose of these lists would again be to exemplify, but not to define, the scope of the general-purpose criterion (WHO, 2004).

Agent lists

The WHO collected a comprehensive list of agents that have been listed in various treaties. This list already shows how much variation there can be in different agent assessments (WHO, 2004). Another list of bio-terroristic agents has been compiled by the CDC (2012). The agents are categorized with respect to their danger. Category A agents of the CDC are mostly for airborne transmission, therefore also Category B agents of the CDC should be included when considering water contamination.

The United States Departments of Health and Human Services and of Agriculture compiled the "Select Agents and Toxins List" which have the potential to pose a severe threat to both human and animal health, to plant health, or to animal and plant products (HHS & USDA, 2008).

Historical data

Purver (1995) summarizes historical data on agents that have been weaponised or stockpiled in recent times or are known to have been used as weapons and McGeorge (1986) and Carus (2001) compiled information on agents that have been used in non-state entities/biocrimes.

Historical records are not complete, however, because former possessor states have not yet made all of the relevant papers available. Nevertheless, the WHO (2004) compiled an extensive list of antipersonnel agents. The information on the actual use of toxic and infective agents for hostile purposes may be even less complete than that on weaponization or stockpiling, not least because of the role of these agents in clandestine warfare, on which official records are often sparse.

Biocrimes

Terrorists and criminals may not use the same agents as those selected by military biological weapons programs (Carus, 2001). Although there are differences between terrorist and criminal uses of biological agents, the biocriminal faces many of the same obstacles as the bioterrorist. Both must acquire, develop, and employ biological weapons, so the technical constrains that appear in criminal cases are likely to apply for terrorist cases. In addition, it is possible that criminal cases will be a leading indicator of possible terrorist interest in biological agents. Nevertheless, the differences between terrorists and

Table 2.1 Short list of pathogens representing a risk to the drinking water system.

Pathogen	Water threat	Stable in water	Infectious dose	Exposure	Chlorine tolerance
Bacteria					
Bacillus anthracis	yes	2 years spores	6000	inh.	Spores resistant
Campylobacter jejuni	yes		400–500		
Escherichia coli – enteropathogenic (EPEC)	yes		for infants the dose is very low. In adults: $>10^6$ total dose		
Escherichia coli – enterotoxigenic (ETEC)	yes		10^8–10^{10}		
Franciscella tularensis	yes	<90 days	100,000,000	ing.	Inactivated, 1 ppm, 5 min
Salmonella spp. (typhoid)	yes	8 days, fresh water	15–20/10,000	ing.	inactivated
Shigella spp. (*S. dysenteriae, S. flexneri*)	yes	2–3 days	10/10,000	ing.	Inactivated, 0.05 ppm, 10 min
Vibrio cholerae	yes	"Survives well"	1000	ing.	"Easily killed"
Vibrio cholerae O1	yes		10^6		
Yersinia pestis	yes	16 days	500	inh.	Unknown
Plesiomonas shigelloides	yes		$>10^6$		
Brucella species (*B. Melitensis, B. Suis*)	probable	20–72 days	10,000	uns.	Unknown
Aeromonas hydrophila and other spp.	possibly		unsuccessful even at high doses		
Chlamydia psittaci	possibly	18–24h, seawater	Unknown		Unknown
Coxiella burnetii	possibly	Unknown	25	uns.	Unknown

Escherichia coli – enteroinvasive (EIEC)	possibly		possibly $\geq$10		
Escherichia coli O157:H7 enterohemorrhagic (EHEC)	possibly		possibly $\geq$10		
Listeria monocytogenes	possibly		$<10^3$		
Vibrio cholerae non-O1	possibly		$>10^6$		
Yersinia enterocolitica and *Yersinia pseudotuberculosis*	possibly		Unknown		
Miscellaneous enterics	possibly		Unknown		
Clostridium perfringens	no/probable	Common in sewage	$5 \times 10^5/\gg 10^8$	ing.	Resistant
Protozoa					
Cryptosporidium parvum	yes	Stable days	1/130	ing.	Oocysts resistant
Entamoeba histolytica	yes		one viable cyst		
Giardia lamblia	yes		one or more cysts		
Viruses					
enteric virus	yes	8–32 days	6	ing.	Readily inactivated (rota virus)
Hepatitis A	yes	Unknown	30/10–100 virus particles		Inactivated, 0.4 ppm, 30 min
Hepatitis E virus	yes		Unknown		
Norwalk virus group	yes		presumed to be low		
Rotavirus	possibly		10–100 infectious viral particles		
Variola	possibly	unknown	10	uns.	Unknown

Source: Data from Burrows and Renner (1999); FDA (2011) and CDC (2012).

criminals suggest that caution needs to be exercised in extrapolating from the experience of criminals to that of terrorists.

Carus (2001) analyzed 180 cases of illicit biological agent use and the number of criminal cases is higher than that of terrorist cases.

Confirmed use and confirmed possession and probable use of agents

In Table 2.2, below, a compiled list of biological agents previously used or found in possession of possible perpetrators.

Bacterial agents: Note that only three of cases involved possession of anthrax. *Yersinia pestis*, *S. typhi*, and *Shigella dysenteriae* strains appear several times, while seven other pathogens appear no more than once.

Viral agents: Few perpetrators have considered viral agents, except when they can use it in a natural state. HIV appears in several cases, including at least four murder cases.

Other pathogens: Only one case involved the use of a parasite. In that case, the perpetrator contaminated food with *Ascaris suum*, a worm that infects pigs and does not normally infect humans.

Table 2.2 Confirmed use and confirmed possession and probable use of agents.

Agent (human pathogen)	Confirmed use	Threatened use (confirmed possession)	Probable or possible use	Total
Salmonella typhi	5	1	6	12
Bacillus anthracis	3	2	2	7
HIV	3		4	7
V. cholera	1		3	4
Yersinia pestis	1	1		2
Shigella	2			2
Salmonella typhimurium	1			1
S. paratyphi	1			1
M. bovis?	1			1
Y. enterocolitica		1		1
C. botulinum, C. tetani		1		1
Yellow fever virus			1	1
Hepatitis A virus			1	1
Giardia			1	1

Source: Data were from Carus (2001).

Naturally occurring microbial water contaminants

Naturally occurring microbes are potential agents that can be used to intentionally contaminate the drinking water system (Ashford *et al.* 2003). National drinking water standards, like those from the EPA (2012c), give an indication on which microorganisms are considered to be a particular threat to the public health. Their occurrence is usually monitored and regulated.

In the U.S.A., the EPA produces the drinking water "Contaminant Candidate List" (EPA, 2012a). The contaminants on the list are known or anticipated to occur in public water systems. However, they are unregulated by existing national primary drinking water regulations. Additional sources of information are the various epidemiological data in relation to water-borne and food-borne outbreaks of infectious diseases. A special issue of the Journal of Water and Health was published in July/August 2006 by EPA (2006) with a series of review articles on endemic gastrointestinal illnesses associated with microbial drinking water exposures.

Other relevant legal frameworks are national regulations for microbial food contaminants and information provided by government agencies on potentially hazardous microorganisms. The US Food and Drug Administration FDA compiled the "FDA Bad Bug Book" (FDA, 2011). It lists and describes microorganisms that are harmful for the food and water supply system. Monroe (2006) reorganized the table and created a rating on the propensity of each pathogen to be waterborne.

Assesment of biological agents with respect to drinking-water attacks

There are several publications that assess which biological agents can be used for attacks on the drinking water supply (Burrows & Renner, 1999; Hickmann, 1999; Khan *et al.* 2001; Field, 2004; Shea & Gottron, 2004; Ottaviani *et al.* 2005; Gleick, 2006; Nuzzo, 2006; EPA, 2007; Winston & Leventhal, 2008; Clark, 2010; Burrows & Birkmire, 2011).

Burrows and Renner (1999) compiled one of the most comprehensive investigations of biological threats to the drinking water system. A very useful summary contains information on: clinical considerations, infective/toxic dose, environmental stability, disinfection efficacy, removal by treatment systems.

2.4.2 Criticality assessment

Although the actual number of known victims and fatalities from bioterrorism and criminal activities has been miniscule compared with many other daily hazards, biological agents have the potential to poison or compromise our food or water supply, to severely damage segments of the economy, to cause human mass casualties and, perhaps most damaging, disrupt our society physically and psychologically.

An unintentional drinking-water contamination event that occurred in Tel Aviv, Israel in July, 2001 showed no matter how minor the contamination event or short-term the disruption of delivery of safe drinking-water, psychological, medical and public health impact could be significant (Winston & Leventhal, 2008).

Thus, water systems have been designated as critical to national security by many governments.

The severity of the attack depends on the number of people affected, the individual exposure and the dose-response function of the contaminant.

2.4.3 Vulnerability assessment

Previous naturally occurring food- and waterborne outbreaks have demonstrated the vulnerability of both the water supply and the public's health to biological contamination of drinking water. Such experiences suggest that a biological attack, or even a credible threat of an attack, on water supplies and water distribution systems represent potential targets for terrorist activity.

The vulnerability is characterized by the ease with which threat agents can be introduced in quantities sufficient to achieve the attacker's purpose once the target has been reached.

Factors determining the vulnerability (Stephenson, 2004; Nuzzo, 2006; EPA, 2002):

- *Accessibility*: How easy is it for someone to enter, operate, collect information, and evade response forces?
- *Availability of information* concerning the target

- *Detectability*: ability to work unobserved,
- *Ease of introducing agents*: does not involve complex steps to weaponise an agent
- *Volumes of contaminant* that can be introduced
- *Ability to uniformly mix* agents into target
- *Time* available for introduction of agents
- *Dilution and treatment* (e.g. chlorination) can limit the effectiveness of water contamination

Detectability

The detectability of an attack can be considerably improved when sensors are in place that can detect contamination of the drinking water supply. The work conducted in the DINAMICS EU collaborative research project was aiming at creating a automated and sensitive detector for bioterroristic attacks with microorganisms. The American Society of Civil Engineers produced a Guidelines for Designing an Online Contaminant Monitoring System (ASCE, 2004). Other projects (Hindson *et al.* 2005) or attempts of offering commercial solutions have been made (e.g. by Early Warning, Inc.).

Point of attack (DHS & FBI, 2008)

Contamination of raw water sources (lakes, reservoirs, rivers, and wells) prior to treatment: Contaminants could be introduced at various points along the transmission line, particularly if aqueducts are open or conduits are above ground and easily accessible. Contaminants also may be pumped directly into the raw water sources prior to transportation or at treatment plant intakes. Water quality monitoring by local water authorities and the effects of fluorination/chlorination and the filtering of surface waters can limit the effects of a potential contaminant.

Contamination of distribution systems or water storage tanks following treatment: Contaminants could be pumped directly into aboveground storage tanks or uncovered storage reservoirs. A contaminant also could be introduced into a drinking water distribution system using access points such as fire hydrants and most types of commercial and residential connections.

Backflow contamination: Contaminants could be pumped back into the water system through output sources such as restroom sinks, toilets, or water tanks, potentially contaminating the water system in a localized area.

Disabling or sabotaging the drinking water system: Terrorists could increase the effect of contaminants by tampering with or disabling treatment equipment through cyber or physical attacks. If equipment is disrupted and raw water is not properly treated, contaminants introduced into a water source could continue through the finished water storage tanks and into the distribution system.

2.5 RISK ESTIMATION

2.5.1 Hierarchical risk estimation

The best estimate of the risk of a bio-terroristic attack on the drinking water system has to incorporate all aspects of the threats. This can reasonably only be done by a staged approach that uses different methods for the different kinds of information and its certainty. A hierarchical approach is proposed that starts at the international and national political level and lists and describes the actual and potential adversaries. The next stage would use this information to get a better characterization of the threats. Particularly, the relation between adversaries, attack mode and biological hazard has to be characterized. An influence diagram, whether formalized or not, will help to clarify the interdependencies between various factors.

Recent literature presents more complex models that try to capture the interdependence of the many factors involved in the risk analysis (Hudson *et al.* 2001; NRC, 2002; Latourrette & Willis, 2007; Benett, 2008; Lindhe, 2008; Ezell *et al.* 2010).

In the final stage, a modified form of QMRA can be applied. The advantage of this staged approach, represented in Figure 2.1, is that at each stage a different panel of experts will be needed and different methods can be used.

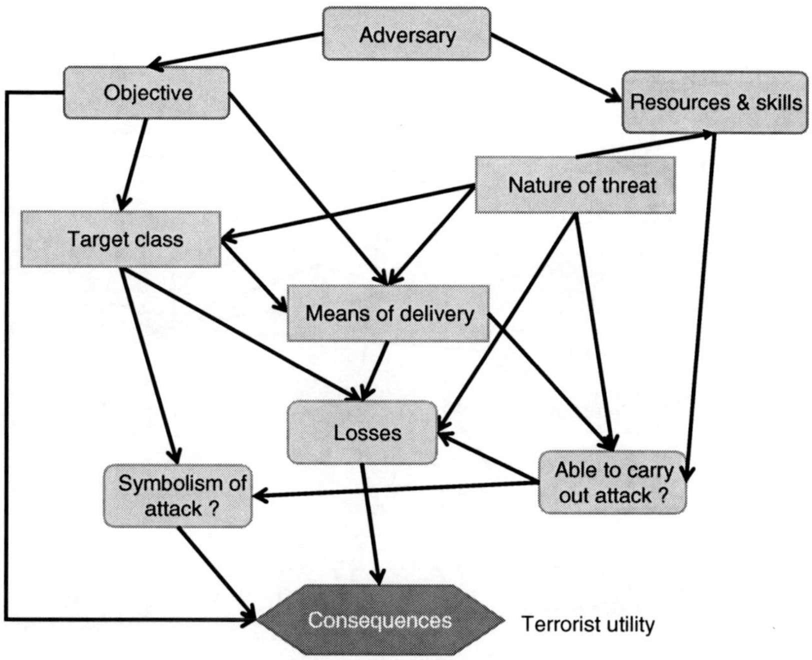

Figure 2.1 Probabilistic threat characterization. This figure is modelled after Paté-Cornell and Guikema (2002).

In the case of bio-terroristic attacks on the drinking water system, "nature of threat" refers to the biological agent. The selection of the agent is influenced by skills and resources available to the adversaries and it determines what losses can be inflicted and what the likelihood is that the attack can actually carried out. This diagram helps to decide which of the pathogens have to be considered for a more detailed analysis.

The graph in Figure 2.2 shows a subset of possible interaction between the various factors influencing the risk of a particular threat scenario. The factors are grouped according to

- Point of attack (vulnerability and technical parameters of a particular distribution system)
- Biological characteristic of the chosen pathogen
- Technological constraints of the chosen pathogen
- Characteristics of a particular adversary

The skills of the adversary have strong influence on the ability to grow larger quantities of an agent, depending on the agent involved. Some pathogens are easy to culture (e.g. bacteria), while others are extremely difficult (e.g. viruses). The assessment of an adversary's skills is conducted on a higher level of risk assessment. The adversary's object determines which amount of pathogen is necessary. Historically perpetrators who adopted objectives that required significant quantities of agent usually selected biological agents easy to culture (Carus, 2001; McGeorge, 1986; Purver, 1995).

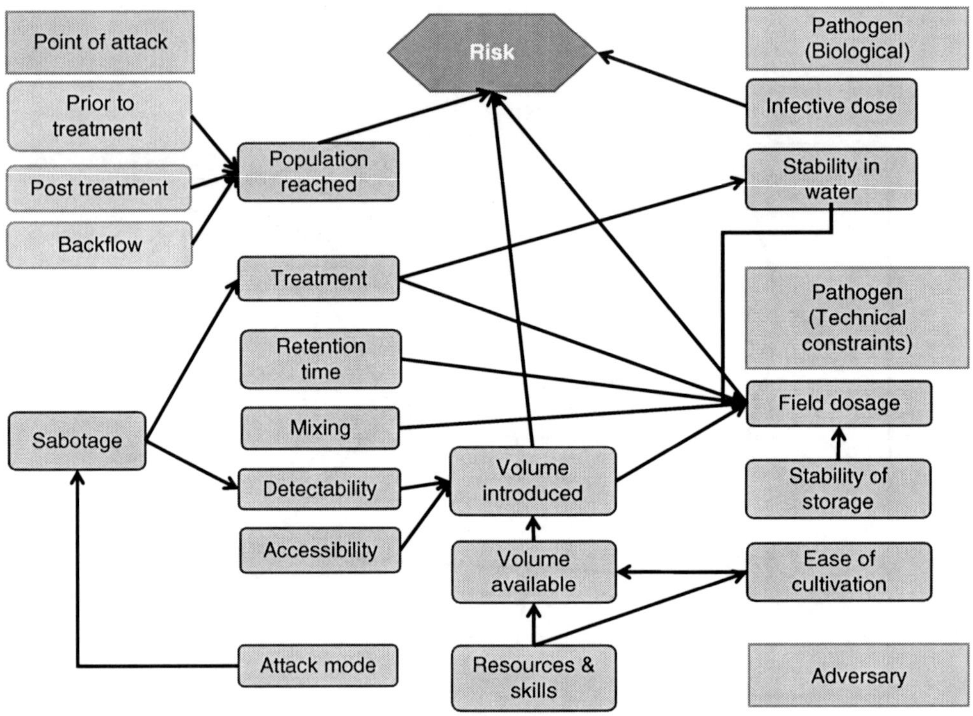

Figure 2.2 Risk Estimation Model showing some of the dependencies between the various influence factors. The rectangular boxes with sharp corners only denote the category of factors.

One of the most important factors for the risk calculation is the *effective field dosage*, which depends on many other factors, like stability of storage, volume introduced, mixing, retention time, treatment and stability in water after release.

Municipal water systems are designed to eliminate impurities by various kinds of treatments, especially pathogens. As part of this process, communities use filters to remove particles from the water and add chlorine to kill any organisms that remain. So, depending on whether the *point of attack* is before or after the water treatment plant, the effective field dose of the pathogen is influenced dramatically. The terrorist could inject pathogenic organisms directly into the mains past the treatment and quality control stations. The exact topology of the drinking-water system and the point of attack determine how many people might be affected by the contamination. On the other hand, residual chlorine might still mitigate the severity of the attack. So it also has to be considered that terrorists could use chlorine resistant agents.

Attacking the reservoir might not seem the most effective strategy for a terrorist when downstream there is a water treatment plant, but when the attack is combined with acts of sabotage stopping the water purification (Hickmann, 1999) this would usually give the largest distribution of the pathogen to the population.

Since the chance that a sabotage is detected is usually higher than the detectability of a water contamination, a possible strategy could be to defeat the antibacterial agents in the water system by overloading their capacity (Clark, 1980). Nevertheless, it has to be considered that only a small fraction of the water drawn from an urban water supply is ingested by the population, by far the largest share is used for personal hygiene, watering lawns, washing clothes and cars, flushing toilets, and so on.

2.5.2 Quantitative microbial risk assessment

The National Academy of Sciences report (NAS, 1983) broke down quantitative microbial risk assessment (QMRA) into four stages:

- *Hazard identification*: covers the nature of a disease or adverse health impact,
- *Dose–response assessment*: covers quantitative studies, often from animal model toxicology experiments that evaluate the frequency of a response *versus* the dose of the toxicant,
- *Exposure assessment* evaluates the potential doses of the toxicant among the receptor population groups of concern, and
- *Risk characterization*: links the quantitative dose-response model predictions from the experimental data to the exposure doses to quantitatively predict the risks of the adverse effect in the target population. Factors to address dose scaling and both inter-species and intra-species differences in susceptibility are usually considered.

This approach was refined through subsequent research (Haas, 1999; Armstrong, 2005; EPA, 2007; Parkin, 2008; EPA, 2009; Schijven *et al.* 2011; USDA & EPA, 2011; Camrawiki, 2012).

Performing a reliable QMRA for a certain pathogen in the drinking water supply and for a given population requires knowledge of the concentrations of a pathogen in the source water, removal or inactivation efficiency of the treatment process and consumption of drinking water. Mathematical modelling can be used to estimate the effects on health of low doses of pathogens in drinking-water.

The QMRA approach quantitatively assesses the likelihood, consequences and scale of effects of the specified scenarios (Teunis *et al.* 2000; Hunter *et al.* 2003; Hörman, 2005; WHO, 2011). The following concepts are relevant:

Problem formulation and hazard identification. All potential hazards, sources and events that can lead to the presence of microbial pathogens (i.e. what can happen and how) should be identified and documented for each component of the drinking-water system.

Exposure assessment. Exposure assessment in the context of drinking-water consumption involves estimation of the number of pathogens to which an individual is exposed, principally through ingestion. Exposure is determined by the concentration of pathogens in drinking-water and the volume of water consumed. The main component of exposure assessment, which is common to all pathogens, is the *volume of unboiled water consumed by the population*, including person-to-person variation in consumption behaviour and especially consumption behaviour of vulnerable subpopulations. For microbial hazards, it is important that the unboiled volume of drinking-water, both consumed directly and used in food preparation, is used in the risk assessment, as heating will rapidly inactivate pathogens. Exposure assessment must account for variability of such factors as concentrations of pathogens over time and volumes ingested. The local variation of drinking-water consumption has to be considered. Exposure can be expressed as a single dose of pathogens that a consumer ingests at a certain point in time or the total amount over several exposures in a certain period of time (e.g. over a year).

Dose–response assessment. The probability of an adverse health effect following exposure to one or more pathogenic organisms is derived from a dose–response model. Available dose–response data have been obtained mainly from studies using healthy adult volunteers. However, adequate data are lacking for vulnerable subpopulations, such as children, the elderly and the immune-compromised, who may suffer more severe disease outcomes. Dose–response model is based on the estimation of the probability of an infection in relation to a specific dose. Infection is a conditional event, since for an infection to occur, one or more viable pathogens must have been ingested and one or more of these ingested pathogens must have survived in the host's body. The concept of the single-hit principle, even a single pathogen may be able to cause infection and disease, supersedes the concept of (minimum) infectious dose that is frequently used in older literature. For natural, well-dispersed water contaminations, pathogena are assumed to be Poisson distributed. The dose–response relationship simplifies to an exponential function, in the case one can assume that the individual probability of any organism surviving and starting infection is the same. More complex models use the beta-Poisson dose–response relationship. In the case of an intentional contamination, this assumption of Poisson distribution might not be true and the distribution of pathogens will depend on volume, concentration as well as on the dispersion function of the system at the point where the pathogen is introduced into the drinking water system.

Risk characterization. Risk characterization combines the data collected on exposure, dose–response and the incidence and severity of disease. The probability of infection can be estimated as the product of the exposure by drinking-water and the probability that exposure to one organism would result in infection. Not all infected individuals will develop clinical illness; asymptomatic infection is common for most pathogens. The percentage of infected persons who will develop clinical illness depends on the pathogen, but also on other factors, such as the immune status of the host and if acquired immunity to that pathogen exists in a portion of the population.

Data requirements. Data collection and documentation is usually the most time-consuming part of the any risk assessment. General issues concerning the quality and relevance of data to risk assessments are addressed in for example, in FAO/WHO (2003, 2008) risk assessment guidelines. There are two basic types of data required for a risk assessment, whether qualitative or quantitative, namely:

- The data used to describe the risk pathway, and thus construct the model framework; and
- The data used to estimate the model input parameters.

The model input parameters must all be numerical for a quantitative risk assessment. In the absence of numerical data, quantified expert opinion or surrogate data are needed to fill the gaps. Furthermore uncertainty or variability of the parameter estimates must be incorporated into the model, generally this as done by means of distributions.

2.5.3 Data sources

The Center for Advancing Microbial Risk Assessment (CAMRA) started the CAMRA Wiki project Camrawiki (2012) to become a central repository for QMRA knowledge and data available to the risk analysis community of scientists. The goal is for each pathogen to have a Pathogen Safety Data Sheet (PSDS) that gives a brief overview of the hazard and its associated risks and to have a completed dose–response model. At the time of compiling this report, the database listed 8 viruses, 12 bacteria and 4 Protozoa.

Pathogen Safety Data Sheet (PSDS) gives morbidity rate, mortality rate, incubation time, recommended best fit dose response parameters and survival information for all agents. The Public Health Agency of Canada (PHAC, 2012) has an on-line repository for Pathogen Safety Data Sheets (previously titled

Material Safety Data Sheets for infectious substances). PSDS are technical documents that describe the hazardous properties of a human pathogen and recommendations for work involving these agents in a laboratory setting.

The EPA Water Contaminant Information Tool (WCIT) contains detailed information describing the substance, its behaviour in water, and potential health effects. WCIT information includes among others: pathogen properties, availability, fate and transport, medical and toxicity information, drinking water and wastewater treatment.

The VITAL FP7 project (ViTAL, 2007) reviewed the literature to find QMRA studies that assessed the infection risks for viruses quantitatively: 10 such studies were retrieved.

The effect of chlorination and other water treatment methods in deactivating and/or eliminating pathogens from drinking water has been investigated and described in several publications and reports (LeChevallíer *et al.* 1981; Hoeger *et al.* 2002; Stanfield *et al.* 2003; Rose *et al.* 2004; WHO, 2004; LeChevallier & Au, 2005; Hörman, 2005; Rose & O'Connell, 2009).

The FP5 Project Microrisk (Mons *et al.* 2005) produced several interesting reports on the "Estimation of the consumption of cold tap water for microbiological risk assessment," the "Efficacy of water treatment processes" and on "Pathogens in drinking water sources" as well as on the "Persistence of pathogens."

Dembek (1997) lists food and waterborne pathogens in "Medical Aspects of Biological Warfare", with information on infective dose and mortality and additional medical details. Beside an epidemiological assessment, the book includes a number of chapters dealing with medical and biological characteristics of individual pathogens.

The Environmental Protection Agency EPA (2009) produced a draft "Protocol for Microbial Risk Assessment" which gives an overview on *Dose–Response Relationships for Waterborne Pathogens*, which has been adapted from McBride *et al.* (2002).

Sinclair *et al.* (2008) published results on the persistence of category A select agents in water.

EPA (2007) surveyed and summarized literature into "A Compendium of Prior and Current Microbial Risk Assessment Methods."

The WHO Guidelines for Drinking-water Quality (WHO, 2011) presents also lists on pathogens that can be transmitted through drinking-water and organisms for which transmission through drinking-water has been suggested but for which evidence is inconclusive.

It also provides microbial fact sheets that include information on human health effects, sources and occurrence, routes of transmission and the significance of drinking water as a source of infection. The Annex contains a description of Treatment methods and their performance.

EPA's Thesaurus of Terms Used in Microbial Risk Assessment EPA (2012b) is a compilation of definitions from many sources within EPA and outside of EPA.

A NRC (2004) publication lists "Emerging and Re-emerging Waterborne Pathogens" and describes the health effects and the mode of transmission for each pathogen.

REFERENCES

Armstrong T. W. (2005). A Quantitative Microbial Risk Assessment Model for Human Inhalation Exposure to Legionella. Ph.D. thesis, Drexel University, USA.

ASCE (2004). Interim Voluntary Guidelines for Designing an Online Contaminant Monitoring System. American Society of Civil Engineers, USA.

Ashford D. A., Kaiser R. M., Bales M. E., Shutt K., Patrawall A., McShan A., Tappero J. W., Perkins B. A. and Dannenberg A. L. (2003). Planning against biological terrorism: lessons from outbreak investigations. *Emerging Infectious Diseases*, **9**(5), 515–519.

Bartram J., Corrales L., Davison A., Deere D., Drury D., Gordon B., Howard G., Rinehold A. and Stevens M. (2009). Water Safety Plan Manual: Step-by-Step Risk Management for Drinking-Water Suppliers. WHO, Geneva.

Benett S. (2008). WMD Terrorism Risk Assessment in the Department of Homeland Security, Presentation, S&T Stakeholder Conference, June 2–5 2008, USA. Www.dtic.mil/ndia/2008homest/benn.pdf (accessed 31 January 2012)

Benford D. J. (2001). Principles of Risk Assessment of Food and Drinking Water Related to Human Health. International Life Sciences Institute, Europe.

Burrows W. D. and Birkmire S. E. (2011). Water supply, vulnerability, and attack specifics In: Encyclopedia of Bioterrorism Defense, Katz Rebecca and A. Raymond, Zilinskas (eds), Hoboken, NJ, USA.

Burrows W. D. and Renner S. E. (1999). Environmental Health Perspectives. Biological warfare against as threats to potable water. U.S. Army Center for Health Promotion and Preventive Medicine, Aberdeen Proving Ground, MD 21010-5403, USA. http://ehp03.niehs.nih.gov/article/fetchArticle.action?articleURI=info:doi/10.1289/ehp.99107975 (accessed 31January 2012)

CAMRA Microbial Risk Assessment Wiki (CAMRAwiki), Center for Advancing Microbial Risk Assessment (CAMRA), jointly at Michigan State University (MSU) and Drexel University, USA. http://wiki.camra.msu.edu. (accessed 31 January 2012)

Carus S. (2001). Bioterrorism and Biocrimes: The Illicit Use of Biological Agents Since 1900, Center for Counterproliferation Research, National Defense University, Washington, D.C., USA. http://www.fas.org/irp/threat/cbw/carus.pdf (accessed 31 January 2012)

Catlin M. and Kautter D. (2007). Carver Plus Shock Method for Food Sector Vulnerability Assessments. FDA, USA.

CDC (2012). Bioterrorism Agents/Diseases by Category, Center of Disease Control, Department of Health and Human Services, USA. http://emergency.cdc.gov/agent/agentlist-category.asp (accessed 31 January 2012)

Clark R. C. (1980). Technological Terrorism. Devin–Adair, Old Greenwich, CT.

Clark R. M. 2010. Potential contamination agents of interest. In: Wiley Handbook of Science and Technology for Homeland Security, John G. (ed), Voeller Wiley-Blackwell, Hoboken, NJ, USA. pp. 1–18.

Copeland C. (2010). Terrorism and Security Issues Facing the Water Infrastructure Sector, Report RL32189, CRS Report for Congress, Congressional Research Service, The Library of Congress, USA. http://www.fas.org/sgp/crs/terror/RL32189.pdf (accessed 31 January 2012)

Dawson D. (2003). Foodborne Protozoan Parasites. ILSI Europe Emerging Pathogen Task Force, International Life Sciences Institute, Brussels, Belgium.

Dechesne M. and Soyeux E. (2007). Assessment of source water pathogen contamination. *Journal of Water and Health*, **5**(Suppl 1), 39–50 .

Dembek Z. F. (1997). Medical Aspects of Biological Warfare, Series: Textbooks of Military Medicine, Office of The Surgeon General. US Army Medical Department Center and School Borden Institute, Washington, DC, USA.

Department of Homeland Security (2008). DHS Risk Lexicon, Risk Steering Committee. US Department of Homeland Security, USA.

DHS and FBI (2008). Potential Terrorist Attack Methods. Homeland Security, Federal Bureau of Investigation, USA. http://publicintelligence.net/dhsfbi-ufouo-potential-terrorist-attack-methods (accessed 31 January 2012).

Dufour S. M., Koster W., Bartram J., Ronchi E. and Fewtrell L. (2003). Assessing Microbial Safety of Drinking Water – Improving Approaches and Methods. WHO Water Quality Series OECD. London, UK.

EPA (2002). Vulnerability Assessment Factsheet, Report EPA 816-F-02-025, Office of Water (4601M), Environmental Protection Agency, USA.

EPA (2006). Waterborne disease research summaries, *Journal of Water and Health*, **4**(Suppl 2). http://www.epa.gov/nheerl/articles/2006/waterborne_disease.html (accessed 31 January 2012)

EPA (2007). A Compendium of Prior and Current Microbial Risk Assessment Methods, Report EPA/600/R-07/129, Threat and Consequence Assessment Division National Homeland Security Research Center United States Environmental Protection Agency, USA.

EPA (2009). Protocol for Microbial Risk Assessment. Report EPA-SAB-10-008, Office of Science and Technology, Office of Water, Washington, DC, and Environmental Protection Agency, USA.

EPA (2012). National Primary Drinking Water Regulations. Environmental Protection Agency, USA. http://water.epa.gov/drink/contaminants/index.cfm#Microorganisms (accessed 31 January 2012)

EPA (2012a). Third Contaminant Candidate List, Environmental Protection Agency, USA. http://water.epa.gov/scitech/drinkingwater/dws/ccl/ (accessed 31 January 2012)

EPA (2012b). EPA's Thesaurus of Terms Used in Microbial Risk Assessment. http://water.epa.gov/scitech/swguidance/standards/criteria/health/microbial/thesaurus_index.cfm (accessed 31 January 2012)

Ezell B. C., Bennett S. P., von Winterfeldt D., Sokolowski J. and Collins A. J. (2010). Probabilistic risk analysis and terrorism risk. *Risk Analysis*, **30**(4), 575–589.

FAO and WHO (2003). Hazard characterization for pathogens in food and water. In: Guidelines. Microbiological Risk Assessment, Jean-Louis Jouve, Jørgen Schlundt (eds), Series No 3. p. 61, Food and Agriculture Organization of the United Nations and World Health Organization, Geneva, Switzerland.

FAO and WHO (2008). Exposure Assessment of Microbiological Hazards in Food. In: Guidelines. Microbiological Risk Assessment, Ezzeddine Boutrif, Jørgen Schlundt (eds.), Series No 7. p. 61, Food and Agriculture Organization of the United Nations and World Health Organization, Geneva, Switzerland.

FDA (2011). Foodborne Pathogenic Microorganisms and Natural Toxins Handbook, The "Bad Bug Book", Food and Drug Adminstration, USA. http://www.fda.gov/food/foodsafety/foodborneillness/foodborneillnessfoodbornepathogensnaturaltoxins/badbugbook/default.htm (accessed 31 January 2012)

Fedra K. (2008). Technological risk assessment and management: can we integrate terrorist attacks? In: Integration of Information for Environmental Security. NATO Science for Peace and Security Series C: Environmental Security, H. G. Coskun, H. K. Cigizoglu and D. Maktav, (eds), Springer, Dordrecht, pp. 353–373, 498.

FEMA (2005). Risk Assessment, A How-To Guide to Mitigate Potential Terrorist Attacks Against Buildings, Report FEMA 452, Federal Emergency Management Agency U.S. Department of Homeland security, USA.

Field M. S. (2004). Assessing the risks to drinking-water supplies from terrorists attacks. In: Water Supply Systems Security, W. M. Larry (ed.), 1st edn, From Water Supply Systems Security, McGraw-Hill Professional, New York, NY, USA.

Fowle J. R. and Dearfield K. L. (2000). Risk Characterization Handbook. Environmental Protection Agency 100-B-00-002, Washington, DC, USA, p. 1–57.

Gleick P. H. (2006). Water and terrorism. *Water Policy*, **8**, 481–503.

Haas C. N. (1983). Estimation of risk due to low doses of microorganisms: a comparison of alternative methodologies. *American Journal of Epidemiology*, **118**(4), 573–582.

Haas C. N., Rose J. B. and Gerba C. P. (1999). Quantitative Microbial Risk Assessment, John Wiley & Sons. Hoboken, NJ, USA.

HHS and USDA (2008). List of Select Agents and Toxins, 7 CFR Part 331, 9 CFR Part 121, and 42 CFR Part 73, U.S. Department of Agriculture's Food Safety and HHS USA. http://www.selectagents.gov/select agents and Toxins list.html (accessed 31 January 2012)

Hickman D. C. (1999). A Chemical and Biological Warfare Threat: USAF Water Systems At Risk, Counterproliferation Paper No. 3, USAF Counterproliferation Center, Air War College, Air University, AL, USA. http://www.au.af.mil/au/awc/awcgate/cpc-pubs/hickman.htm (accessed 31 January 2012)

Hindson B. J., Makarewicz A. J., Setlur U. S., Henderer B. D., McBride M. T. and Dzenitis J. M. (2005). APDS: the autonomous pathogen detection system. *Biosensors and Bioelectronics*, **20**(10), S1925–S1931.

Hoeger S. J., Dietrich D. R. and Hitzfeld B. C. (2002). Effect of ozonation on the removal of cyanobacterial toxins during drinking water treatment. *Environ. Health Perspect*, **110**(11), 1127–1132.

Hokstad P., Røstum J., Sklet S., Rosén L., Pettersson T., Linde A., Sturm S., Beuken R., Kirchner D. and Niewersch C. (2009). Methods for risk analysis of drinking water systems from source to tap, Report TECHNEAU D4.2.4, TECHNEAU is an Integrated Project Funded by the European Commission under the Sixth Framework Programme (contractnumber 018320).

Hörman A. (2005). Assessment of the Microbial Safety of Drinking Water Produced from Surface Water Under Field Conditions, PhD thesis, Department of Food and Environmental Hygiene, University of Helsinki, Helsinki, Finland.

Hudson L. D., Ware B. S., Laskey K. B. and Mahoney S. M. (2001). An Application of Bayesian Networks to Antiterrorism Risk Management for Military Planners. Technical Report, Department of Systems and Engineering and Operation Research, George Mason University. http://www.dsbox.com/docs/UAI_Antiterrorism_Paper.pdf (accessed 31 January 2012)

Hunter P. R., Payment P., Ashbolt N. and Bartram J. (2003). Assessment of risk. In: Assessing Microbial Safety of Drinking Water – Improving Approaches and Methods, Chapter 3, WSH/DOC 2003, Water Quality Series, OECD, A. Dufour, M. Snozzi, W. Koster, J. Bartram, E. Ronchi and L. Fewtrell (eds.), IWA Publishing, London, UK, pp. 79–109.

ILSI (2008). Considering water quality for use in the food industry. Report ILSI Europe Environment and Health Task Force, ILSI, Europe.

IRM (2002). A Structured Approach to Enterprise Risk Management (ERM) and the Requirements of ISO 31000, Institute of Risk Management, UK. http://www.theirm.org/publications/documents/ARMS_2002_IRM.pdf (accessed 31 January 2012)

Khan A. S., Swerdlow D. L. and Juranek D. D. (2001). Precautions against Biological and Chemical Terrorism Directed at Food and Water Supplies. Public Health Reports, Volume 16.

Koopmans M. and Duizer E. (2003). Foodborne viruses: an emerging problem. *International Journal of Food Microbiology*, **90**(2004), 23–41.

Larson E., Aiello A. E. and Arbor A. (2006). Systematic risk assessment methods for the infection control professional. *AJIC Practice Forum*, **34**, 323–326.

Latourrette T. and Willis H. H. (2007). Using probabilistic terrorism risk modeling for regulatory benefit-cost analysis. *RAND*, Report WR-487-IEC, Working Paper, http://www.rand.org/content/dam/rand/pubs/working_papers/2007/RAND_WR487.pdf, 1–46.

LeChevallier M. W. and Au K. 2004. Water Treatment and Pathogen Control: Process Efficiency in Achieving Safe Drinking Water. IWA Publishing, London, UK.

LeChevallíer M. W., Evans T. M. and Seidler R. J. (1981). Effect of turbidity on chlorination efficiency and bacterial persistence in drinking water. *Applied and Environmental Microbiology*, July 1981, 159–167.

Leson J. (2005). Assessing and Managing the Terrorism Threat, Report NCJ 210680, Bureau of Justice Assistance, Office of Justice Programs, USA.

Lindhe A. (2008). Integrated and Probabilistic Risk Analysis of Drinking Water Systems. PhD thesis. Department of Civil and Environmental Engineering, Chalmers University, Göteborg, Sweden 1–241.

Masse T., O'Neil S. and Rollins J. (2007). The Department of Homeland Security's Risk Assessment Methodology: Evolution, Issues and Options for Congress. CRS Web, Washington, DC, USA, pp. 1–33.

McBride G., Till D., Ryan T., Ball A., Lewis G., Palmer S. and Weinstein P. (2002). Pathogen Occurrence and Human Health Risk Assessment Analysis, Microbiology Research Programme Technical Report TR122, Ministry for the Environment, NZ. http://www.mfe.govt.nz/publications/water/freshwater-microbiology-nov02/freshwater-microbiology-nov02.pdf (accessed 31 January 2012)

McGeorge H. J. II (1986). The deadly mixture: bugs, gas, and terrorists. *NBC Defense & Technology International*, **1**(2), 56–61.

Monroe W.-S. (2004), Waterborne rating, Civil and Environmental Engineering, Cornell University, USA. http://ceeserver.cee.cornell.edu/mw24/cee454/Readings/badbugs.htm (accessed 31 January 2012)

Mons M., Blokker M., van der Wielen J., Medema G., Sinclair M., Hulshof K., Dangendorf F. and Hunter P. R. (2005). Estimation of the consumption of cold tap water for microbiological risk assessment. FP5 MICRORISK Project Nr. CG020114. http://www.microrisk.com/uploads/microrisk_tap_water_consumption.pdf (accessed 31 January 2012)

National Academy of Sciences (1983). Risk Assessment in the Federal Government: Managing the Process. National Academy Press, Washington, DC, USA.

National Research Council (2002). Making the Nation Safer: The Role of Science and Technology in Countering Terrorism. The National Academies Press, Washington, DC, 2002. 1. Print. http://www.nap.edu/openbook.php?record_id=10415 (accessed 31 January 2012)

National Research Council (2004). Indicators for Waterborne Pathogens. The National Academies Press, Washington, DC, 2004. 1. Print. http://www.nap.edu/openbook.php?record_id=11010 (accessed 31 January 2012)

Nuzzo J. B. (2006). The biological threat to U.S. water supplies: toward a National Water Security Policy. *Biosecurity and Bioterrorism*, **4**(2), 147–159.

Ottaviani M., Drusiani R., Lucentini L., Ferretti E. and Bonadonna L. (2005). Misure di prevenzione e di sicurezza dei sistemi acquedottistici nei confronti di possibili atti terroristici (Preventive and Security Measures Towards Potential Terrorist Attacks to Water Systems.). Rapporti ISTISAN 05/4, Ministero della Salute, Istituto Superiore di Sanità and Federgasacqua, Italy.

Parkin R. T. (2008). Foundations and Frameworks for Human Microbial Risk Assessment. Center for Risk Science and Public Health, School of Public Health and Health Services, The George Washington University Medical Center, Washington, DC, 1–129.

Paté-Cornell E. and Guikema S. (2002). Probabilistic modeling of terrorist threats: a systems analysis approach to setting priorities among countermeasures. *Military Operations Research*, **7**(4), 1–42.

Public Health Agency of Canada (2012). Pathogen Safety Data Sheets and Risk Assessment. CA. http://www.phac-aspc.gc.ca/lab-bio/res/psds-ftss/index-eng.php (accessed 31 January 2012)

Purver R. (1995). Chemical and Biological Terrorism: The Threat According to the Open Literature. http://www.csis-scrs.gc.ca/pblctns/thr/cbtrrrsm01-eng.asp (accessed 31 January 2012)

Riha J. (2009), Semi-quantitative risk assessment of water structures. International Symposium on Water Management and Hydraulic Engineering, Ohrid, Macedonia, Paper: A28, pp. 81–88.

Rios J. (2010). Adverserial risk analysis for counterterrorism modelling, Workshop on Adverserial Decision Making, DIMACS 2010, Piscataway, NJ, USA.

Rose L. J. and O'Connell H. (2009). UV light inactivation of bacterial biothreat agents. *Applied and Environmental Microbiology*, **75**(9), 2987–2990.

Rose L. J., Rice E. W., Jensen B., Murga R., Peterson A., Donlan R. M. and Arduino M. J. (2005). Chlorine inactivation of bacterial bioterrorism agents. *Applied and Environmental Microbiology*, **71**(1), 566–568.

Rosén L., Hokstad P., Lindhe A., Sklet S. and Røstum J. (2007). Generic Framework and Methods for Integrated Risk Management in Water Safety Plans, Report TECHNEAU D4.1.3, TECHNEAU is an Integrated Project Funded by the European Commission under the Sixth Framework Programme (contractnumber 018320).

Schijven J. F., Teunis P. F. M., Rutjes S. A., Bouwknegt M. and de Roda Husman A. M. (2011). QMRAspot: a tool for quantitative microbial risk assessment from surface water to potable water. *Water Research*, **45**, 5564–5576.

Schmid D., Stüger H.-P., Pet D., Pongratz P., Brüller W. and Allerberger F. (2008). Bioterrorism: A review of Potential Bioterrorism Agents. Austrian Agency for Health and Food Safety (Ages), Austria. http://www.ages.at/uploads/media/Bioterrorism_110509_07.pdf (accessed 31 January 2012)

Schroeder C. M., Jensen E., Miliotis M. D., Dennis S. B. and Morgan K. M. (2007). Microbial risk assessment. In: Infectious Disease: Foodborne Diseases, S. Simjee (ed.), Humana Press, Totowa NJ, USA, pp. 435–455.

Shea D. A. and Gottron F. (2004). Small-scale Terrorist Attacks Using Chemical and Biological Agents: An Assessment Framework and Preliminary Comparisons. CRS Web, Washington, DC, USA, pp. 1–81.

Sinclair R., Boone S. A., Greenberg D., Keim P. and Gerba C. P. (2008). Persistence of category a select agents in the environment. *Applied and Environmental Microbiology*, **74**(3), 555–563.

Stanfield G., LeChevallier M. and Snozzi M. (2003). Treatment efficiency. In: Assessing Microbial Safety of Drinking Water: Improving Approaches and Methods, A. Dufour, M. Snozzi, W. Koster, J. Bartram, E. Ronchi and L. Fewtrell, (eds), IWA Publishing, London, UK, pp. 159–178.

Stephenson J. B. (2004). Experts' views on how future federal funding can best be spent to improve security. Report GAO-04-1098T, United States General Accounting Office, USA.

Teunis P. F. and Havelaar A. H. (2000). The Beta Poisson dose-response model is not a single-hit model. *Risk Analysis*, **20**(4), 513–520.

Thoeye C., Eyck K. V., Bixio D., Weemaes M. and Gueldre G. D. (2003). Methods used for health risk assessment. In: State of the Art Report Health Risks in Aquifer Recharge Using Reclaimed Water, R. Aertgeerts and A. Angelakis (eds), World Health Organization, Copenhagen, Denmark, 123–152.

Tuduk T. (2004). Homeland security quantitative risk analysis. *The ISSA Journal*, **2**(1), 14–18.

US Army (1998). Biological Warfare Agents as Potable Water Threats, Medical issues Information paper No. IP-31-017, US Army Center for Health Promotion and Preventive Medicine, Aberdeen Providing Ground, MD, USA. https://www.ncjrs.gov/App/Publications/abstract.aspx?ID=191802 (accessed 31 January 2012)

USDA and EPA (2011). Microbial Risk Assessment Guideline: Pathogenic Microorganisms with Focus on Food and Water, Report DRAFT Version 5.3, Interagency Microbiological Risk Assessment Guideline Workgroup, U.S. Department of Agriculture's Food Safety and U.S. Environmental Protection Agency, USA.

ViTAL (2007). Integrated Monitoring and Control of Foodborne Viruses in European Food Supply Chains), FP projekt KBBE-2007-2-4-03. www.eurovital.org (accessed 31 January 2012)

WHO (2004). Public Health Response to Biological and Chemical Weapons: WHO guidance. 2nd edn, World Health Organization, Geneva, Switzerland. http://www.who.int/csr/delibepidemics/biochemguide/en/ (accessed 31 January 2012).

WHO (2011). Guidelines for Drinking-Water Quality. 4th edn, World Health Organization, Geneva, Switzerland.

Willis H. H., Morral A. R., Kelly T. K. and Medby J. J. (2005). Estimating Terrorism Risk. RAND, Santa Monica, CA, USA, pp. 1–66.

Winston G. and Leventhal A. J. (2008). Unintentional drinking-water contamination events of unknown origin: surrogate for terrorism preparedness. *Water Health*, **6**(Suppl 1), 11–9.

Woo G. (2008). Terrorism risk. In: Wiley Handbook of Science and Technology for Homeland Security, John G. Voeller (ed.). Wiley-Blackwell, Hoboken, NJ, USA, 1–17.

Chapter 3

Sample collection procedures for Online Contaminant Monitoring System

Miloslava Prokšová, Marianna Cíchová and Lívia Tóthová

3.1 INTRODUCTION

A reliable supply of safe and good drinking water quality is an important requirement for well-being and economic development of society. One of the main water utilities concerns is to protect water supply systems from damages caused by deliberate or accidental interventions. Due to the unpredictability of such events water companies can only hardly prevent them and find appropriate strategy for elimination of their consequences. To manage such accidents it is necessary to have all available information on hazards identification and subsequent remedy.

A bioterrorism attack is the deliberate release of viruses, bacteria, or the other germs (agents) used to cause illness or death in people, animals, or plants. These agents are typically found in nature, but it is possible that they could be changed to increase their ability to cause disease, make them resistant to current medicines, or to increase their ability to be spread into environment. Biological agents could be spread through the air, through water, or introduced in food. Terrorists may use biological agents because they can be extremely difficult to detect and do not cause illness for several hours to several days. Some bioterrorism agents, like the smallpox virus, can be spread from person to person and some, like anthrax, can not (http://www.cdc.gov/; Khan *et al.* 2001; Arendt, 2003).

Water control technologies are indispensable for the production of safe drinking water. They allow for the surveillance of source water quality and the detection of biological and chemical threats, thus defining the boundary conditions for the subsequent treatment and providing earlywarning in case of unexpected contaminations. They are mandatory for the permanent control of the treatment process and the efficacy of each single treatment step, and they safeguard the high quality of finished water.

Methods of microbiological contamination monitoring rely on periodic water sampling and analysing that typically take hours or days to evaluate. These methods are usually sufficient for compliance monitoring but are inadequate for early warning systems because when the results are known a considerable amount of the contaminated water could be consumed. Managers of water systems need to have very quick information on the type and extend of contamination so proper protective actions could be implemented. Therefore a ready and reliable device for identification of biological threads could be a useful tool in managing such events.

3.2 MICROBIAL MONITORING OF DRINKING WATER

Water supply utilities are responsible for delivering safe and good drinking water that does not have adverse health effect to consumers. In the case of a terrorist attack water supply operators need to have very quick information on the type and extend of contamination so proper protective actions can be implemented. Therefore a time consuming process of microorganisms identification should be completed by a supporting system that immediately indicates disruption of a water supply system physical integrity and tampering with a water quality (on-line monitoring of surrogate parameters).

The assessment of the microbiological quality of water is the key priority for both water suppliers and surveillance agencies. Microbiological quality is of principal concern because of the acute risk to health posed by viruses, bacteria, protozoa and helminths in drinking-water (Burrows & Renner, 1999). Therefore, monitoring and assessment of drinking-water is primarily a health-based activity which emphasises the protection of public health through ensuring that the water supplied is of a good quality.

It is important to know about the quality of the drinking-water supplied to a community to be sure that it is safe to drink. According to Council Directive 98/83/EC of 1998 on the quality of water intended for human consumption Member States shall take the measures necessary to ensure that water intended for human consumption is wholesome and clean. For the purposes of the minimum requirements of this Directive, water intended for human consumption shall be wholesome and clean if it:

(a) Is free from any micro-organisms and parasites and from any substances which, in numbers or concentrations, constitute a potential danger to human health, and
(b) Meets the minimum requirements set out in Table 3.1 in view of microbiological quality.

Table 3.1 Microbial parameters and parametric values (Anon, 1998).

Parameters	Parametric value (number/100 ml)
Escherichia coli (*E. coli*)	0
Enterococci	0

Samples should be taken so that they are representative of the quality of the water consumed throughout the year. Member States shall ensure that additional monitoring is carried out on a case-by-case basis of micro-organisms for which no parametric value has been set, if there is reason to suspect that they may be present in amounts or numbers which constitute a potential danger to human health.

Water suppliers have to be aware of the contaminants that could be in a water source before the water is treated. Monitoring of water during and after treatment will provide information about the effectiveness of any treatment processes. Monitoring the water will also show whether the water complies with drinking water standards.

Microorganisms are living organisms. In addition, when they are introduced into water, they do not form a perfect solution, but a suspension with an inherent degree of variability. And so, appropriate sampling is essential to provide representative samples to the laboratory in charge of testing. Sampling features depend on the objective of sampling, but also on the nature of the sample.

International standard ISO 19458 provides guidance on planning water sampling regimes, on the selection of sampling locations and the collection, on sampling procedures for microbiological analysis and on transport, handling and storage of samples until analysis begins. It focuses on sampling for microbiological investigations. Effective monitoring of drinking water requires collaboration between sampling programme designers, water treatment plant and distribution system operators, sample collectors, laboratory analysts and data users. Specific sampling protocols can vary widely in accordance with different purposes and different analytical methods.

Examples of sampling purposes include:

(a) Checking of drinking water to ensure compliance with national and/or international regulations (e.g. WHO Guidelines for Drinking Water Quality and the EU Drinking Water Directive);
(b) Determination of the efficiency of a drinking water treatment plant or components there of (e.g. disinfection);
(c) Quality monitoring of the water leaving the treatment plant;
(d) Quality monitoring of the water within the distribution system (including distribution within large buildings);
(e) Search for the cause of contamination of the distribution system (e.g. in response to customer complaints);
(f) Monitoring of the corrosive potential of drinking water to plumbing;
(g) Assessment of the effects of materials in contact with water on the water quality (chemical and biological);
(h) Monitoring of the influent water and the various processing stages in a food or beverage processing plant, including necessary treatment steps.

The sampling points for monitoring shall be determined by the competent authorities. Member States should take samples at the points of determined for control to ensure that water intended for human consumption meets the requirements of the Directive 98/83/EC. However, in the case of a distribution network, a Member State may take samples within the supply zone or at the treatment works for particular parameters if it can be demonstrated that there would be no adverse change to the measured value of the parameters concerned.

The number of microbiological samples taken and their frequency will vary depending on resources and population size, but as far as possible samples should be taken at least quarterly by the supplier and at least annually by the surveillance agency. The number of samples taken is largely dependent on population supplied, time available, analytical resources and type of distribution network. However, the more samples that are taken mean the more representative the results. In Table 3.2 is a very crude guide for the minimum number of samples that should be taken:

Table 3.2 Minimum frequency of sampling and analyses for water intended for human consumption supplied from a distribution network or from a tanker or used in a food-production undertaking (Anon, 1998).

Volume of water distributed or produced each day within a supply zone (Notes 1 and 2) m^3		**Check monitoring number of samples per year (Notes 3, 4 and 5)**	**Audit monitoring number of samples per year (Notes 3 and 5)**
	$\leq$100	Note 6	Note 6
>100	$\leq$1,000	4	1
>1,000	$\leq$10,000		1 + 1 for each 3,300 m^3/d and part thereof of the total volume

(Continued)

Table 3.2 Minimum frequency of sampling and analyses for water intended for human consumption supplied from a distribution network or from a tanker or used in a food-production undertaking (Anon, 1998) (*Continued*).

Volume of water distributed or produced each day within a supply zone (Notes 1 and 2) m^3		**Check monitoring number of samples per year (Notes 3, 4 and 5)**	**Audit monitoring number of samples per year (Notes 3 and 5)**
>10,000	100,000	$4+3$ for each 1,000 m^3/d and part thereof of the total volume	$3+1$ for each 10,000 m^3/d and part thereof of the total volume
>100,000			$10+1$ for each 25,000 m^3/d and part thereof of the total volume

Note 1: A supply zone is a geographically defined area within which water intended for human consumption comes from one or more sources and within which water quality may be considered as being approximately uniform.
Note 2: The volumes are calculated as averages taken over a calendar year. A Member State may use the number of inhabitants in a supply zone instead of the volume of water to determine the minimum frequency, assuming a water consumption of 200 l/day/capita.
Note 3: In the event of intermittent short-term supply the monitoring frequency of water distributed by tankers is to be decided by the Member State concerned.
Note 4: For the different parameters in Annex I, a Member State may reduce the number of samples specified in the table if:
(a) The values of the results obtained from samples taken during a period of at least two successive years are constant and significantly better than the limits laid down in Annex I, and
(b) No factor is likely to cause a deterioration of the quality of the water.
The lowest frequency applied must not be less than 50% of the number of samples specified in the table except in the particular case of note 6.
Note 5: As far as possible, the number of samples should be distributed equally in time and location.
Note 6: The frequency is to be decided by the Member State concerned.

3.3 SAMPLING PLAN

Sampling plan is one of the most important part of monitoring proposal. Before any sampling programme is devised, it is very important that the objectives of the programme are carefully established since they are the major factors in determining the position of sampling sites, frequency of sampling, duration of sampling, sampling procedures, subsequent treatment of samples, and analytical requirements. The degree of accuracy and precision necessary for the estimation of water quality concentrations sought should also be taken into account. The sampling programme should be designed to be capable of estimating the error in such values as affected by statistical sampling error and errors in analysis.

It is important to take into account all relevant data from previous programmes at the same or similar locations and other information on local conditions. Previous personal experience of similar programmes or situations can also be very valuable when setting up a new programme for the first time or in the cases of crisis situation.

Sampling plan is depend on analytical methods as well. Another plan is built if common methods are used in laboratory or if used on line monitoring system.

If a biohazard sensor is supported by a surrogate parameters monitoring, types of instruments and their locations are determined by parameters to be monitored, detection limits and accuracy of the data needed. Other important issues that have to be considered are: installation and operational requirements (periodic *versus* continuous sampling, data collection, communications, maintenance requirements, etc.), integration of the instruments with existing water quality monitoring systems, number of instruments that may be installed, and so on.

Local site conditions and supply system considerations that should be considered as well. They include easy access to the device site by authorized personnel because all instruments require periodic maintenance. This factor is directly related to the operations, maintenance and upgrading of the system. At the same time, the site should be secure against access of unauthorized persons. There should be available space for the instruments and auxiliary equipment. There are some other requirements to be considered, for example suitability of candidate device or sample collection method for the sampling site, including the discharge of waste stream, access to electricity power, data transfer and telecommunication equipment, physical security of the site to guard against unauthorized access or tampering. Hydraulic conditions at sampling sites are among the most important factors for the installation of instruments because turbulence in the pipe might affect sample collection or measurement. Good candidates for installing additional sampling instruments may be existing sampling sites for a basic or compliance monitoring.

3.3.1 What should the sampling plan include?

Generally for both approaches, the plan should include the following elements.

- Brief description of the water system that includes source, treatment, storage, distribution system maintenance, pressure zones, number of connections, population, and so on.
- Map of the distribution system with the routine and repeat sampling sites identified, distribution piping locations, entry points, and so on.
- Sample siting plan that includes sample site addresses, the minimum number of samples collected, rotation schedule of sample sites, chlorine residual monitoring, contact person and phone number, sampling procedure or protocol, and so on. In the event that a routine site sample tests positive for some pathogens, the plan should list repeat sites for each routine site and should include a written procedure of what steps the water system will follow to investigate a positive sample.

There are several system-wide and topological factors important for sensor locations selection, including:

- Localities with the highest potential of contamination entry (such as reservoirs, blow off valves, pump stations) due to the lack of physical security and ease insertion of contaminants;
- Probable types of contamination (according the risk analysis or educated guess);
- Contaminant transport time and concentration that also influence where and how many sensors need to be installed. Likely contaminant transport rates in the network (due to flow, dilution and decay), changes in contaminant properties due to bulk water properties, wall effects (pipe material, turbulence, biofilm) and mixing all affect the time required for the contaminant to reach consumers at a certain concentration;
- Instrument accuracy and detection limits;
- Vulnerable populations (such as the elderly, ill or children) at different parts of the network;
- Relative water demand and associated flow characteristics because of the temporal and physical characteristics of the network. Temporal factors include diurnal (e.g. morning *vs.* noon), daily (e.g. weekday *vs.* weekend) and seasonal (e.g. summer *vs.* winter) variations. Physical factors include pipe length, size, condition, material, accessories, bends, T-s, and so on.
- Frequency of sampling (periodic *vs.* continuous) and the amount of data collected and analysed.

Programmes for investigation of causes of contamination should be designed to determine the characterization of polluting discharges of unknown origin. They are generally based on knowledge of the nature or natures of the contaminants, and the coincidence of the periodicity of the appearance of contamination and of sampling. These criteria necessitate that the sampling, in contrast with that carried out for water quality management and quality characterization, should be carried out with a fairly high frequency in relation to the frequency of appearance of contamination. Inventory sampling from a large number of locations is often found to be useful in locating undocumented sources of contaminants.

The knowledge about whole distribution network is needed before preparation of sampling plan. There are information such as water treatment condition, source of drinking water, water tretment construction. Under the terms these information the important part belong to flow conditions in view of bioterroristic attack.

Flow measurements for water quality purposes

Contamination loads cannot be assessed without flow measurements. This subclause indicates the flow principles that should be taken into account when setting up a sampling programme.

There are five aspects of flow that need to be measured, namely,

(a) Flow direction,
(b) Flow velocity,
(c) Discharge rate,
(d) Flow structure,
(e) Cross-sectional area.

Flow direction

The flow direction is self-evident in most inland watercourses, but in navigation canals and drainage channels. The flow direction can vary with time and there is a possibility of reversal of direction and even counter flow situations. Rivers can also show reverse flow in eddies or in other circumstances. Knowledge of the pattern of groundwater flow within an aquifer is of primary importance for assessing the consequences of aquifer contamination and for sites selection for sampling boreholes. The pattern of water movement in tanks affects the mixing of the contents in treatment processes and the settling of suspended matter should be taken into account to ensure that representative samples are collected.

Flow velocity

Current velocity is important for

- Calculating the discharge rate,
- Calculating the mean velocity or time of travel which, for water quality purposes, is the time required for a given body of water to move through a given distance,
- Assessing the effect of turbulence and the mixing of a water body produced by velocity.

Discharge rate

The discharge rate is the volume of liquid that passes a given point per unit time. Information on the mean and on extreme rates of discharge is essential for the design and operation of water treatment plants.

Flow structure

The structure of the flow can strongly influence the rate of mixing vertically and laterally. Care should be taken to assess whether flow is in one confined channel, in several channels (i.e. braided) and whether or not eddies are present. Ideally, samples should be collected from a single, well-mixed channel; observations of flow structure in multiple channels and eddies, for example, suggest that samples might not be representative.

Cross-sectional area

Sampling cross sections can range from being approximately rectangular to having a deep channel at one edge, from shallow and wide to narrow and deep. These features affect both mixing and erosion, and they can change over time in natural streams and man-made channels.

Sampling sites

The sampling sites shall provide representative characteristics and account for any vertical, horizontal and temporal variations and shall be identified precisely following the general recommendations of ISO 5667-1 and ISO 5667-5, taking into account additional aspects specific to microbiology. The sampling sites should provide adequate coverage of the distribution network and pressure zones. It is also important to select sampling sites that provide the least amount of negative influence on the water sample. Many water systems utilize dedicated sampling stations in the distribution system that are used exclusively for sampling purposes, eliminating many outside influences that may potentially impact water samples.

Sampling points where conditions are unstable should be avoided, and the heterogeneity of the hydraulic system shall be taken into consideration. In studies on the efficacy of disinfection, the sampling point shall be chosen to ensure that the reaction is complete. All the points of a network are not equivalent, as there may be dead ends and sections where the flow is reduced, particularly if the network is fed from two sources. The quality at the outlet of a well-mixed tank is generally the same as in the body of water, but can be quite different from the inlet.

Sampling equipment

The sample equipment should be designed to preserve the composition of the sample from losses due to adsorption and volatilization, or from contamination by foreign substances.

The sample equipment used to collect and store the sample should be chosen after considering, for example, resistance to temperature extremes, resistance to breakage, ease of good sealing and reopening, size, shape, mass, availability, cost, potential for cleaning and re-use, and so on.

In addition, the sample equipment used to collect and store the samples should be selected by taking into account the following predominant criteria:

(a) Minimalization of contamination of the water sample by the material of which the container or its stopper is made, for example, leaching of inorganic constituents from glass (especially soft glass) and organic compounds and metals from plastics and elastomers (plasticized vinyl cap liners, neoprene jackets);
(b) Ability to clean and treat the walls of the containers, to reduce surface contamination by trace constituents such as heavy metals or radionuclides;
(c) Chemical and biological inertness of the material of which the container is made, in order to prevent or minimize reaction between constituents of the sample and the container;

(d) Sample containers which can also cause errors by adsorption of chemical determinants. Trace metals are particularly liable to this effect, but other determinants (e.g. detergents, pesticides, phosphate) can also be subject to error.

Sampling pipes are generally used in automatic sampling to supply samples to continuous analysers or monitors. During the residence time within the pipe, the sample can be considered as being stored in a container having the composition of the sampling line. Guidelines for the selection of materials for sample containers also, therefore, apply to sampling pipes.

Guidance on sample containers for microbiological examination is detailed in ISO 5667-16 and ISO 19458. Sample containers should be able to withstand the high temperatures that occur during sterilization. During sterilization or sample storage, the materials should not produce or release chemicals that could inhibit microbiological viability, release toxic chemicals or encourage growth. The samples should remain sealed until opened in the laboratory, and should be covered to prevent contamination.

Contamination prevention (Sampling contamination sources)

Contamination prevention during sampling is essential. All possible sources of contamination should be taken into account and the appropriate control applied if necessary.

Potential sources of contamination can include the following:

(a) The residue of earlier samples remaining on sampling containers, funnels, scoops, spatulas and other equipment;
(b) Contamination from the sampling site during sampling;
(c) Residual water in or on ropes, chains or extension handles;
(d) Contamination of funnels from preserved samples;
(e) Contamination of bottle caps or tops by dust or water;
(f) Contamination of the barrel of syringes and through filter medium;
(g) Contamination from hands, fingers, gloves and general handling;
(h) Contamination from internal combustion exhaust;
(i) Inappropriate sampling devices, bottles and filtration devices;
(j) Degraded reagents.

3.4 NEW SECURITY APPROACHES FOR DRINKING WATER

Generally, water supply systems are considered as very vulnerable to deliberate chemical and/or biological contamination. One of the reasons is a large scale of water systems where protection measures should cover large areas of water sources, raw water mains, treatment plants and distribution systems. The major protection against sabotage acts, including destruction of infrastructure of a water supply system, is a system's robust construction, so that it is able to operate in the case of damage or natural disaster. It is especially important in large supply zones (big cities), where water supply should provide a multiple protection of reliable water quality within the entire system – from water treatment plants, through water reservoirs to interconnected distribution network.

Drinking water can become contaminated at water source, during treatment and in the distribution system. Dealing with an accidental or deliberate contamination of water requires complex knowledge on contamination risks at different parts of water supply system and efficient methods of the risk management. Recently new management method was introduced for drinking water quality assurance and reliable water supply that is based upon holistic approach to the water system from the catchment area to the consumer. This approach is called *Water Safety Plan* (WSP) in the WHO (2004). In documents published by the European Commission a term *risk assessment and risk management* or HACCP (*hazard analyzis and critical control points*) is used for the similar approach.

To apply such an approach managers of a water supply system should have a good understanding of a system to carry out a risk analysis of the whole system (catchment area – water source – treatment – distribution) and identify points and conditions of the system that could affect drinking water quality. Based on this analysis a water safety plan is prepared with identification of risk points, ways of water quality protection and control as well as required prevention and remedial measures. The main objective of WSPs is an introducing of good water supply practice by protection of water sources against contamination, reduction or elimination of contamination in treatment processes and prevention of water contamination during accumulation and distribution. Integral parts of WSP are management procedures and measures for normal operating conditions and for conditions when quality of water could be endangered, including accidental and/or deliberate contamination. Therefore WSP is considered as a useful basic tool for any hazard management including biohazard (Deininger & Meier, 1998; Geldriech, 1998; Sekheta, 2006).

The identification and implementation of control measures should be based on the multiple-barrier principle. The strength of this approach is that a failure of one barrier may be compensated by effective operation of the remaining barriers, thus minimizing the likelihood of contaminants passing through the entire system and being present in sufficient amounts to cause harm to consumers.

The vulnerability of an individual water supply system depends on specific local conditions and protection measures should be tailored according the real situation. Nevertheless, there are some general features of water supply security that are shortly discussed below focusing on protection against biological threats.

3.4.1 Water sources

A security of water sources depends on the case of access to the source and on the type of subsequent water treatment. There are two types of water sources: surface water sources (lakes, reservoirs, river intakes) and groundwater sources (springs, boreholes, wells). Surface water sources are more vulnerable to a contamination (both biological and chemical) due to the direct access to the water body. However, the dilution effect of a large volume of water is the reason, why contamination of reservoirs would not likely produce a large risk to public health. Furthermore, surface water used to be treated by more or less complex treatment processes that could serve as an effective barrier against many types of deliberate contamination. On the other hand, ground water from shallow catchment areas and wells where treatment is not provided could present in the case of contamination a more serious human health threats.

Resource and source protection can be considered as a first barrier in protection of drinking-water quality. Effective catchment and source protection management has many benefits. By decreasing the risk of the source water contamination, the amount of treatment required is reduced. This may reduce the production of treatment by-products and minimize operational costs. However, introducing effective protection measures against intentional contamination can be at water sources difficult, if not impossible. For example, prevention of unauthorized persons access by increasing physical security (guards and fences) or on-line water quality monitoring are not likely to be cost-effective.

3.4.2 Raw water transport

Pipes delivering raw water to the treatment plant can be relatively easy targets to terrorist attacks including deliberate biological contamination. However, subsequent water treatment steps serve as a multiple barrier against contamination intrusion into the distribution system and reduce (but not eliminate) public health threats.

Measures to minimize risks of raw water pipes contamination are similar as at water sources: protection against unauthorized persons access, assurance of pipes physical integrity.

3.4.3 Treatment plants

Water treatment plants have a crucial role in the security of the whole water supply systems. Well operating and flexible water treatment could serve as one of effective measures in reducing up-stream water contamination. On the other hand, failures in protecting water safety in treatment plants can be only hardly remedied in subsequent parts of supply systems. Various filtration processes are used in drinking-water treatment, including granular, slow sand, precoat and membrane filtration (microfiltration, ultrafiltration, nanofiltration and reverse osmosis). With proper design and operation, filtration can act as a consistent and effective barrier for microbial pathogens and may in some cases be the only treatment barrier (e.g. for removing *Cryptosporidium* oocysts by direct filtration when chlorine is used as the sole disinfectant) (Lindquist *et al.* 2007).

Application of an adequate level of disinfection is an essential element for most treatment systems to achieve the necessary level of microbial risk reduction. Taking into account the level of microbial inactivation required for the more resistant microbial pathogens through the application of the Ct concept (Ct-product of disinfectant concentration and contact time) for a particular pH and temperature ensures that other more sensitive microbes are also effectively controlled.

Considering bioterrorism, disinfection step could provide an effective barrier. The most commonly used disinfection is chlorination. To a lower extent, ozonation, UV irradiation, chloramination and application of chlorine dioxide are also used. Chlorination is effective against many pathogenic biological agents. However, some of very dangerous pathogens are chlorine-resistant (e.g. sporulating bacteria). Ozonation is generally more effective against biological contamination, but does not provide any residual protection of a distribution system as it is by chlorination. These methods are very effective in killing bacteria and can be reasonably effective in inactivating viruses (depending on type) and many protozoa, including *Giardia* and *Cryptosporidium*. For effective removal or inactivation of protozoal cysts and oocysts, filtration with the aid of coagulation/flocculation (to reduce particles and turbidity) followed by disinfection (by one or a combination of disinfectants) is the most practical method (Polaczyk *et al.* 2008; Hill *et al.* 2005; Rajal *et al.* 2007).

3.4.4 Service reservoirs and distribution

Maintaining good water quality in the distribution system depends on the supply system design and operation and on maintenance and survey procedures to prevent contamination. Because of the nature of the distribution system that usually include many kilometres of pipes, numbers of storage tanks and interconnections with users, there is a great potential for tampering and vandalism and consequently opportunities for microbial and chemical contamination.

If contaminated water contains pathogens or hazardous chemicals, it is likely that consumers will be exposed to them. Even where disinfectant residual is used to limit microbial re-grows in the distribution system, it may be inadequate to overcome the deliberate contamination or may be ineffective against some or all of the pathogen types introduced. As a result, pathogens may occur in concentrations that could cause infection and illness of water consumers.

Protection of the physical integrity of distribution system is essential for providing safe drinking water. Water distribution systems should be fully enclosed, and storage reservoirs and tanks should be securely roofed with external drainage to prevent contamination. Control of short-circuiting and prevention of stagnation in both storage and distribution contribute to prevention of microbial growth. A number of strategies can be adopted to maintain the quality of water within the distribution system, including use of backflow prevention devices, maintaining positive pressure throughout the system and implementation of efficient maintenance procedures. It is also important that appropriate security measures are put in place to prevent unauthorized access to or interference with the drinking-water system infrastructure.

3.5 NEW APPROACH OF ONLINE CONTAMINATION MONITORING DEVICE (OCMD)

Common cultivation methods of microbiological contamination monitoring rely on periodic water sampling and analysing that take hours or days to evaluate. These methods are usually sufficient for compliance monitoring but are inadequate for early warning systems because by the time the results are known a considerable portion of the contaminated water could be consumed. New technologies of pathogen detection differ significantly from those currently used for conventional water quality monitoring. In developing pathogen detection systems, efforts have mainly focused on the detection of pathogen genome with high sensitivity and accuracy. Polymerase chain reaction (PCR)-based methods are applied most frequently to detect and identify pathogens according to unique genomic DNA. However, the conventional PCR approach typically involves a minimum of 6 h of tedious labour in carrying out the sample processing, DNA extraction, DNA amplification and agarose gel electrophoresis. In the past decade, the development of micro-fabrication technology has prompted significant advances in many fields, particularly in the miniaturization of chemical and genomic analysis devices. The micromachined analytical system, which integrates sample collection, sample concentration and pretreatment with the DNA extraction, amplification, hybridization and detection, can be realized by combining functional microfluidic components. Many such devices for different types of environmental samples have been reported in the literature, including micro-PCR chips (Northrup *et al.* 1993), micro-DNA chips (Fan *et al.* 1999), micro-DNA biosensor (Kwakye *et al.* 2006) and so on.

Technical specifications of new OCMDs should be specifically designed for the detection water contamination during treatment and distribution to consumers. Some important factors should be considered in connection with preparatory procedures, water sampling or water intake, filtration, purging of captured microorganisms, lysing and DNA extraction methods, detection of microorganisms, measurement and signal evaluation by a sensor.

Service reservoirs or water towers are regarded as the most vulnerable parts of water supply systems. It can be reasonably assumed that they would be the most probable targets of a terrorist attack. Design of a water sampling part of the OCMD should take into consideration specific construction and operating conditions of service reservoirs/water towers (or other sampling points). Two possibilities could be considered for water sampling from reservoirs, sampling from under the water level, or sampling on the discharge from the reservoir.

Usually, at neither of these sampling points there is a pressure sufficient for the filtration of required water volume in reasonable time through membrane filters (or an other chosen filtration device). Therefore, it would be necessary to use a pressurized sampling device.

The requirements for the filtration arise from the size differences of used microorganisms (viruses, bacteria, protozoa). The filtration could be made by three ways depending on the necessity to separate different types of microorganisms:

- Set of three separate membrane modules for each filtration membrane with a specific pore size;
- One filtration module with a membrane with the smallest pore size;
- Microfiltration or ultrafiltration membrane units for separation of particles from 0.10 μm and 0.005 μm, respectively.

If mutual separation of captured microorganisms is required a one-step filtration can be sufficient that would be optimal for the minimization of a handheld device. A membrane with the pore size 0.22 μm or a microfiltration or ultrafiltration membrane unit can be used for the filtration (Clancy *et al.* 1998; Schaub *et al.* 1993).

It is important to know that microfiltration and ultrafiltration technique requires higher operating pressures (0.03–0.3 MPa and 0.1–0.6 MPa, respectively).

3.5.1 Location of OCMD in water system

Contamination events, whether accidental or deliberate, can affect any part of a water system. OCMD should provide coverage of all parts of the system in which contamination presents a risk. OCMD should provide detection of contamination events, location information and necessary laboratory analyses to identify and measure specific contaminants. An ideal but very expensive solution (number of samples, contaminants, equipment of instruments) would be to locate sampling sites at all important nodes of a distribution system vulnerable to contamination. Parts of a water supply system where a monitoring device can be located are presented in the Table 3.3.

Table 3.3 Parts of a water supply for potential locations of monitoring devices.

Location	Threat of intentional contamination	Time available for warning and response	Request for locating of device
Source waters	Relatively low	Large quantities of contaminant needed – therefore it is easy to detect	Covers large part or all of system
Raw water transport	Slightly higher than for sources		Covers large part or all of system
Treatment plant	Slightly higher than for source or transport; insider treads higher		
Finished water reservoirs	Considerably higher		Many of locations require coverage
Early distribution system	Moderate, particularly at sites to which access can be gained (valves, pumps and check points)	Relatively long	Several locations to cover entire system

(*Continued*)

Table 3.3 Parts of a water supply for potential locations of monitoring devices (*Continued*).

Location	Threat of intentional contamination	Time available for warning and response	Request for locating of device
Mid distribution system	Higher, covers many of the likely contamination entry points (valves, pumps, inspection and sampling ports)	Moderate to little	Multiple locations to get full coverage
Entry pipes for likely targeted customers	Higher risk area; except better cooperation from such customers	Very little	–

There are several locations on a water supply system that have a higher risk of a terrorist attack and are more proper for the location of monitoring device, among them especially:

- Incoming water supply line where there is a greater certainty of impact and economy of use of a likely contamination (typically a high security government facility or a target of high iconic value);
- Pump station discharges where there is the greatest assurance of having the contaminant effectively and broadly distributed. Pump stations have also some advantages as places for monitoring device location in terms of power supply, telecommunication and other infrastructure needed to support a device;
- Storage facilities (service reservoirs and/or water towers) where water is usually not under pressure and contaminant introduction would be easier. However, its effect is less certain, particularly if the water exchange from the storage facility is slow.

In selecting sites for online contamination monitoring device location it is useful to identify as many candidate sites as practical. It includes identification of points with the highest vulnerability to contamination using the risk analysis methodology. The assessment should include evaluation of accessibility of the insertion point as well as the sensitivity of the affected population. If a utility is not restricted in its resources, best locations can be selected by a hydraulic analysis of a water supply system. A functional and well-calibrated network model can be useful for the determination of dominant contaminant pathways. Alternatively, when a model is not available, knowledge of the system and educated guesses on dominant pathways and device locations might be used.

An assessment of the parameters discussed above leads to the identification of the following main factors for consideration in selection of type of instrument and its location:

- Minimize contamination detection time for a given number of sensors, *versus* minimize the number of sensors for a specified time of detection;
- Maximize monitoring coverage for all consumers versus maximize coverage for vulnerable (at risk) consumers such as populations at schools, nursing homes, hospitals, and so on;
- Continuous monitoring versus periodic monitoring;
- Automated sampling versus manual sampling;
- Instrument life cycle costs;
- Instrument ease of use and maintenance by utility.

REFERENCES

Anon (1998). European Council Directive 98/83/EC of 3 November 1998 on the quality of water intended for human consumption.

Arendt N. (2003). Bioterrorism, cyberterrorism and water supplies. *Wisconsin Water Well Association Journal*, January 15. http://www.sehinc.com/news/company/seh-news/2003/jan/15/bioterrorism-cyberterrorism-and-water-supplies (accessed 15 August 2011)

Clancy J. L. and Fricker C. (1998). Control of cryptosporidium – how effective is drinking water treatment? *Water Qual. Int.*, **1**, 37–41.

Burrows D. W. and Renner S. E. (1999). Biological warfare agents as threats to potable. *Water Environ. Health Perspect.*, **107**, 975–984.

Deininger R. A. and Meier P. G. (1998). Sabotage of public water supply systems. *Proceedings of the NATO Advanced Research Workshop on Security of Water Supply*, Tihany, Hungary.

Fan Z. H., Mangru S., Granzow R., Heaney P., Ho W., Dong Q. and Kumar R. (1999). Dynamic DNA hybridization on a chip using paramagnetic beads. *Anal. Chem.*, **71**, 4851–4859.

Geldriech E. D. (1998). Microbial quality issues for drinking water. *Proceedings of the NATO Advanced Research Workshop on Security of Water Supply*, Tihany, Hungary.

Hill V. R., Polaczyk A. L., Hahn D., Narayanan J., Cromeans T. L., Roberts J. M. and Amburgey J. E. (2005). Development of a rapid method for simultaneous recovery of diverse microbes in drinking water by ultrafiltration with sodium polyphosphate and surfactants. *Appl. Environ. Microbiol.*, **71**, 6878–6884. http://www.bt.cdc.gov/ (accesseed 15 August 2011)

ISO 19458 (2006). Water quality – Sampling for microbiological analysis.

ISO 5667-1 (2006). Water quality. Sampling. Part 1: Guidance on the desing of sampling programmes and sampling techniques.

ISO 5667-5 (2006). Water quality. Sampling. Part 5: Guidance on sampling of drinking water from treatment works and piped distribution systems.

ISO 5667-16 (1998). Water quality – Sampling – Part 16: Guidance on biotesting of samples.

Khan A. S., Swerdlow D. L. and Juranek D. D. (2001). Precaution against Biological and Chemical Terrorism Directed at Food and Water Supplies. Public Health Reports. Vol. 116, January-February, pp. 3–13.

Kwakye S., Goral V. N. and Baeumner A. J. (2006). Electrochemical microfluidic biosensor for nucleic acid detection with integrated minipotentiostat. *Biosens. Bioelectron.*, **21**, 2217–2223.

Lindquist H. D. A., Harris S., Lucas S., Hartzel M., Riner D., Rochele P. and DeLeon R. (2007). Using ultrafiltration to concentrate and detect *Bacillus anthracis, Bacillus atrophaeus* subspecies *globigii*, and *Cryptosporidium parvum* in 100-liter water samples. *J. Microbiol. Methods*, **70**, 484–492.

Northrup M. A., Ching M. T., White R. M. and Watson R. T. (1993). DNA amplification with a microfabricated reaction chamber. In: Digest of Technical Papers: Transducers (*Proc. 7th mt. Conf. on Solidstate Sensors and Actuators*). Institute of Electrical and Electronic Engineers, New York, pp. 924–926.

Polaczyk A., Narayanan J., Cromeans T., Hahn D., Roberts J., Amburgey J. and Hill V. (2008). Ultrafiltration-based techniques for rapid and simultaneous concentration of multiple microbe classes from 100-L tap water samples. *J. Microbiol. Methods*, **73**, 92–99.

Rajal V. B., McSwain B. S., Thompson D. E., Leutenegger Ch. M., Kildare B. J. and Wuertz S. (2007). Validation of hollow fiber ultrafiltration and real-time PCR using bacteriophage PP7 as surrogate for the quantification of viruses from water samples. *Water Res.*, **41**, 1411–1422.

Sekheta M. A. F., Sahtout A. H., Sekheta F. N., Pantovic N. and Al Omari A. T. (2006). Terrorist threats to food and water supplies and the role of HACCP implementation a one of the major effective and preventive measures. *Internet J. Food Saf.*, **8**, 30–34.

Schaub S. A., Hargett H. T., Schmidt M. O. and Burrows W. D. (1993). Reverse Osmosis Water Purification Unit: Efficacy of Cartridge Filters for Removal of Bacteria and Protozoan Cysts When RO Elements are Bypassed. Rpt no TR9207, AD A266879. Biomedical Research and Development Laboratory, Ft. Detrick.

WHO and OECD (2003). Assessing Microbial Safety of Drinking Water. Improving Approaches and Methods.

WHO (2004) Guidelines for Drinking water Quality, Volume 1: Recommendations. 3rd edn, World Health Organization, Geneva.

Chapter 4

A device to extract highly diluted specimens out of large volumes of water for analysis in lab-on-a-chip detection systems

Christoph Zeis

4.1 THE NEED FOR A MACRO-TO-MICRO-FLUIDIC INTERFACE

4.1.1 The DINAMICS EU research project

In the context of the "Diagnostic Nanotech and Microtech Sensors" (DINAMICS) FP6 EU collaborative research project, an integrated system for microbiological water testing was proposed. Such automatic system must be able to autonomously sample a defined amount of drinking water and test for the presence of pathogens by searching for pathogen-specific nucleic acids. For the sake of significance of sampling and for increasing the chance of finding even relatively diluted pathogens, a large volume of water must be collected and concentrated in order to prepare a specimen to be tested in the nucleic-acids biosensor device which is part of the system.

This chapter deals with the technological choices and solutions available for the implementation of the first step of the DINAMICS pipeline, that dealing with the water concentration. This step aims to reduce a large volume of water (50–100 l) to the range of millilitres needed for further separation (see Chapter 6), lysis (see Chapter 5) and analysis methods (see Chapter 7).

4.1.2 Device needed for generating input for analytics

To generate representative water samples as input into the microfluidic diagnostic system, a special apparatus is needed. This apparatus, here defined as Inline Separator or Preconcentrator, needs to be able to increase the concentration of suspended material, viruses, bacteria, protozoa in a water sample to a minimum concentration above the detection limit of the employed biosensor. In this respect, it works as a so called macro-to-micro-fluidic interface. The required increase of concentration depends on the specified input volume of the detection device, the infectious dose of the most critical target organism, the detection limit of the sensor (or the minimum number of units to be input into the PCR, respectively) and the specified processing time. It is crucial to hit the optimum balance between those parameters and thus the preconcentrator is an essential module in the system architecture of the microfluidic detection device.

The following requirements were identified for the preconcentration device:

(1) Fast (30 minutes or less for concentration process)
(2) Concentration increase at least 1500 times

(3) Robust and Reliable (also working with highly turbid waters)
(4) Reusable (no need for parts exchange unless an alarm occurred)
(5) Eligible for Cleaning in Place (no dead volumes in fluid path, all wetted parts resistant to aggressive media)
(6) Covering most pathogen sizes from virus to protozoa
(7) Portable (at least by car to process surface water sources)
(8) Cost effective (<10,000 Euro)
(9) Sustainable (little or no chemical amendments)

The overall DINAMICS system is designed as an early warning system for potential water contamination and thus time is a crucial factor. The time for the entire detection process must be kept as low as possible. This results in strict requirements regarding the effectiveness of each process step. As an indication, the overall DINAMICS process is divided into 5 discrete steps.

(1) Preconcentration
(2) Lysis
(3) DNA Isolation
(4) PCR
(5) On chip detection

The most time consuming process in this chain could be PCR which takes about 30 minutes. All other steps will have to fit into this timeframe of 30 minutes.

Since many microbiological species are likely to be attached to suspended solids (sediment), discarding the sediment is not currently considered as an option before the preconcentration step. The chosen separation principle thus must be robust enough to also cope with waters with high turbidity. Several principles seem to be qualified for the separation: the following chapter will present some of these.

4.2 PRINCIPLES OF SEPARATION

The macro-micro-fluidic interface is that part of the system that reduces the volume of a water sample from several litres (up to 100 litres) to a few millilitres while retaining suspended pathogens, thus increasing their concentration.

Several approaches to concentration of large volumes of water can be found in literature resources none of these methods is *per se* qualified to be fully automated. However successful rapid concentration of bacteria, viruses and protozoan pathogens from large volumes of water is known from sundry sources of literature. Methods described comprise dead-end filtration (henceforward referred to as Normal Filtration NM), cross-flow filtration (Tangential Flow Filtration TTF) and also centrifugation or spin columns.

To assess the identified concentration methods as far as their eligibility for the macro-micro-fluidic interface is concerned, parameters have been introduced to describe and evaluate each method in a distinguished and comprehensive way. Focus has to be set on rapid response (low filtering and downstream processing time), reliable recovery of highly diluted pathogens (ability to collect and process large volumes of sample water) and reliable recovery of all kinds of pathogens (ability to collect different sizes simultaneously).

4.2.1 Survey of concentration methods

Among the concentration methods mentioned above the filtration methods employing Ultrafiltration with Hollow Fibre Filters are the most promising. At first a continuous centrifugation method was also taken into consideration but had to be discarded due to high cost, high weight and large dimensions.

Filtration with Hollow Fibre Filters can be done in two different setups:

- Dead End Filtration (Normal Flow)
- Cross Flow Filtration (Tangential Flow)

Dead end filtration/normal flow

A hollow-fibre dead-end ultrafilter can be used to filter water directly from the tap, driving the system with pressure also taken from the tap. Sample recovery is obtained in a separate step by a pump-driven back-flow. Kearns *et al.* (2008) report that such a system proves useful for concentrating Bacillus atrophaeus spores. The microorganisms were concentrated directly from the unmodified water stream, that is, no dechlorinating, surfactant or dispersing agents were added, while instead previous studies used finite sample volumes of water amended with chemical additives.

In the first step water is taken from the tap and enters the filter through the feed valve while the valve to the sample outlet is closed. Since the second end of the hollow fibre filter module is closed the water is driven through the filter membrane by tap pressure and the permeate is directed into waste. In the second step the feed and waste valves are closed and the organisms and particles retained in the filter membrane are eluted by using a pump to drive a backflush buffer in reverse direction through the filter and toward the sample outlet. A scheme is in Figure 4.1.

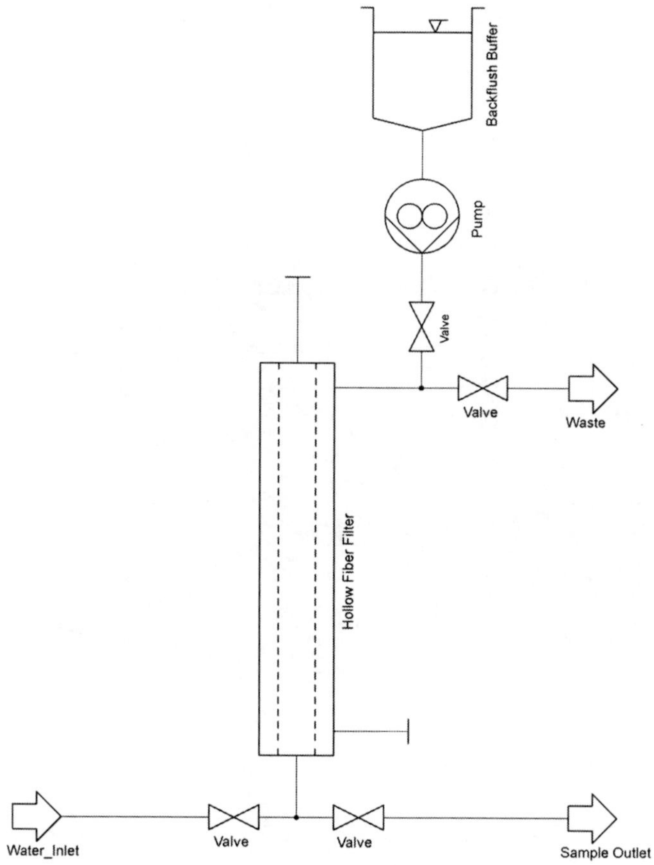

Figure 4.1 Schematic of a dead-end concentration system.

The advantages of this principle are its moderate hardware requirements. Since the pressure to drive the water through the membrane is directly taken from the tap there is no need for feed or circulation pumps as it is the case with the latter mentioned Tangential Flow Filtration. The only pump needed is the backflush pump but using a smart fluidic setup this could also be eliminated.

To verify this principle an experimental setup has been realized and evaluated (see Figure 4.2). The filter module employed was a Inge dizzer® S 0.9 MB 2.5 with pore size of 20 nm, a membrane area of 2.5 m^2 and an internal fibre diameter of 0.9 mm.

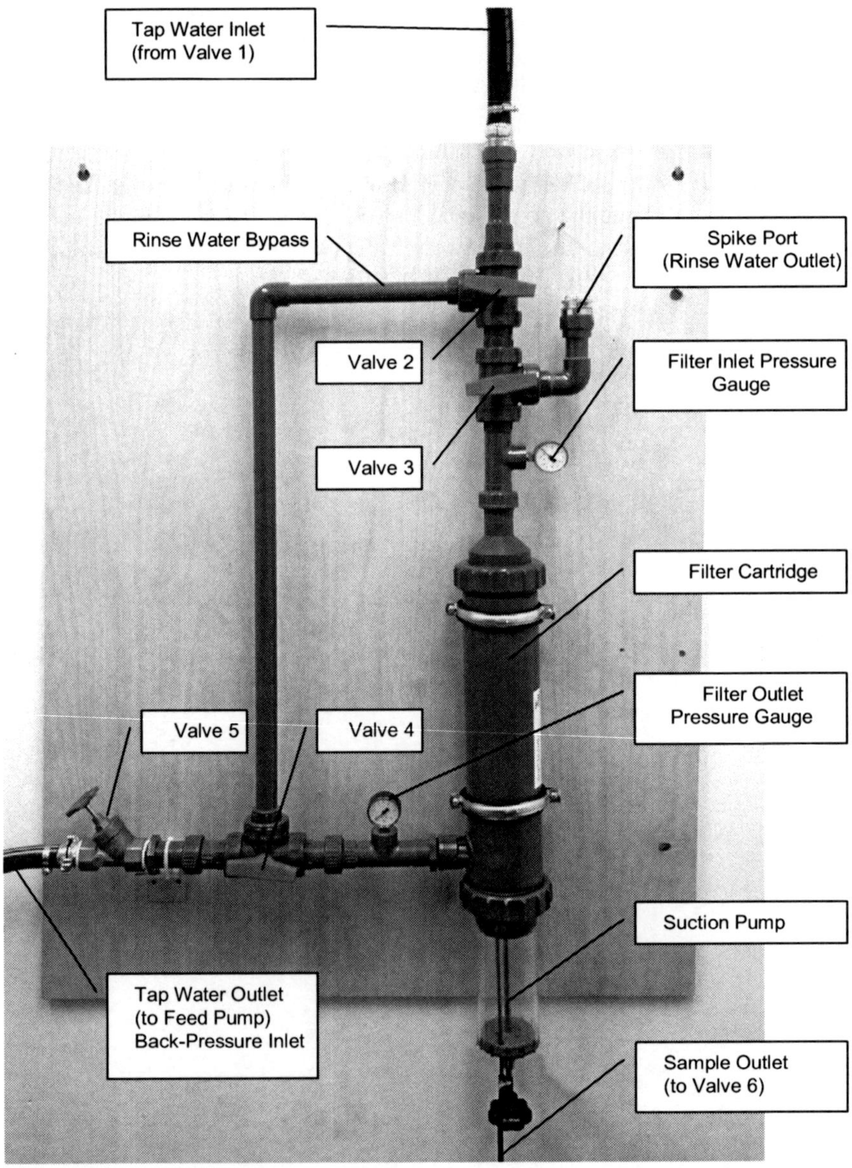

Figure 4.2 Dead end filtration: experimental setup.

As it turned out the disadvantage of this principle is that the membrane is very likely to clog. Since – following Bernoulli's law – the transmembrane pressure linearly decreases along each fibre from its feed side to its terminated end the highest flow through the membrane occurs at the bottom end of the filter module while the remote end only contributes little to the overall flowrate. Thus the membrane clogs beginning at the bottom while the remote end remain almost clean. When backflushing this effect is reversed. Most of the backflush flow crosses the membrane at its clean areas while there is almost no flow at the clogged parts. This makes backflushing very ineffective and leaves most retained particles and organisms stuck in the membrane.

Tests were conducted using live bacteria (*Escherichia coli*) which were spiked into the system at a given concentration. The eluted sample was analysed by cell culture and PCR. Although the simplicity of this filter setup was tempting the recovery results remained poor and thus this principle was discarded. However a prototype had already been drafted and will be presented later.

Cross flow filtration/tangential flow

The cross flow filtration follows a slightly different setup (see the scheme in Figure 4.3). Unlike dead end filtration the opposite end of the hollow fibre module remains open and the sample water is permanently circulated through the fibres by a pump. When closing the throttle at the filter outlet to a certain extent some back pressure is exerted and a small portion of water (approx. 5% of the circulation flow rate) is forced through the membrane leaving the system at the permeate outlet and going to waste while simultaneously being replaced by the same amount of water taken from the sample reservoir. Membrane clogging is prevented by the high velocities of the circulating water (1 m/s) which invokes the Pinch Effect. While each volume leaving the system as clear permeate water is being replaced by a volume of sample water taken from the sample reservoir the concentration of suspended particles and organism within the loop keeps increasing over time.

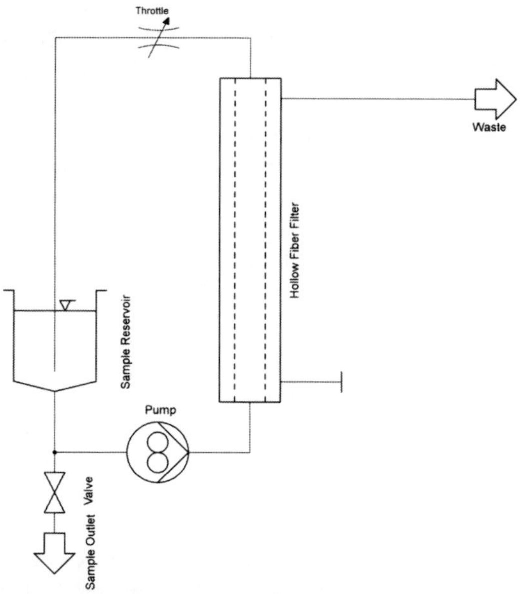

Figure 4.3 Scheme of a 10-liter ultrafiltration experimental setup.

After the desired amount of sample water has been fed into and through the system (or if the sample reservoir is empty) the pump is turned off and the entire water volume is drained through the sample outlet valve.

The major advantage of this principle is that the filter membrane is prevented from clogging due to the Pinch Effect which means that due to its high velocity along the membrane the circulating water cleans the surface while sedimented particles and organismns are most likely localized in an area between the center axis of each fibre and its wall. There is no need for a dedicated backflush mechanism. Instead the remaining water can be drained by gravity. The recovery rate of this setup is superior to the dead end filtration.

However, the disadvantage of this setup is that – compared to the earlier mentioned dead end filtration – it is quite energy consuming due the need for the circulation pump. Also the need for a sample reservoir which may easily exceed the volume of 100 liters must be considered problematic as far as system dimensions and portability are concerned.

As mentioned in many literature resources the recovery results can effectively be improved by using various chemical reagents as amendments to the sample water (i.e. Tween80, calf serum, fetal bovine serum, sodium thiosulfate, glycine, etc.). For an automated system, this necessity for various reagents appears to be disadvantageous and contradicts the requirements of easy fluidic handling and of sustainability. It also may be detrimental for some applications downstream of the preconcentrator. Hill and co-workers (2005) mitigated those disadvantages by using only one reagent (sodium polyphosphate). This reagent is able to fulfil several functions within this filtration setup, for example filter blocking, detaching bacteria from suspended particles, dispersing Cryptosporidium oocysts and Giardia cysts and keeping particles in suspesion. Last but not least it does not interfere with PCR unlike other potential amendments.

Continuous flow centrifugation (CFC)

As mentioned before Continuous Flow Centrifugation had also been taken into consideration for the Preconcentration Device. Zuckerman and co-workers (2006) survey the use of continuous flow centrifugation for monitoring waterborne protozoa from large volumes of various matrices. According also to apparatuses manufacturers, such as Beckmann-Coulter (Dorin, 2004), continuous flow centrifugation is most suitable for sedimentation of bacteria, larger subcellular particles and cell debris.

The most significant parameter for the CFC is the specific sedimentation coefficient of each particle or organism to be concentrated. This parameter establishes a correlation between the flowrate of sample through the centrifuge and the rotor speed. The lighter a particle is the higher the rotor speed has to be to force it to the centrifuge cushion or it takes more time respectively. While this still works fine with larger organisms like bacteria and protozoa it becomes more and more difficult when entering the virus domain. Since the Preconcentration Device is required to also concentrate viruses in a limited time frame, this principle had to be excluded.

4.3 THE NEEDLE IN THE HAYSTACK

The main challenge in designing and dimensioning a general purpose water preconcentrator is the vast discrepancy between the potentially infectious dose of pathogens and the limit of detection of a detector chip (see Table 4.1). Translated into a worst-case scenario, this results in numbers like: 10 pathogens ingested at once might already cause a threat whereas the detector needs at least 1000 units to create a reliable signal. To translate these absolute numbers into concentrations, we used the average amount of water a person uses (or should use) per day: 2 litres. This means that 5 pathogens per litre is the concentration the DINAMICS detector would have to respond to. We call that "the needle in the haystack."

Table 4.1 Infectious dose *vs.* detection limit of target organisms. Infectious doses from Burrows and Renner (1999) and Corlett and Stier (1990). If for the infectious dose a range was given, the table shows the more critical limit of this range (lower value). Detection limit presents an indicative value taken as a guideline for the DINAMICS research project.

Kingdom	Organism	Type	Infectious dose	Detection limit	Ratio
Prokaryotes	*Yersinia pestis*	Gram (−)	500	1000	2.00
Prokaryotes	*Clostridium perfringens*	Spore-forming	500.000	100000	0.20
Prokaryotes	*Salmonella* spp.	Gram (−)	15	1000	67
Prokaryotes	*Shigella* spp.	Gram (−)	10	10	1.00
Prokaryotes	*Vibrio cholerae*	Gram (−)	1000	100000	100.00
Prokaryotes	*Campylobacter* spp.		400	1000	2.5
Prokaryotes	*Listeria monocytogenes*	Gram (+)	1000	1000	1.00
Virus	variola major	dsDNA virus	10	10	0.00
Virus	Norovirus, Norwalk like calciviruses (NLV)	ssRNA virus	presumed to be low	10	0.00
Virus	Rotavirus	dsRNA virus	10	10	0.10
Virus	Hepatitis A	ssRNA virus	10	10	0.10
Virus	Hepatitis E	ssRNA virus	unknown	10	0.00
Eukaryotes	*Cryptosporidium parvum*		1	10	10.00
Eukaryotes	*Giardia lamblia*		1	10	10.00

So how much water would an apparatus have to process to deliver a representative specimen to the microfluidic analytic device? The maximum input volume for the first process in the chain, the cell lysis, was given with 1 ml. Without preconcentration of the water sample not even one pathogen would be captured.

4.4 TECHNICAL DESCRIPTION OF THE DINAMICS CONCENTRATION APPARATUS

The method of concentration is strongly dependent on the estimated amount of water to be processed. For a reliable result, all subsequent process steps of the detection device have to be taken into account. Without high concentration gain in the Inline Separator and without improvement of the subsequent analytical process, the required water volumes to be processed become unrealistically large. This can be worsened in case it is impossible to use the entire volume coming from a unit to feed the next one, so a fraction of the volume and its analyte content is lost.

Within the DINAMICS EU research project, the author (along with other personnel at Provenion Engineering Company) took care of the system architecture for the automatic pathogen detection system, the project plan and the coordination and execution of system integration. These tasks opened the opportunity to review and modify the system architecture in order to optimise its yield in collecting pathogens. The small and effective Inline Separator device presented below became possible as a result of one fundamental decision. It is the postulation that for each process step 100% of the particles/pathogens are to be delivered to the subsequent ones thus avoiding those aliquoting effects. On the other hand, as many discrete process steps as possible should be combined into one continuous step which also decreases the chance of losing target pathogens and simultaneously saving process time which is crucial for an early warning system. Only after these process parameters were adapted to these requirements it was possible to determine the needs for the Inline Separator and Macro Microfluidic Interface.

The required volume of water sample was reduced by magnitudes by four means:

(1) adding a second filtration stage that further concentrates the output of the first stage
(2) Increasing sample volume to be lysed from 1 ml to 20 ml
(3) Realizing continuous flow between Inline Separator and Lysis Chamber
(4) Realizing continuous flow between Lysis Chamber and DNA-Isolator

The internal volumes and efficiencies of those two stages were calculated in that way that 100% of the output of the second stage can be processed by the subsequent Lysis Chamber thus fulfilling the demand mentioned above.

The required volume of water sample could then be calculated to 32 litres which would be reduced to 20 ml thus increasing the concentration by factor 1600.

4.4.1 First draft of the dead-end filtration system

Since dead-end-filtration is very tempting as far as fast process time and automation potential is concerned experiments were conducted and a first prototype was designed simultaneously.

Efficiency calculation

To adapt the Inline Separator's output volume as close as possible to the volume processed by the Lysis Chamber it was necessary to consider several parameters for the filter cartridges. Hence, the corresponding calculations were conducted starting with the determined lysis input volume of 20 ml. Tables 4.2 and 4.3 show the estimated filter efficiencies per filter stage based on the parameters of the two filter cartridges identified to best match the requirements:

Table 4.2 Efficiency calculation stage 1.

Microza Module Polyacrylonitrile Membrane ACP-2053		
Parameter	**Value**	**Unit**
Residual volume	0.31	l
Filter area	0.60	m^2
Permissive filter load	300	l/m^2h
Time	24	min
Volume of water input	36	l
Flow-rate	1.50	l/min
Factor total water volume elution	3.00	times
Dead volume of filter end cap	50.00	ml
Dead volume between filter and valve	15.00	ml
Total dead volume	65.00	ml
Total eluted volume	1.00	l
Gain factor	36	

Table 4.3 Efficiency calculation stage 2.

Microza Module Polyacrylonitrile Membrane ACP-0013		
Parameter	**Value**	**Unit**
Residual volume	9,0	ml
Filter area	0,002	m^2
Permissive filter load	300	l/m^2h
Time	24	min
Volume of water input	1000	ml
Permeate flow-rate	43,0	ml/min
Permissible flow-rate	8,5	ml/min
Required flow-rate	41,7	ml/min
Factor total water volume elution	2,00	times
Dead volume of filter	2,00	ml
Dead volume between filter and valve	0,50	ml
Total dead volume	2,50	ml
Total eluted volume	20,50	ml
Gain	49	
Total gain	**1765**	

Draft

Figure 4.4 shows the first draft of a tap pressure driven dead end filtration device. Each stage comprises a hollow fibre ultrafiltration cartridge and a reservoir for its elution. The dimensions were calculated to fulfil each filtration step within 24 minutes which was supposed to be the cycle time for the entire detection device. The advantage of such a system would be that no energy is consumed to drive the fluid through the filter membrane except the pressure supplied through the water pipe itself.

Figure 4.4 First CAD draft of the two staged inline separator (dead end filtration).

Backflush and recovery issues employing dead end filtration

As promising as the dead-end-filtration might have been, experiments showed that there is an issue regarding the recovery of the retentate. Several attempts to recover the material that was retained by the filter membrane showed only little success. Thus dead-end filtration finally had to be discarded.

4.4.2 Tangential (cross-flow) filtration system

As the results of the first prototype were not satisfying, the design was changed towards tangential flow filtration, also referred to as cross-flow filtration. The underlying principle of this method is that the filter cartridge – and thus each of its hollow fibres – is open at both ends allowing the filtrate to circulate from the filter feed through the filter cartridge to its outlet port, back to the batch tank and again into the filter. By adding slight counter pressure to the outlet, a small portion of the filtrate is forced through the membrane thus leaving the system as permeate while the main portion of the filtrate keeps circulation along the membrane. This process persists until the batch tank is empty and only the circulation path is still filled with sample water, referred to as retentate.

One of the advantages of this principle is that there is no need for a filter backflush step since the particles remain in the circulation system. This also opens the chance to operate the system in a continuous mode. The main advantage is that there is a negligible risk of filter clogging. In case the fluid velocity along the filter membrane is high enough (> 1 m/s) the so called Pinch Effect guarantees that particles in the fluid will concentrate in an area between the pipe wall and the pipe centre axis. Hence, there is only little likelihood that particles come close enough to the membrane to clog it. On the other hand this principle is much more energy consuming than dead end filtration since the high fluid velocity along the filter membrane requires a circulation pump. In the case of the first filtration stage, the required velocity yields a flow rate of 30 litres/minute at a pressure of 1.8 bar(g). The pump employed by Provenion's prototype consumes 1 kW at these ratings. In Figure 4.5, the CAD drawing of the two-stage tangential flow filtration.

4.4.3 Steps towards a continuous working device

As mentioned above, the performance of the entire detection device can be improved by making processes continuous instead of batch. This paradigm was implemented by combining the second filtration stage of the Inline Separator with the Lysis Chamber and the first step of the DNA-Isolation into one single step.

The Inline Separator (Tangential Flow Version) itself started with a batch design. This means that at the beginning of each sample a batch vessel (max. volume 50 liters) would have to be filled with tap water which is then circulated through the filter as described above. However, this would contradict the postulation of a continuous working system. As a consequence, the first filtration stage was one last time redesigned to evaluate the continuous mode. Instead of using a 50 litre batch tank the water is directly taken from the tap and enters the system at the same flow rate as the permeate water leaves (1.8 litres/minute). This eliminates the need for a giant batch tank and thus results in a smaller design.

To emphasize the purpose of the Inline Separator acting as Macro-Micro Fluidic Interface, the two stages were built employing different technologies. The first (macro) filtration stage is mainly built with components also used in pharmaceutical industries (hygienic flanges, stainless steel, etc.). The second (micro) filtration stage comprises components known from medical or diagnostic devices (silicone tubing, pinch valves). Since the main purpose of the prototype described was to act as a demonstrator, it was set up as a flat and easy visible design. All components were mounted in one plane to increase transparency. The same components set up in a more sophisticated design should result in a device not larger than half the size of a dishwasher. Photographs of details of the system are presented in Figures 4.6, 4.7 and 4.8.

Figure 4.5 CAD image of the final two staged inline separator (tangential flow filtration).

Inline separator specifications

- Input
  - Volume: 36 liters
  - Minimum required concentration of target pathogen: 5 units/liter
- Filtering
  - Permeate Rate Stage 1: 90 liters/h (1.5 liters/min)
  - Retentate Volume Stage 1: 1 liter
  - Permeate Rate Stage 2: 43 ml/min
  - Retentate Volume Stage 2: 20 ml

Figure 4.6 1st filtration stage.

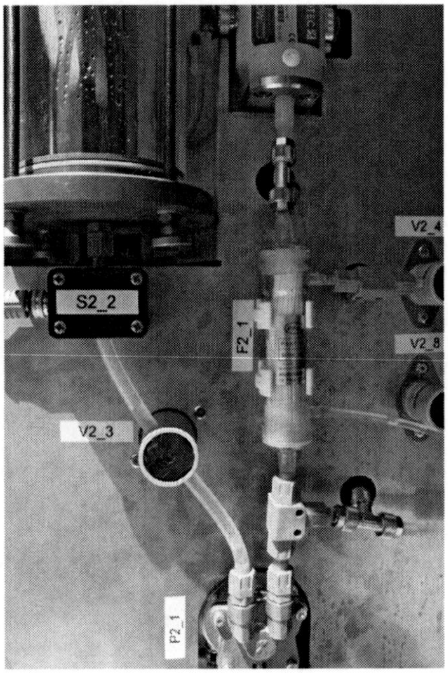

Figure 4.7 2nd filtration stage.

Figure 4.8 Circulation pump of 1st stage and vacuum pump for draining 1st stage and transferring to 2nd stage.

- Gain after 24 minutes of Cross-Flow-Filtering
  - Factor 36 in Stage 1
  - Factor 50 in Stage 2 (due to a better ratio of surface to internal volume)
  - Total gain 1800

- Output
  - o Volume: 20 ml at 1 ml/min
  - o Concentration: greater 8 units/ml (estimated required target concentration at Lysis Chamber Input for successful detection of target is 4 units/ml)

4.5 CONCLUSIONS

The device described here, referred to as the Inline Separator or as the Macro Micro Fluidic Interface, is a first development step towards an automated device for concentrating the contents of large water volumes to an amount processable in microfluidic diagnostic systems. As there is no such system purchasable on the market, the effort of development was significantly high.

The presented Prototype Inline Separator offers several opportunities to vary parameters and chemical water amendments to improve the process performance. For future designs, the benefits of an entirely continuous system should be taken into consideration. Doing so might decrease the required material and device dimensions and thus might finally open the door to a fully portable detection device.

REFERENCES

Burrows W. D. and Renner S. E. (1999). Biological warfare agents as threats to potable water. *Environ. Health Perspect.*, **107**(12), 975–984.

Corlett D. A. and Stier R. F. (1990). Foodborne Pathogenic Microorganisms & Natural Toxins Handbook (The "Bad Bug Book"). United States Food & Drug Administration.

Dorin M. and Cummings J. (2004) Principles of Continuous Flow Centrifugation. Beckman Coulter, Technical Information T-1780.

Hill V. R., Polaczyk A. L., Hahn D., Narayanan J., Cromeans T. L., Roberts J. M. and Amburgey J. E. (2005). Development of a rapid method for simultaneous recovery of diverse microbes in drinking water by ultrafiltration with sodium polyphosphate and surfactants. *Appl. Environ. Microbiol.*, **71**, 6878–6884.

Kearns E. A., Magaña S. and Lim D. V. (2008). Automated concentration and recovery of micro-organisms from drinking water using dead-end ultrafiltration. *J. Appl. Microbiol.*, **105**, 432–442.

Zuckerman U. and Tzipori S. (2006). Portable continuous flow centrifugation and method 1623 for monitoring of waterborne protozoa from large volumes of various water matrices. *J. Appl. Microbiol.*, **100**, 1220–1227.

Chapter 5

Sustainable DNA/RNA release methods for in-line waterborne pathogen screening devices

Hunor Sántha

DEFINITIONS AND GLOSSARY

DNA: Deoxy-ribo Nucleic Acid
RNA: Ribo Nucleic Acid
Nucleotides; nucleic acids: DNA or RNA content; DNA or RNA molecules
DNA/RNA release: A process in which the nucleotide content of a pathogen becomes accessible for the reagents of a molecular diagnostic assay.

(*Note#1:* The scientific literature uses typically the following expressions for the process: "cell lysis", "cell disruption", "DNA/RNA extraction", *Note#2:* this latter expression is not absolutely correct, because extraction means (or includes also) the purification step of nucleotides from other cell/viral residues e.g. proteins, lipids. *Note#3:* since DNA/RNA release is necessary also in the case of viruses, though viruses have only a protein capsid around their nucleotide content, it is not correct to mention any "cell" in the case of revealing the DNA or RNA content of a virus, as viruses do not fulfill the requirements of being/having a cell).

5.1 INTRODUCTION

Detection and identification of pathogenic microorganisms (bacteria, viruses, spores, fungi, protozoa) in samples under test can be based either on the selective detection of "structural building parts" of a certain pathogen (typically protein-like molecules), which can serve as antigens for immunological cells, thus, are suitable for being the target of an antibody molecule (which must be the spatial 3D complement of a certain portion of the antigen molecule to a sufficient extent), or on the selective detection of the presence of the "design code" (the nucleotide content of the certain pathogen) of the "structural building parts".

The former method is implemented in immunosensor type biosensors and the latter in nucleotide sensor type biosensors. The selectivity of biosensors is ensured by a so called "key-lock" or "host-guest" principle which occurs between the molecules of target analyte and the selective molecular recognition elements ("bioreceptors") immobilised into/onto the biosensor. Immunosensors and nucleotide-sensors together represent the affinity-biosensor family among the biosensors, of which the focus is on the nucleotide biosensors and the related sample preprocessing of a nucleotide-biosensor based device in this chapter. A rough classification method of biosensors [which type of sensors count to be part of the chemosensors according to the definitions of the IUPAC (Thévenot *et al.*, 2001)] is shown in Figure 5.1, taken from our previous work (Sántha, 2009). More about biosensors and challenges in design/manufacturing of biosensor based devices can be read for example from Harsányi, 2000 and Bonyár *et al.* 2010 and in other chapters of this book dedicated to biosensing techniques. In this chapter the emphasis is much more on the fact, that affinity-biosensors inherently demand a well-tailored/optimised sample handling concept and sample handling structures for a well working biosensor based device. This is in contrast to the reactive-biosensor based devices, because latter ones are called and can be considered as "probe-type" devices (Pickup, 2007) while the former ones always need a controlled incubation process. During this process the analyte molecules (if present in the sample) are brought together with the capturing molecules (named also as probe/receptor layer) for a certain time. The parameters of incubation (time, temperature, volumes, diffusion lengths etc.) will determine the overall performance of the biosensor based device with the same importance as the quality of the molecular building-parts responsible for the affinity-biosensing phenomenon (i.e. the capturing molecules) and its detection performance (i.e., the signal transducer technique and the related instrumentation).

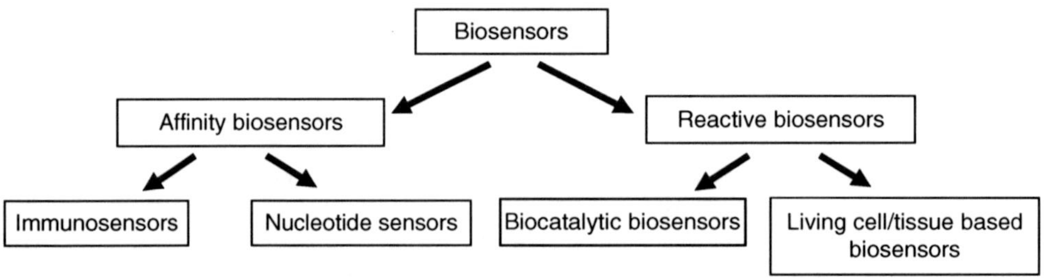

Figure 5.1 A classification method of biosensors.

In bioanalytical process-chains and assays often not the sensing principle/detection method itself is the challenge but the adequate preprocessing methods of the sample. Design challenges are emphasized many times among stakeholders of the biosensor, the µTAS- (micro-Total-Analysis-Systems), the LoC (Lab-on-a-chip) and the BioMEMS community for example (Sántha, 2009; Di Carlo *et al.* 2005; Huang *et al.* 2002) because the pre-processing of the sample has to be reproducible, effective, scalable, cost efficient, compatible with later assay steps also in their novel miniaturised/Point-of-Care (PoC)/single use disposable formats.

This is especially true for the DNA/RNA detection based bioassays, where with the invention of the PCR (Polymerase Chain Reaction) and the several subsequently developed nucleotide amplification techniques (other similarly thermocycled variants or isothermal ones) the detection limit (LoD) has been pushed down to few copies (theoretically/potentially 1, practically speaking few 10s, 100s) of target entity (bacteria, virus etc.) per sample under test (Laureyn *et al.* 2007).

Hereby we emphasize further the significance and difficulties related to a proper sample handling. Pathogen detection techniques which are based on capturing and analysis of nucleotide content are more universalizable, than those based on capturing of "structural building parts" since every pathogen even the smallest viruses have "design codes" which follow similar molecular structures, namely as nucleotides, and the sequence of these structures holds the information. Thus, the presence of nucleotide based "design codes" is such a conserved concomitant sign of life, which must be present in the sample under test if the pathogen is/was there. An additional advantageous possibility is the deployment of any of the existing nucleotide amplification techniques in the process-chain so even a less sophisticated and less advanced detection/signal transducer may fulfil the requirements of the application. Namely, in terms of accuracy, precision etc. orders of magnitude lower performance may fit for the purpose in comparison to immuno-biosensors.

However, compared to immuno-biosensors based methods a great disadvantage is that the target molecules to be captured – the nucleotide content – are never on the surface of a pathogen, thus, under any circumstances these typically protected/packaged DNA/RNA molecules have to be released first from the "body" of the pathogen before any capturing can happen.

The "body" of the pathogen can show significantly different levels of robustness/durability beginning from a simple protein cover layer (e.g. a virus capsid) across a combination of phospholipide cell membranes with glycopolysaccharide cell-walls (e.g. Gram-positive bacteria) and ending up with thick and dense, multilayered protein based spore-walls withstanding even few hours in boiling water or even a part of the less rigorous disinfection procedures.

Thus, one has to conclude that among the sample handling R&D challenges of research of biosensor based devices (e.g. LoC, μTAS applications) the sample handling/preprocessing issues related to nucleotide-biosensor based devices are the most complex ones, since the release of nucleotide content is inevitable, thus, the number of necessary process steps have to be increased.

The scope of this book chapter is to introduce a successfully realised prototype device and method resulting from the "DINAMICS" FP6 project funded by the European Commission. This prototype was suitable for multi-pathogen DNA/RNA release and the authors reveal the most important considerations leading to the final solution.

5.2 SURVEY ON PATHOGEN-LYSIS/CELL-DISRUPTION METHODS

The scope of our literature and market survey was to find candidates for a "one-method-fits-all" solution. The main requirement was to find methods which can guarantee an acceptable yield of DNA/RNA release in the case of any of the targeted pathogens of our project. The gold standard and most general method for pathogen cell-lysis is the boiling of the sample for a predetermined time, for example 10 minutes. In order for assessing the yield artificially spiked test samples can be used and the efficacy of release of nucleotide content (i.e. the yield) can be measured by subsequent DNA/RNA detection and quantitation, as the quantity of pathogens in the artificially prepared sample is known in such cases.

As a first benchmark, best practice methods known for human/mammalian cells in order for make their nucleic acid content available for further assaying were considered, as these are used and studied extensively, especially since the success of the Human Genome Project. The nucleic acid content of such cells may have significant relevance both clinically and in medical/biological research, thus, there are commercially available reagent kits for this (probably very well known for members of microbiology lab staff), and there are examples also for simpler (Kulinski *et al.* 2009) and more sophisticated LoC or μTAS type prototypes (Easley *et al.* 2006) which successfully solved a part or many of the above mentioned sample handling/preparatory R&D challenges. However, the methods that perform well and reproducibly in the case of, for example, the osmotically vulnerable human white- or red-blood cells (Laureyn *et al.* 2007; Di Carlo *et al.* 2005) may be very far from being efficient for all the targeted pathogens of the DINAMICS project (see the targets in a subsequent section "Considerations …").

Anyway, even in these "simpler methods" fitting for white- or red-blood cells significant obstacles have to be handled some way. The incompatibility of chemical reagents deployed typically for cell disruption and nucleic acid extraction (guanidine and isopropanol) with the subsequent PCR process was found to be a major challenge associated with integrating sample treatment steps into a microfluidic process flow. Fluidic isolation of the cell disruption and nucleic acid extraction from the PCR domains can be accomplished by differential flow–resistances of channels, elastomeric valves, or laminar flow. Problems arising from the incompatibility of the poly(dimethyl siloxane) (PDMS) valves with organic solvents can be circumvented by putting water layers/columns as inter-spacers, because the water can effectively serve as a barrier to organic solvents.

After considering the methods for mammalian/human cells, the survey work was focussed only on pathogen microbes. The three major types of cell membranes/cell walls in pathogen microbes to be considered: 1) Gram-negative type, 2) Gram-positive type and 3) chitinous. The first two are typical to bacteria distinguished by their behaviour when stained with Gram stain (WIKI #1, 2012) the latter is typical to fungi. Viruses have only a protein encapsulation (called 'capsid') around their RNA or DNA content, thus, they represent the less problematic case for releasing the nucleotide content, however, the fact that many of them have not DNA inside but RNA for storing the "design codes", and in several nucleotide sensing methods RNA molecules first have to be converted (copied) to DNA before the detection can occur, they represent additional sample handling/postprocessing challenges in the subsequent part of the analysis.

Gram-negative bacteria have a relatively thin cell wall consisting of a few layers of peptidoglycan surrounded by a second lipid membrane containing lipopolysaccharides and lipoproteins. In fact this membrane can not be called a cell wall, since it is a thin polymer layer (WIKI #2, 2012).

This type of bacteria is the easiest to disrupt, thus, methods suitable for human/mammalian cells may be suitable for these as well. The Gram-positive cell wall structure shows more resistance to environmental stress. It possesses a thick cell wall containing many layers of peptidoglycan and teichoic acids. Peptidoglycan is a polymer consisting of sugars and amino acids that forms a mesh-like layer, it gives the cells massive integrity. The third type of microbes we are going to regard are the fungi. Many species of this kingdom are very hazardous and resistive. Their cell wall's integrity is mainly provided by the polymer consisting of chitin monomers. It builds up a massive layer around the cell.

The following impacts may be exploited in pathogen disruption for DNA/RNA release:

- (1) Physical impacts
  - (a) Elevated temperature
  - (b) High electrical field
    - (i) AC
    - (ii) DC
  - (c) Mechanical
    - (i) Real mechanical impact
    - (ii) Sonic pressure waves
    - (iii) Osmotic pressure
- (2) Chemical impacts
  - (a) Detergents
  - (b) Chaotropic salts
  - (c) pH shift
- (3) Biological impacts
  - (a) Cell-wall digesting enzymes (e.g. lysozyme or proteinase K)

As mentioned above, the boiling of the sample in order for releasing its intracellular content is considered as the gold standard. It has established traditions in microbiology, thus, we consider this as the relevant example for the elevated temperature impact among the physical ones. For the rest of the potential methods we have followed the above written bullet points to reveal the state-of-the-art of pathogen-disruption methods and devices.

5.2.1 Examples for devices utilizing electric field

In the first approach under the aegis of the DINAMICS project a "reagentless cell lysis and DNA/RNA release method" and prototype module was planned to be elaborated in order to spare with addition of extra reagents. We aimed at a simplified usability, thus, deploying only pure electrical impacts seemed to be of first preferred choice. The critical value which can successfully destroy the membrane of a cell corresponds to a transmembrane potential of approximately 1 V (Lee *et al.* 1998) as compared to 0.07 V under normal conditions. This value must be definitely higher, if the cell is surrounded not only with a phospholipid bilayer membrane but with an additional cell-wall too.

Our first efforts were targeting on the on-market available devices for electrical cell-lysis and electroporation. It is important to mention the difference between electroporation and lysis. While the previous concept means the opening of the cell membrane/cell wall in order to be able to introduce foreign molecules into the cells inner plasma but the cell can regenerate later; the latter is the cells full disruption, at which process the cells inner contents exit to the outer environment and the cell dies. Unfortunately we were not able to identify commercially available devices for electrical cell-lysis only some for electroporation, but these latter did not provide enough practical guidelines for our work, since our goal is to lyse the cells to a sufficient extent, in order for the DNA/RNA content inside becomes accessible for subsequent analysis, thus, knowledge about parameters/impacts temporarily opening the cell-membranes of cells/microbes not reinforced with a cell-wall is not enough.

However some important issues were revealed. Many authors from the research domain reporting on electrical cell lysis set-ups/prototypes faced the problem of over-lysing when multi-target compatible protocols were to be established. Over-lysing occurs if parameters of the impact are not kept in the "process window" compatible with certain target pathogens/cells of the sample, and the intracellular content of those will be conceivably damaged. In one example the conclusion was drawn, that the efficacy of electrical lysis was successfully increased from 8%, to 30% on leukocyte cells when the dimensions of the electrodes were protruded in a 3D shape instead of 2D planar ones, furthermore it could have been increased to 90% with simple measures in design (i.e. increasing the number of subsequent electrode rows from 3 to 4–5), but further efforts for a total lysis could jeopardise the quality of lysate (Lu *et al.* 2006). As the here cited experimental work was performed on leukocytes, which are considered to be easy subjects for electrical cell-lysis or electroporation, one has to assign even higher attention to the risk of electrical over-lysis in the case of more resistant pathogens, for example Gram-positive bacteria and spores.

Among our most relevant findings two sources (Lu *et al.* 2006; Wang *et al.* 2006) were informative enough for considering their concept during the prototype development efforts of our project and two review type articles (Huang *et al.* 2002; Sun *et al.* 2006) have been found too, which are suggested for further reading to those who are interested in a more comprehensive analysis of state-of-art sample preprocessing advancements. One of the considered concepts is based on DC electric field, and a quite simple structure, while the other deploys AC waveforms or pulsed DC and a more unique layout.

A DC electric field based cell-lysis device

This device (Wang *et al.* 2006) utilizes DC electric field between two collateral electrodes. The layout has been set in such way that the electric field between the electrodes is amplified by the geometric design of the surface.

It has been shown that this device lysed *E. coli* bacteria with a very good efficiency, in the best setup >95%. Among 3 realised and investigated layouts the authors measured <12 μA in 2 of them and <50 μA in a 3rd one, and they concluded that the lowering in threshold field strength in devices according to the 3rd configuration happened possibly due to the increased Joule heating. Thus, this method can be considered partly among the elevated temperature based methods at last.

The cells move from the sample reservoir (biased to GND) to the receiving reservoir (biased with + voltage) through a channel with varying width. This design ensures far larger field strength, than the diameter and the voltage would suggest. The following equations explain how it works:

$$E_1 = \frac{V}{L\left(2 + \frac{W_1}{W_2}\right)}, \qquad E_2 = \frac{V}{L\left(1 + 2 \cdot \frac{W_2}{W_1}\right)}, \qquad \frac{E_2}{E_1} = \frac{W_1}{W_2}$$

where W_1 and W_2 are the widths of the channel, E_1 and E_2 are the electric field strengths between the collateral electrodes, L is the length of the channels, and V stands for the voltage given to the electrodes. Three configurations with the parameters in Table 5.1 were investigated.

Table 5.1 Dimensions realised and tested by Wang *et al.* (2006).

Configuration	W_1 (μm)	W_2 (μm)	W_1/W_2	L (mm)
A	203	25	8,12	5
B	212	33	6,42	2,5
C	219	115	1,9	2,5

This module has the advantage that it is easy to implement and the supplementary circuits are easy to build. Though, it has to be considered that the voltage used is relatively high, so care has to be taken that no bubbles arise between the electrodes as it could seriously corrupt the efficiency of the device. Furthermore the low flow-throughput of the individual channel(s) may be a bottleneck in the system design, even if multiplied in a final quasi continuous screening device.

An AC electric field based cell-lysis device

This idea is based on a more complex electrode design placed in the bottom of a 30 μm high reaction chamber. Yeast (*Saccharomyces cerevisiae*) protoplasts and *Escherichia coli* (strain ATCC 25922) were tested for lysis. Cabbage and radish protoplasts were also examined with a different type of electrode. According to the test-results, the device was able to successfully disrupt Yeast protoplasts up to 80% efficacy, which fact was measured by optical microscopy counting of intact protoplasts after treatment. Here we note that Yeast cells are considered to be quite resistant against electrical lysis, thus, the 80% looks to be a great success, however, the protoplast state of the microbes under test means that their cell-wall was removed enzymatically before the pulsed electrical field treatment.

The electrode structure is based on Cr/Au interdigitated electrodes having triangle shaped teeth on the fingers on both sides and a Parylene C (4 μm thick poly-p-xylylene) coating on the backbone of the fingers. The complete surface of the module can be coated with a 500 nm thick Teflon (AF 1601S) too, since it ensures protection against oxidation, though, such a Teflon overlayer increases the voltage needed. The specific waveforms deployed in order to reach high lysis efficiency ratio at low voltage levels consist of repeated few hundred kHz or few MHz AC burst periods and pauses in between.

This concept is apparently somewhat more suitable for high flow-throughput designs, but its AC electric field or pulsed DC electric field based lysis efficay proved to be promising (i.e. 80%) only in the case of protoplast forms of microbes (i.e. which had no cell-walls after an enzymatic pretreatment).

5.2.2 Examples for devices harnessing mechanical impact

Direct mechanical impact based devices

Theoretical considerations suggest that the "reagentless cell lysis and DNA/RNA release method" and prototype module targeted in the DINAMICS project could be realised also by deploying certain mechanical impacts on the pathogens. And these may be standalone techniques but also complementary and integrated to the formerly investigated electrical impacts if needed.

The most straightforward method of deploying mechanical impact on microbes (i.e. shear forces and tension) is to force/collide them to rigid structures for example by bead milling. However such macroscaled devices and methods are more typical to analytical laboratories, thus, were not considered in our in-line, quasi-continuous system, which should respect requirements for automation.

In line with these requirements two papers which deploy special microfluidic structures with adequate fabrication methods have been identified and considered among the direct mechanical impact based concepts. One of them (Di Carlo *et al.* 2003) exploits the inherent phenomenon related to Deep Reactive Ion Etching (DRIE) of silicon, namely that the high aspect ratio, almost perfectly perpendicular-to-plane columns/crests remaining as sidewalls after the etching of trenches/channels in between them will get a longitudinal micro-blades covered surface according to the cyclic repetitive character of the removal of material during the penetration into deeper and deeper regions of the silicon substrate. The authors optimised/modified their DRIE process in order to exaggerate this side effect, for an improved cell-disruption capability, when the sidewalls are deployed in certain few micron wide cell filtering structures.

Although the idea is very creative, the absolute lysis efficacy tested on HL-60 human leukaemia cells was unacceptably low. Within the filter region total protein and haemoglobin accessibilities of 4.8% and 7.5% were observed respectively as compared to 1.9% and 3.2% for a filter without nanostructured barbs as compared to a chemically lysed sample (as a benchmark of 100%).

A more efficient method among the microfluidics, LoC or μTAS friendly solutions is the deployment of porous structures, through which cells/pathogens can be forced through in a pressure driven way. A quite recent example of such systems (Kulinski *et al.* 2009) was capable to perform lysis and extraction of DNA from bacteria infected human urine samples in the presence of human whole blood contamination. The established proof-of-concept for their sample preparation microfluidic device that uses shear and frictional forces generated with a ca. 10 bar overpressure through a polymer monolith with small pores (ca. 80% of pores below 2 μm diameter) was as efficient or more efficient than the positive control (i.e. a commercially available DNA extraction kit of Qiagen Inc.) in isolation of bacterial DNA from simulated Urinary Tract Infection samples (urine spiked with *E. Coli* and whole blood) in a range of concentrations, 10^5–10^1 CFU/mL. The integrated sample preparation channel processed a 100 μl sample with one wash in less than 40 min.

Reducing the elution volume used there and increasing the channel volumes could lead to more time savings. With further design development, this system could be suitable for integration with in-line amplification and detection technologies even in thermoplastic (thus affordable/cost efficient) platforms.

Sonic pessure waves based lysis devices

As indicated at the introduction of this chapter, there are numerous techniques and instruments used for cell disruption and extraction of cell DNA/RNA for further analysis: boiling, use of enzymes capable to digest the compact cell-wall, bead-mill method (use of glass or ceramic beads), sonication, breaking the lipid barrier with detergents, treatment with solvents, high pressure gradient, high-shear mechanical methods and treatment with electric field. Some of these methods favour the implementation in on-field deployable device formats, like sonication and electroporation/lysis with electric field. The intent of this subchapter is to discuss sonication techniques, to introduce the related phenomena causing cell lysis/release of nucleotide content and comparison of these processes. Similarly to the case of electrical methods discussed above, here again our first efforts were targeting on the on-market available devices. But before presenting the most relevant results found, a short explanation of ultrasonication is necessary.

Ultrasound is an oscillating pressure wave with a frequency of approximately 20 kHz or higher. Since ultrasound is a mechanical phenomenon it requires a medium for propagation. The energy coupled into the medium will cause cyclic compressions (and subsequent relaxations), thus, certain heat will be generated under any circumstances. With 1–3 W/cm^2 intensities in physiotherapy applications local temperature rise can be caused in the patient's tissue (Elsner *et al.* 1989). In aqueous solutions, the application of ultrasound is accompanied by cavitation too. Cavitation occurs when gases dissolved in liquid are induced to form cavities or microbubbles. The radius of these bubbles can range from 100 to 250 μm. There are two types of cavitation that are said to impact on pathogens and biomacromolecules in a solution in a direct or indirect manner:

- *Stable cavitation* (or gas body activation) and
- *Transient* (or vaporous or inertial) cavitation

Stable cavitation occurs in solutions when the application of low intensities of ultrasound (ca. $1W/cm^2$) generates a current of microbubbles. During the negative half-pressure cycle of the ultrasound wave, the microbubbles increase in size; during the positive half-pressure cycle, the bubbles decrease in size. Transient cavitation occurs during the application of higher intensities of ultrasound (greater than $1W/cm^2$). In this case, microbubbles may begin to oscillate in size but at some point during sonication, they reach a critical size and collapse. The extreme heat generated at the point of this collapse, several thousand Kelvin, is enough to cause the formation of free radicals from water. The collapses that occur due to transient cavitation are considered as high energy vents and can be very destructive. In fact, they can be responsible for erosion, cell disruption, luminescence and shearing of biological molecules.

This oscillation in the microbubble scale causes shearing stress as microstreaming occurs at the bubble surface and bubbles interact with each other, with the solution and the vessel walls. Mechanical stress is responsible for causing cell lysis and is suspected to be responsible for most of the DNA degradation in solution. Ultrasonic degradation of DNA in solution occurs by breaking hydrogen bonds and by single-strand and double-strand ruptures of the DNA helix. Increasing the intensity of the ultrasound above 2 W/cm^2 is followed by increases in single-strand ruptures due to the creation of free radicals by transient cavitation. Following sonication, the distribution of the resulting DNA fragments approaches a lower size limit of 100–500 bp.

When cellular components are exposed to ultrasound, there are two mechanisms that are primarily responsible for the observed effects on biological molecules. The first is direct mechanical damage via bulk heating. The second is by indirect interaction by chemical reaction as free radicals attack molecules.

Regarding the timescale of the necessary impact, we refer to a study performed with L5178Y mouse lymphoma cell cultures at 1 MHz ultrasound frequency and 1–5 W/cm^2 intensities. The cell disintegration efficacy dropped to zero for pulses shorter than 1 ms, and the optimum pulse duration was 30 ms. Regarding the intensity 1 W/cm^2 was the cell disintegration threshold and 5 W/cm^2 was the optimum under conditions providing stabilization of the cavitation field (Clarke *et al.* 1970).

In another paper the optimization of duration of ultrasonication and its parameters is discussed in the case of *Bacillus Cereus* (Gabig-Ciminska *et al.* 2005). These bacteria are Gram-positive – which means, they have a thicker multilayered cell-wall – when young, but may become Gram-negative as they age.

A suspension of vegetative bacterial cells was continuously sonicated (disrupted) with sampling after 30 s, 2.5 min, 5min, 10 min, and 13 min, and at 10 min an optimum of the yield was found. The ultrasonic power output was 100W @ 30 kHz operating frequency with a microtip of 1 mm in diameter. During ultrasound cavitations, the samples were cooled in an ice water bath until completion of the procedure. After a heat treatment (95°C, 10 min) and removal of the solid particles by centrifugation (5000 g, 10 min), the lysates were subjected directly to the assay. An early endpoint PCR analysis was used to characterize the DNA fragmentation as a function of ultrasonication time.

The first minutes of sonication increased the signal due to both increased DNA release and increased DNA fragmentation. The latter is assumed to increase the signal due to improved diffusion and faster hybridization of the target DNA. Too long (>10 min) sonication decreased the signal, presumably due to loss of hybridization sites on the targets caused by extensive DNA fragmentation. The results were intended to form a basis for rational design of an ultrasound cell disruption system integrated with analysis on chip that will move nucleic acid-based detection through real-time analysis closer to reality.

There are three main types of ultrasonic cell disruptor devices:

- Probe-type (horn-type) sonicators
- Ultrasonic baths
- Minisonicators

Probe-type sonicators

Probe-type sonicators are the easiest to purchase and implement by their nature. They are the most commonly used types by chemists. These instruments are applied for treating greater amounts of liquids: homogenizing food ingredients (like juices), mechanical cleaning or sanitization of tools, cell permeabilization and disruption of various cell types. Probe-type sonicators typically consist of two elements: a transducer (which turns the electric signals into mechanical motions, typically piezoceramics) and a tip (often called probes or horns or sonotrodes).

In Figure 5.2 the photo of such a device is presented. The device works the following way: the ultrasonic power supply converts line voltage to high frequency electrical energy. This electrical energy is transmitted to the probe where it is converted to mechanical energy. The vibrations from the probe are coupled to and intensified by the tip. The probe vibrates in a longitudinal direction and transmits this motion to the titanium tip immersed in the solution.

To prevent excessive heating, ultrasonic treatment is applied in multiple short bursts to a sample immersed in an ice bath. Sonication with probes (horns) is best suited for volumes <100 ml.

Main advantages of this type of devices are that it is simple to manufacture, to maintain and to use. One can vary the tips of the probes (small probes for smaller amount of samples, for the use of higher frequency etc.) and the intensity of sonication depending on the application to be performed. This is illustrated by Figure 5.3. Processing volumes range from 0.3 ml to 2000 ml in some cases (depending mostly on the usage of different probes).

Ultrasonic baths

In an ultrasonic bath (WIKI #3, 2012) the object to be treated (typically cleaned) is placed in a chamber containing a suitable ultrasound conducting fluid (an aqueous or organic solvent, depending on the application). In aqueous cleaners, the chemical added is a surfactant which breaks down the surface tension of the water base. An ultrasound generating transducer is built into the chamber, or may be lowered into the fluid. It is electronically activated to produce ultrasonic waves in the fluid. The main mechanism of cleaning action comes again by the energy released from the creation and collapse of microscopic cavitation bubbles, which break up and lift off dirt and contaminants from the surface to be cleaned. The higher the frequency, the smaller the nodes between the cavitation points which allows for more precise cleaning. The bubbles created can be as hot as 10000 K and 3000 bar, but are so small that cleaning and removal of dirt is the main result.

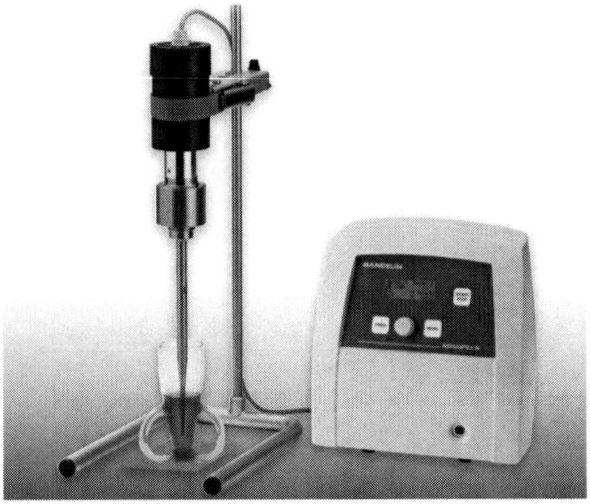

Figure 5.2 Typical appearance of a probe-type sonicator (SONOPLUS HD 3100 system) with the electrical power supply and driver unit (right side) and the transducer & tip structure (left side) [reprinted with permission from BANDELIN (2005)].

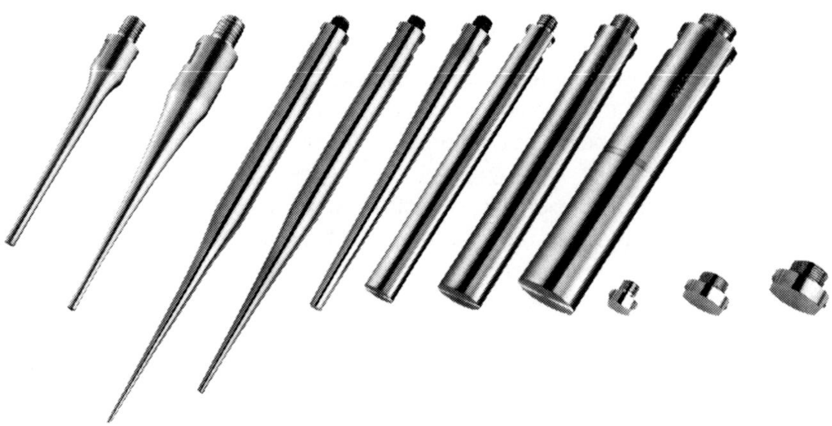

Figure 5.3 Various microsonicator tips currently in use [reprinted with permission from BANDELIN (2005)].

The great difference between bath type sonication and probe type sonication is, that the latter brings the ultrasonic power output in direct contact with the sample with customisable (exchangeable) structures (i.e. the sonotrodes). It means that adjustments to the sample volume can be done easily, and furthermore the former method has inherent losses in terms of energy efficacy compared to the latter. But the sonotrodes have also some drawbacks: 1) the probe comes directly in contact with the biological sample, thus, contamination between different samples is frequently experienced, 2) as the sonication energy depends on the depth of the sonication probe in the liquid, it is disadvantageous on the reproducibility, 3) the probe system is tedious to work with, produces foam, and only one sample can be treated at a time, 4) additionally, the probe system generates aerosols, which are hazardous by biosafety rules.

A very innovative system on the market combines the advantages from both concepts and overcomes the disadvantages (Diagenode Ltd., 2012). The so called *Bioruptor* System is based on a water bath with high power ultrasound generating elements located below the tank. With this system 6 to 12 closed tubes can be sonicated together and the continuous rotation of tubes allows even distribution of the energy. With a probe sonicator, the microstreaming phenomenon would be limited to the vicinity of the probe, whereas for the *Bioruptor*, the whole volume of water present in the tank is exposed to ultrasound energy.

In its 15 ml or 50 ml tubes, a central metallic bar inside is in contact with the sample and facilitates the transfer of the ultrasound inside the tubes. This metallic bar is not a probe but "reflects" the ultrasound originated from the water bath and improves the sample sonication efficiency by a patented resonance system. Produced in stainless steel it is not prone to corrosion.

Anyway, innovative or not and efficient or not, the system is too bulky to be integrated in an in-line lab-on-a-chip system, thus, the survey had to go further.

Minisonicators

Ultrasound microbial cell disrupters operating at around 20 kHz are often physically large and, due to significant heating, can be unsuitable for small sample volumes where biochemical integrity of the extracted product is required. This type of sonicators is basically a lead-zirconate-titanate (PZT) ceramic structure, positioned in a specifically designed vessel, according to the expected effect of sonication (to form standing waves, minimize heat dissipation on walls etc.)

Minisonicators are devices consisting of a small horn transducer and a microfluidic cartridge. The purpose of the minisonicator is to rapidly disrupt bacterial spores for further DNA analysis. The method usually requires a sonication time of 60 seconds. At this type of cell lysis device the treated microbes and spores have to be prepared for further PCR analysis. As in electric field lysis, in this case real-time PCR seems to be an appropriate method to check the efficiency of the method. A microelectromechanical system (MEMS) based piezo-electric microfluidic minisonicator has been realised still in 1999 (Belgrader *et al.* 1999). It operates in the frequency range around 380 MHz. The device lyses cells in the absence of chemical, biologic or microparticle agents in a continuous way as the sample flows through a channel with a diameter of 50 μm.

In another paper (Borthwick *et al.* 2005) the development of a compact device based on a 63.5-mm diameter, 6.5-mm thick tubular transducer for rapid cell disruption in small-volume samples in a high-intensity acoustic cavitation field with minimal temperature rises is described. Suspensions of *Saccharomyces cerevisiae* – which is a fungi, thus, no sense to talk about Gram positivity or negativity – were exposed to cavitation for various times in the compact device of the authors and a 20-kHz probe sonicator. Cell disruption was assessed by protein release and by staining. Yeast cell disruption was greater in the novel 267-kHz compact sonicator than in the 20-kHz compact probe sonicator for the same exposure time.

With the explanations above and these two examples of minisonicators we tried to justify the idea that ultrasound can be a rapid method to release and fragment DNA for use in biosensor based devices. It can be applied directly to cellular samples.

Salt concentration, exposure time, power, and temperature can be manipulated to control the length and potentially the form of fragment desired (single- or double-stranded). With careful control of sonication and hybridization conditions, this sample processing method has the potential to move nucleic acid biosensor technology towards "real-time" analysis (Mann *et al.* 2004).

5.2.3 Chemical impact based methods (i.e. no Cell-Wall digestive enzymes added)

While the above discussed physical DNA/RNA release methods may enable a "reagentless" lab-on-a-chip process flow, the chemical or the biological impacts deployed typically in cell lysis/nucleotide release need the addition(s) of reagent(s). In the case of chemical methods, the impact can be addition of for example detergents or dramatic shift of ionic strengths or shift of pH. for example for Human Papilloma Virus (HPV) or other viruses it is very typical to use only guanidium a chaotropic salt (see below explanation) and SDS, since the nucleotide content of HPV is packaged only in a protein capsid.

The most widespread lysis buffers typically contain a detergent (e.g. Sodium-Dodecyl-Sulfate (SDS), Tween 20, Triton X-100, Nonidet P-40) and proteinase K, an enzyme which is capable to digest/decompose even the multilayered cell-walls of certain microbes. However, this latter agent is of biological nature, thus, should be classified among the biological impact based methods, which were not subject of our work at all, since one of the original objectives of the DINAMICS project was to eliminate the lengthy (~30 min.) biological sample pretreatment steps.

Among molecular biology techniques the "salting out" procedure is beeing widely used also apart from cell lysis and DNA/RNA release methods. The principle is, that the addition of a so called "chaotropic salt" (for example 6M guanidine thiocyanate, or 6M sodium chloride during or after a cell lysis step) disrupts protein structures by interfering with hydrogen bonding, Van der Waals interactions and hydrophilic/hydrophobic interactions, because the high concentration of salt competes with the proteins and other macromolecules for the aqueous solvent, effectively dehydrating the protein/macromolecules. As cellular proteins become largely insoluble in the presence of the chaotropic agent, they can be removed by centrifugation or filtration easily in a subsequent step if the assay requires a'purified' sample matrix (Goodwin *et al.* 2011).

Beside the detergents and the chaotropic agents discussed above the shift of pH is the third chemical impact which can cause release of nucleotide content of a microbe. We present an example from the literature (Di Carlo *et al.* 2005) in which the minimum concentration of sodium hydroxide which is capable to cause the lysis of a trapped HeLa cell (a human tumor cell line) was tested.

One of the key findings of the experiment was, that 20 mM NaOH (pH = 11.70 in PBS) was still an effective concentration, but 10 mM (pH = 11.20 in PBS) did not cause cell lysis. From this result the authors were able to deduct the necessary OH^- ion generating capacity around the cathode of a microfluidic cell-lysis reaction chamber for local OH^- generation by electrochemical water decomposition.

5.3 CONSIDERATIONS FOR WATER SAMPLES AND IN-LINE (QUASI CONTINUOUS) OPERATION

In the title of this chapter the word "sustainable" refers to the intention, that preferably and ideally no consumables should be used for the operation of the DINAMICS pathogen detection system, thus, this is true for the nucleotide release submodule too. If one considers, that in order for being selective in the detection our system deploys biosensors, which type of sensors always have only a limited lifetime/usability, the "no consumables" requirement is immediately handled with flexibility. Practically speaking the exchange interval of the biosensor elements should be maximised, and the usage of single use/disposable consumables (e.g. filters, reagents etc.) should be minimised.

The requirements related to the "in-line" operation of the whole assay (and also of the sample preprocessing i.e. the nucleotide release) means, that the water supply chain should be monitored continuously (or at least quasi continuously, with a controlled delay) without human intervention. Alone from theoretical considerations one can deduce, that only the quasi continuous approach may work, since it is an axiom, that the affinity type biosensors all need a certain (tailorable but predetermined) incubation time for exact concentration measurements. It means, that depending on the complete assay time vs. the acceptable maximum delay in detection of a pathogen hazard, multiple "pipelines" have to be operated simultaneously with shifted cycle-starts.

Under the aegis of the DINAMICS project a "reagentless cell lysis and DNA/RNA release method" and prototype module was targeted as the part of the final prototype device. However at last the meaning of "reagentless" became limited to anything other than the enzymatic cell wall/cell membrane digestion based methods, which would be too time consuming. Two partners with microbiological background – MikroMikoMed Ltd. (Budapest, Hungary) and Water Research Institute, (Bratislava, Slovakia) – and one partner with hardware and software technological background – Budapest University of Technology and Economics – Dept. for Electronics Technology (Budapest, Hungary) – were involved in this RTD work.

After the above described general technical considerations the target organisms for the final prototype have been identified from groups of bacteria, viruses and parasites covering more than ten taxons. The outcome is presented in Table 5.2 below. The reasoning behind this set of choice can be found in a report of the consortiums work entitled "Target Contaminant Selection". Also included in the table and among the functional requirements are the detection limits (or device sensitivity) for each organism, described in terms of *the number of organisms in the daily consumed quantity of water*, estimated at 2 litres per person per day.

As discussed in more detail in the "User Requirements Specification" report of the consortium, the detection limits provided in this table are a simplified representation of a complex issue. Their expression in terms of water consumed downstream as distinct from water sampled upstream implies the need to estimate dispersion.

It worth noting, that the method of filtering/concentration of pathogen microbes from the water (i.e. centrifugation or filtering) has strong implications on the applicable DNA/RNA release methods too. The initial hypothesis was that a one- or two-stage filtering will be necessary either way, because the microbiological standards rely on filters with different pore sizes. The main question was if the cell lysis/nucleotid release should take place already *in-situ* in the filter (Concept **B**)) or only in a subsequent reaction chamber with freely dispensed/floating microbes after a wash-out from the filter (Concept **A**)).

Concept (A)

A separate microfluidic module impacts on a highly concentrated solution of freely moving cells which are dispersed in a buffer solution after removal of pathogen cells from the filter(s). The procedure and its results supposed to be observable through optical microscope. A sketch of the method can be seen in Figure 5.4.

Concept (B)

Lysis *in situ* in the filter membranes by means of an electrode pair pressed on the two sides of the filter membrane (the filter has to be immersed in a reagent before DNA/RNA content can be forwarded into the PCR module). This concept is based on a filter membrane with a pore diameter smaller than the sample cell diameter, but larger than the obsolete components of the sample reservoir. Thus, the method consists of multiple subsequent steps. In the first step filters with a pore diameter appropriate for the largest cells should be applied, then a smaller pore sized filter, and so on. This way, in every step separate types of microbes can be disrupted, and their DNA/RNA prepared for further analysis. Figure 5.5 shows, how it works.

Table 5.2 Target organisms and detection limits for the prototype aimed at in the DINAMICS project.

Category[1]	Kingdom	Organism	Type[4]	Priority[2]	Safety level[3]	Infectious Dose	Detection Limit
A	Prokaryotes	Yersinia pestis	Gram (−)	1	3,2*	>100 organisms	10^3
B	Prokaryotes	Clostridium perfringens	Spore-forming	1	2	$>10^8$ organisms	10^5
B	Prokaryotes	*Salmonella* spp.	Gram (−)	1	2,3**	15–20 vs (10^2–10^6) organisms	10^3
B	Prokaryotes	*Shigella* spp.	Gram (−)	1	2,3***	10 organisms	10
B	Prokaryotes	Vibrio cholerae	Gram (−)	1	2	10^3–10^6 organisms	10^5
Other	Prokaryotes	*Campylobacter* spp.		2	2	About 400–500 organisms	10^3
Other	Prokaryotes	Listeria monocytogenes	Gram (+)	2	2	$<10^3$ organisms	10^3
A	Virus	Variola major	dsDNA virus	2		N/A	10
Other	Virus	Norovirus, Norwalk like calciviruses (NLV)	ssRNA virus	1	2	Presumed to below	10
Other	Virus	Rotavirus	dsRNA virus	1	2	10–100 viral particles	10
Other	Virus	Hepatitis A	ssRNA virus	1		10–100 virus particles	10
Other	Virus	Hepatitis E	ssRNA virus	2		Unknown	10
B	Eukaryotes	Cryptosporidium parvum		1	2	One organism	10
Other	Eukaryotes	Giardia lamblia		1	2	One or more cysts	10

*Y. Enterolytica, Y. Pseudotuberculosis
**S. Typhii
***Sh. Dyenteriae Typ1
[1]These categories have been taken from the U.S. Centers for Disease Control and Prevention (CDC)
[2]Priority refers to the priority within the DINAMICS project because of limited resources
[3]"Safety level" refers to CDC specified levels of precautions for Biological agent (1–4 = lowest – highest)
[4]ss = single stranded, ds = double stranded

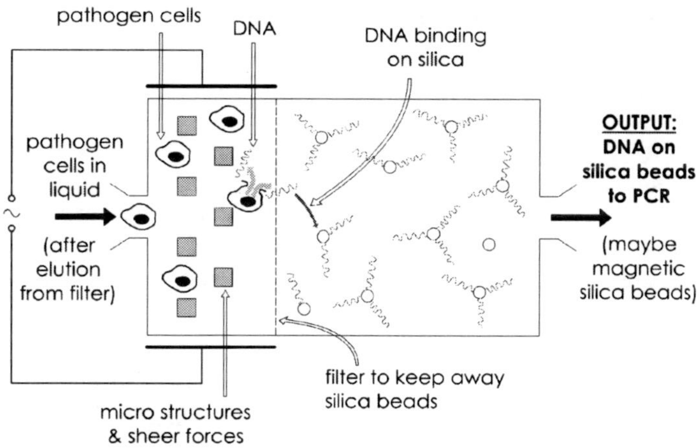

Figure 5.4 The schematic plan to implement a concept'A' lysis device.

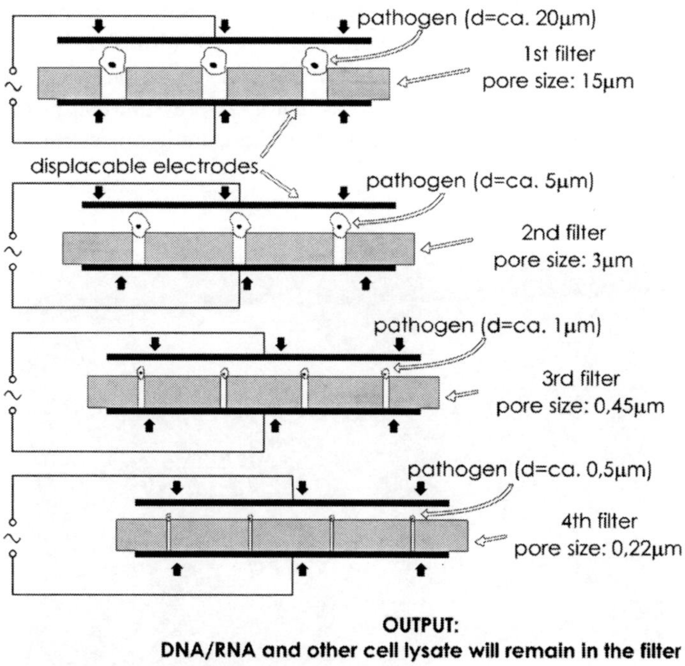

Figure 5.5 The schematic of a concept'B' cell lysis device.

At last filtering proved to be a less integrable manner because of the huge challenge in automation of filter exchanges, thus Concept **B)** was fallen off. Concept **A)** relying exclusively on physical impacts became a first choice, but the possibilities were very diverse and none of the ways was really well established considering that the target pathogen list of table 2 contains more than 10 very-very different microbes, and the project demanded a 1) sustainable, 2) in-line, 3) "one-method-fits-all" solution.

Through various sources of information we concluded, that no commonly accepted guidelines exist in this area, thus, after a literature search revealing the state-of-the-art, certain in-house experiments, several rounds of conceptualization → hardware and software design a series of testing → evaluation → modification have been implemented.

The concept and the proof-of-concept experimental set-up has evolved from a simple purely DC electric field based cell lysis device across a triple-impact (i.e. simultaneous #1 heat, #2 ultrasonic waves, #3 AC electric field) concept towards the final, more industrial, viable and robust solution to be detailed in the last section of this chapter.

5.4 THE REALISED CASE IN THE DINAMICS PROJECT

Based on the comparison of listed lysis methods the heating of the sample to a previously determined temperature in the presence of detergent treatment was realized. This method seemed to be the most practical to be integrated into an automated system where an in-line lysis is necessary with relatively high throughput. An in-house designed electrical circuitry incorporating a microcontroller and communicating with the PC of the user via a USB link was manufactured and assembled in-house on Printed Circuit Board technology and has been linked with 2 MultiPhaser™ NE-501 programmable syringe pumps (OEM product of ProSense, Netherlands) and with an also in-house designed and manufactured lysis chamber holder and temperature actuator. A pressure controller unit is built into the system in order to decrease bubble or foam formation in the sample-buffer mix due to the heating. In Figure 5.6 the whole system is presented.

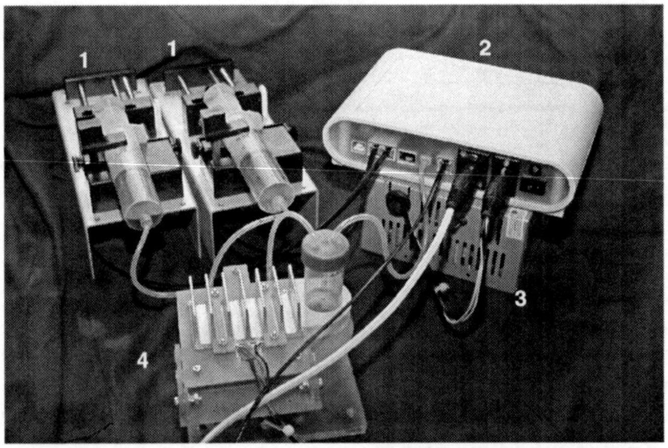

Figure 5.6 Lysis control hardware 1. syringe pumps, 2. liquid handling and temperature and pressure control unit, 3. power supply, 4. lysis chamber and Falcon tube and temperature actuator unit.

To avoid contamination single use lysis chambers made of PDMS were designed. Our preparatory calculations considering the aimed continuous flow lysis protocol, the sensitivity of the final lab-on-a-chip type detection method, the inherent losses between sample handling steps, the probable dangerous initial pathogen concentrations of water samples indicated that a lysis chamber volume of 5 ml is the optimum. Thus microfluidic channel structure has been designed with 2 inlet and 1 outlet orifices and 5 ml internal volume. The maximal outer dimensions are heights: 6 mm, widths: 72 mm, lengths: 65 mm. Earlier a very similar pilot lysis chamber mould was used. The volume of the first version was just 1.5ml which in the light of subsequent calculation did not fit the system requirements. One of each fabricated pieces is shown in Figure 5.7.

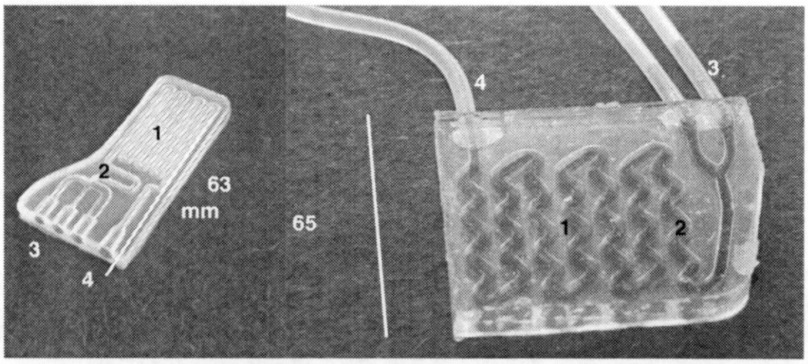

Figure 5.7 The earlier version of lysis chamber (left side) and one of the fabricated pieces of the single use lysis chamber (right side) filled up with 5 ml coloured liquid (both made of PDMS). 1. heating zone, 2. mixing zone, 3. inlets, 4. outlet.

Autodesk Inventor 2010 software was used for designing the objects. Raw PDMS was prepared by adding Sylgard 184 curing agent to Sylgard 184 silicone elastomer in 1:10 m/m ratio. The freshly prepared raw PDMS was casted in a 3D RPT fabricated mold form in a homemade casting workstation consisting of a vacuum exsiccator, a water stream based vacuum pump and tubing or in a vacuum chamber with an oil based vacuum pump. PDMS was purchased from Dow Corning Corp. (USA). For 3D RPT printing we applied an Objet Geometries (Israel) Eden 250 printer with FullCure 720 base material and FullCure 705 support material. FullCure 720 base material and FullCure 705 support material were purchased from Varinex Inc. (Hungary). During the 10 min vacuum exposition all the visible air bubbles left the PDMS body (pressure below 5 kPa). For binding two separately casted PDMS parts together we applied corona treatment surface activation with an Electro-Technic Products Inc. BD-20AC instrument for 5 minutes. This laboratory corona treater works with three different shapes of electrodes with an output voltage between 10–48 kV and 4–5 MHz frequency range.

In the validation procedure the device was used in standalone mode thus the chambers were completed with two 60 ml syringes (sample and buffer source) and with a Falcone tube (as holder for lysate). The chamber was heated and held in its position with a fold-clamping. The mechanical parts were constructed of FR-4 plates the standard yellow base material of Printed Circuit Boards prepared of glass fibre reinforced epoxy. The temperature actuators are two Peltier units (Thermoelectric cooler, TEC) which have 50 W cooling power each and are located at either side of the chamber. The temperature on both sides was measured continuously by two sensor IC's with at least 0.5°C accuracy (Texas Instruments TMP175). This holder is shown in Figure 5.8.

Two different softwares have been written for the lysis control hardware. Both of them are running on PC. The first one is a Graphical User Interface (GUI) and the other is an executable with parameterized inputs. The GUI is useful at the validation process and if standalone mode is used. The layout of the program window is shown in Figure 5.9. This application allows to set all process parameters individually. The process parameters are the temperature of the lysis chamber, the pressure in the closed fluid system, the volume and the flow rate of the sample or the buffer solution. The software connects to the device automatically, additional settings are not required. The parameters to be applied in a certain lysis protocol can be pre-set by the user via a PC, and the process can be initiated by clicking on the "START" button on the screen. After the set amount of fluid flow through the chamber into the Falcon tube the device stops automatically and steps into idle state. The process can be stopped immediately by pressing the "STOP" button at any time. The other software is created for the integrated operation.

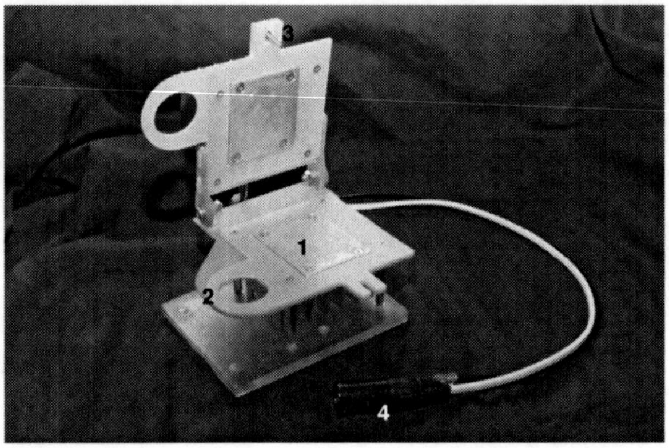

Figure 5.8 Lysis chamber holder and temperature actuator unit (open status) 1. heating zone, 2. Falcon tube holder, 3. closing screw, 4. connector.

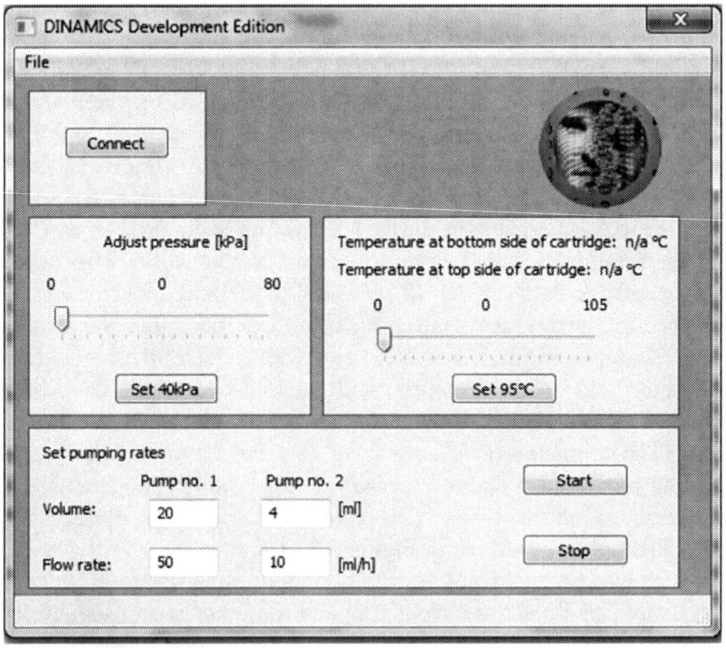

Figure 5.9 The Graphical User Interface.

In the reported experiments the ratio of the lysis buffer and the sample volume were set to 1:5, namely with 10 ml/h and 50 ml/h flow respectively, thus, the residence time of any part of the liquid column subjected to the continuous lysis treatment was 5 minutes in the temperature controlled zone. Considering the whole process the bottom and upper surface of the single use PDMS lysis chambers were heated and kept on 95°C (±2°C) and additionally at 40 kPa (±2%). Table 5.3 contains the characteristics of the control device.

Table 5.3 Physical characteristics.

Parameter	Value		Unit
	Min.	Max.	
Temperature	Ambient temperature	105	°C
Overpressure	0	80	kPa
Volume	–	60	ml
Flow rate	1699	10^{-6}	ml/h

REFERENCES

BANDELIN Ltd. (2005). URL: http://www.lkb.com.au/dl/ultra_sonicator_flyer.pdf (last time accessed 15 March 2012)

Belgrader P., Hansford D., Kovacs G. T. A., Venkateswaran K., Mariella R. Jr., Milanovich F., Nasarabadi S., Okuzumi M., Pourahmadi F. and Northup M. A. (1999). A minisonicator to rapidly disrupt bacterial spores for DNA analysis. *Analytical Chemistry*, **71**(19), 4232–4236, doi: 10.1021/ac990347o.

Bonyár A. and Harsányi G. (2010). Typical problems and solutions in electrochemical measurement cell development. In: *Proc of the 33rd International Spring Seminar on Electronics Technology*. Warsaw, Poland, 2010.05.12-2010.05.16. (IEEE) pp. 433–438, doi: 10.1109/ISSE.2010.5547353.

Borthwick K. A. J. *et al.* (2005). Development of a novel compact sonicator for cell disruption. *Journal of Microbiological Methods*, **60**, 207–216, doi:10.1016/j.mimet. 2004.09.012.

Clarke P. R. and Hill C. R. (1970). Physical and chemical aspects of ultrasonic disruption of cells. *The Journal of the Acoustical Society of America*, **47**(2 Part 2), 649–653.

Diagenode Ltd. (2012). URL: http://ipmb.sinica.edu.tw/microarray/index.files/Bioruptor%20Manual.pdf (last time accessed 15 March/2012)

Di Carlo D., Jeong K.-H. and Lee L. P. (2003). Reagentless mechanical cell lysis by nanoscale barbs in microchannels for sample preparation. *Lab Chip*, **3**, 287–291, doi: 10.1039/b305162e.

Di Carlo D., Ionescu-Zanetti C., Zhang Y., Hung P. and Lee L. P. (2005). On-chip cell lysis by local hydroxide generation. *Lab Chip*, **5**(2), 171–178, doi: 10.1039/b413139h.

Easley C. J., Karlinsey J. M., Bienvenue J. M., Legendre L. A., Roper M. G., Feldman S. H., Huges M. A., Hewlett E. L., Merkel T. J., Ferrance J. P. and Landers J. P. (2006). A fully integrated microfluidic genteic analysis system with sample-in-answer-out capability. *PNAS*, **103**(51), 19272–19277, doi: 10.1073/pnas.0604663103.

Elsner H. and Lindblad E. (1989). Ultrasonic degradation of DNA. *DNA*, **8**, 697–701.

Gabig-Ciminska M., Liu Y. and Enfors S.-O. (2005). Gene-based identification of bacterial colonies with an electric chip. *Analytical Biochemistry*, **345**, 270–276, doi:10.1016/j.ab.2005.07.024.

Goodwin W., Linacre A. and Hadi S. (2011). An Introduction to Forensic Genetics. John Wiley and Sons.

Harsányi G. (2000). Sensors in Biomedical Applications. Technomic Publishing Co., Lancaster, USA/Basel, Switzerland.

Huang Y., Mather E. L., Bell J. L. and Madou M. (2002). MEMS-based sample preparation for molecular diagnostics (review). *Analytical and Bioanalytical Chemistry*, **372**(1), 49–65, doi: 10.1007/s00216-001-1191-9.

Kulinski M. D., Mahalanabis M., Gillers S., Zhang J. Y., Singh S. and Klapperich C. M. (2009). Sample preparation module for bacterial lysis and isolation of DNA from human urine. *Biomedical Microdevices*, **11**(3), 671–678, doi: 10.1007/s10544-008-9277–1.

Laureyn W., Stakenborg T. and Jacobs P. (2007). Genetic and other DNA-based biosensor applications. In: Handbook of Biosensors and Biochips, R. S. Marks, D. C. Cullen, I. Karube, C. R. Lowe and H. H. Weetall (Eds.), John Wiley & Sons Ltd., Chichester, England, pp. 1035–1053.

Lee S.-W. and Tai Y.-C. (1998). A micro cell lysis device. Micro Electro Mechanical Systems, 1998. *MEMS 98 Proceedings*. Print ISBN: 0-7803-4412-X, Heidelberg, Germany, pp. 443–447, doi: 10.1109/MEMSYS.1998.659798.

Lu K.-Y., Wo A. M., Lo Y.-J., Chen K.-C., Lin C.-M. and Yang C.-R. (2006). Three dimensional electrode array for cell lysis via electroporation. *Biosensors and Bioelectronics*, **22**(2006), 568–574, doi: 10.1016/j.bios.2006.08.009.

Mann T. L. and Krull U. J. (2004). The application of ultrasound as a rapid method to provide DNA fragments suitable for detection by DNA biosensors. *Biosensors and Bioelectronics*, **20**, 945–955, doi: 10.1016/j.bios.2004.06.021.

Pickup J. C. (2007). Biosensors for monitoring metabolites in clinical medicine. In: Handbook of Biosensors and Biochips, R. S. Marks, D. C. Cullen, I. Karube, C. R. Lowe and H. H. Weetall (Eds.), John Wiley & Sons Ltd., Chichester, England, pp. 1069–1075.

Sántha H. (2009). *Új, változtatható felépítésü, elektrokémiai bioérzékelő eszközök konstrukciója és alkalmazásuk* (Construction and Application of Novel Electrochemical Biosensor Devices of Variable Structure). PhD thesis, Budapest University of Technology and Economic, Faculty of Electrical Engineering and Informatics, Budapest, Hungary.

Sun Y. and Kwok Y.-C. (2006). Polymeric microfluidic system for DNA analysis (review). *Analytica Chimica Acta*, **556**, 80–96, doi:10.1016/j.aca.2005.09.035.

Thévenot D. R., Tóth K., Durst R. A. and Wilson G. S. (2001). Technical report – Electrochemical biosensors: recommended definitions and classification. *Biosensors and Bioelectronics*, **16**(1–2), 121–131.

Wang H.-Y., Bhunia A.-K. and Lu C. (2006). A microfluidic flow-through device for high througput electrical lysis of bacterial cells based on continuous dc voltage. *Biosensors and Bioelectronics*, **22**(5), 582–588, doi: 0.1016/j.bios.2006.01.032.

WIKI #1 URL: http://en.wikipedia.org/wiki/Gram-stain (last time accessed 15 March 2012)

WIKI #2 URL: http://en.wikipedia.org/wiki/Gram-negative (last time accessed 15 March 2012)

WIKI #3 URL: http://en.wikipedia.org/wiki/Ultrasonic_bath (last time accessed on 15 March 2012)

Chapter 6

The microsystem based core of the DINAMICS water testing system: Design considerations and realization of the chip units

Theo T. Veenstra

6.1 MICROFLUIDICS IN DINAMICS

This chapter gives some insight in the design process of a complex microfluidic system for the DINAMICS project (DINAMICS). The project aimed for the detection of low-concentration pathogens in drinking water. As the envisioned final system would be integrated onto one single microfluidic chip, the development of all functional blocks has to be done taking into account a common fabrication technology for the chips to ensure compatibility in the final stage of integration.

The sensing techniques applied within the DINAMICS project require the handling and manipulation of microlitre volumes. The contents of these few microlitres originate from several tens of litres of (tap) water. Through different stages the tens of litres of water are worked up to a mere 30 microlitres. The first two stages (hollow fibre filtration, lysis) are described in other chapters in this book.

6.2 LAYOUT OF THE FLUIDIC PARTS OF DINAMICS

From the envisioned task of the DINAMICS detection system, a number of functional building blocks can be distinguished (see Figure 6.1).

The first step in the process is the filtration of sample water. A sample of some 10 litres is taken from the main flow. Most of the water of this 10 liter volume is discarded in a two-stage filtration process. About 10 ml of water is finally produced containing all bacteria which were originally present in the 10 litres.

The 10 ml of bacteria-containing sample is then transferred to the Lysis unit for disrupting all cells. The sample is mixed with a chemical lysis solution and heated for a short period of time. As the full 10 ml + lysis solution is processed, the output volume is over 10 ml's.

From the 10+ ml of solution with all cell-debris, the DNA molecules are filtered out in the Isolation chip. Within this chip the DNA is extracted from its solution by binding to a glass surface. After all cell-debris has been flushed through the chip towards waste, the DNA material is re-eluted into a second buffer solution with a volume of only 50 microlitres. Now all DNA material of all cells originally present in the 10 litres of sample are contained in only 50 microlitres.

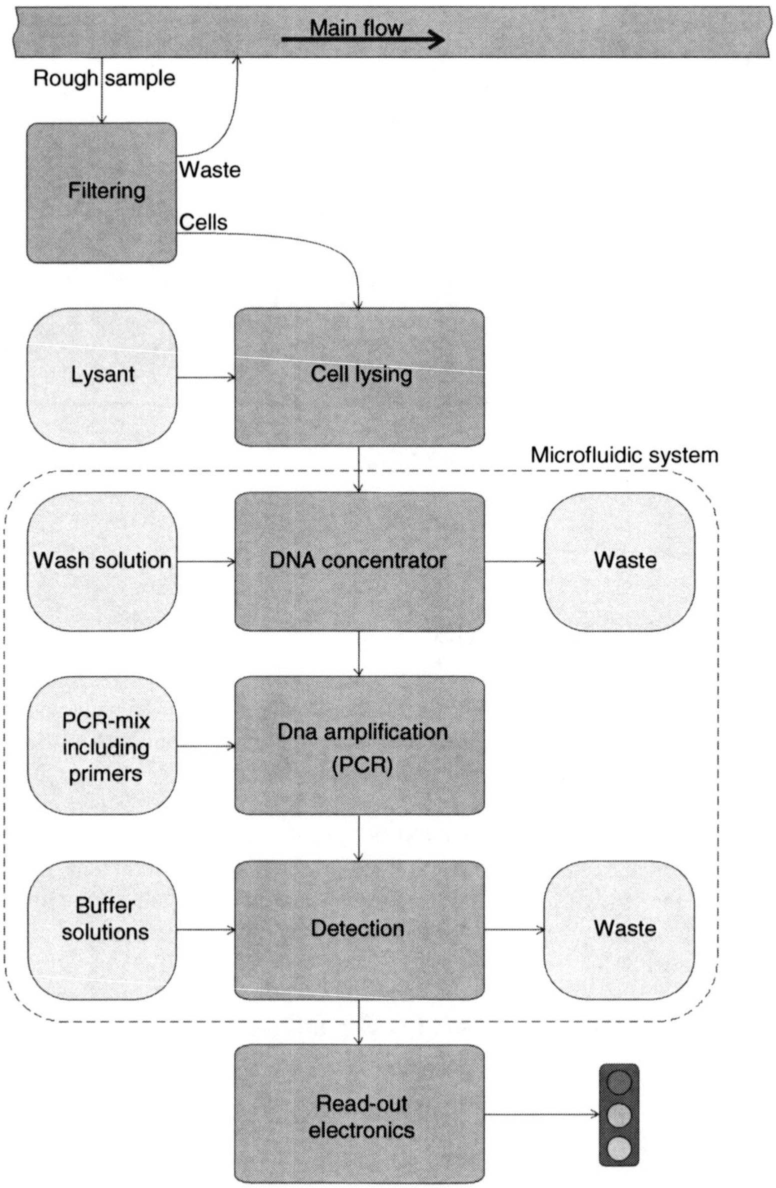

Figure 6.1 Schematic of all fluidic functions in the DINAMICS system indicating also the microfluidic part.

The final sample-preparation step is to multiply the DNA of interest present in the sample solution to such concentration levels that the sensing unit can – with some margin – detect the presence of the pathogens the system is looking for. This multiplication is done by a classic PCR scheme. The cycles of denaturation, annealing and elongation are repeated a fixed number of times, each time doubling the available DNA strands of interest. The output of the PCR-chip is 30 microlitres of sample with copied DNA strands from all pathogens present.

This 30 microlitres are new transported towards the sensing unit. Though within the DINAMICS project two sensing platforms were developed, only one was implemented into the microfluidic platform. This sensor measured a change of capacitance between a set of two electrodes as target DNA was captured onto the surface of these electrodes. This sensing principle is explained in more detail elsewhere in this book. Important to notice here is the fact that the binding events are enabled by a scheme of preparation and flushing buffers.

As this chapter deals with the design of only the microfluidic parts of the DINAMICS system, the filtration and lysis units are not described in the following sections.

6.3 MICROFLUIDIC COMPONENTS IN DINAMICS

From the previous section, it was seen that the DNA-isolation function, the PCR function and the detection were designed with a microfluidic mindset.

Most of DINAMICS building blocks need additional in-/out-puts apart from the ones for the sample-liquid itself. Depending on the function of the building block, this calls for either the capability of switching fluid-streams or for combining two fluid-streams into one single fluid-stream. These basic fluidic functions raise the need for the design/development of valves and mixers, which then can be applied within the functional building blocks.

So the total list of components to be developed consists of the following items:

- Valve (On/Off, combination of valves to create e.g. three-way valves)
- Mixer
- DNA isolation chamber
- PCR chamber
- Detection chamber

These components were developed separately, but as they are to be integrated into a single chip, the fabrication technologies of all separate chips/components have to be compatible with each other. The way this is done is shown in the next paragraph.

6.3.1 The valve component

For the DINAMICS project a valve was developed with minimal dead and displaced volume. As the valve will be used for example in the PCR chip it has to be able to withstand at least 1 Bar of pressure (as the temperature within the PCR chip can be as high as 99°C). Pressures higher than this are not expected in the system at any given time.

The valve design is shown in Figure 6.2. A channel in the chip is covered with a flexible membrane (either Viton or PDMS are ok for this). The channel ends with a through hole to the other side of the chip. Around the entrance of the through hole, a ridge is placed which blocks the flow path when the membrane is forced down. This ridge is the valve-seat. Without any actuation force on the membrane, build-up of pressure will lift the membrane opening a flow path.

A typical test of the performance of the valve is shown in Figure 6.3. During this test run a fixed flow rate was pushed towards the valve. The pressure just in front of the valve is measured with a pressure sensor. The connecting tube to the pressure sensor contains an air bubble, which gives some compliance to the system. At some point during the pressure build-up in time, the valve starts to leak (Figure 6.3). It can be seen that the pressure at which the valve starts to leak is about 8 Bars, after which the valve is able to hold 7 Bars, which is sufficient for the DINAMICS system.

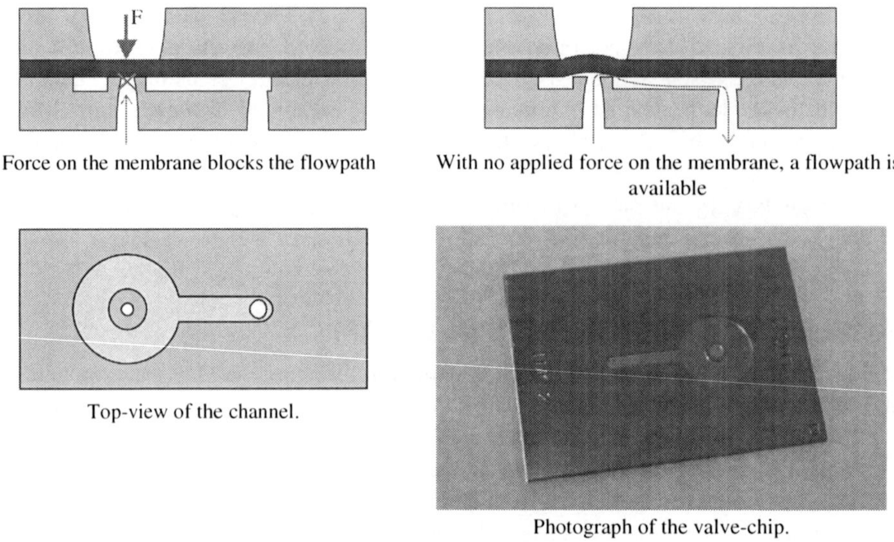

Figure 6.2 Schematics and photograph of the "DINAMICS" valve.

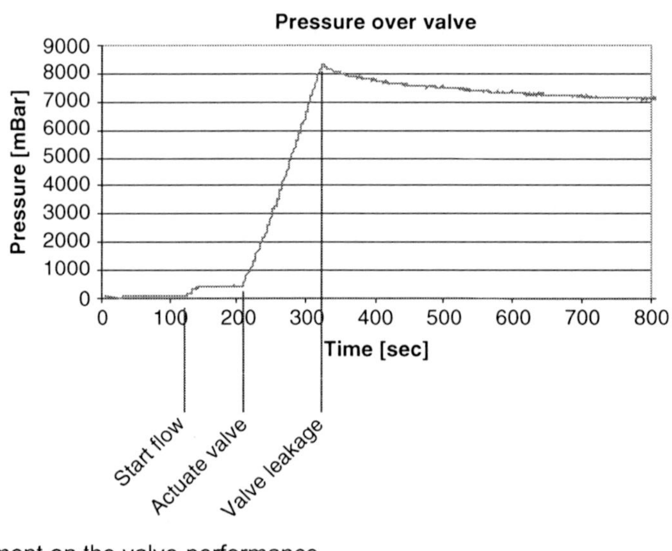

Figure 6.3 Measurement on the valve-performance.

6.3.2 The mixer component

The mixing component of the DINAMICS system is the next basic fluidic component that needs to be developed.

In microfluidics all fluidic flow is laminar. Mixing effects due to turbulence can not be expected, which leaves only diffusion as the driving force for mixing. This means that a trick has to be employed which brings down the diffusion distance. In the literature, several approaches to this end are described (Hessel, 2005) The solution we have chosen is the "lamination mixer," the two fluid streams A and B are brought together such that a distribution ABABA … ABA is constructed over the width of the channel (Figure 6.4). This is done in a wide portion of the channel. After narrowing the channel, the different sections of A and B are only in the order of 1 μm wide, ensuring fast diffusion-based mixing of the compounds in A and B.

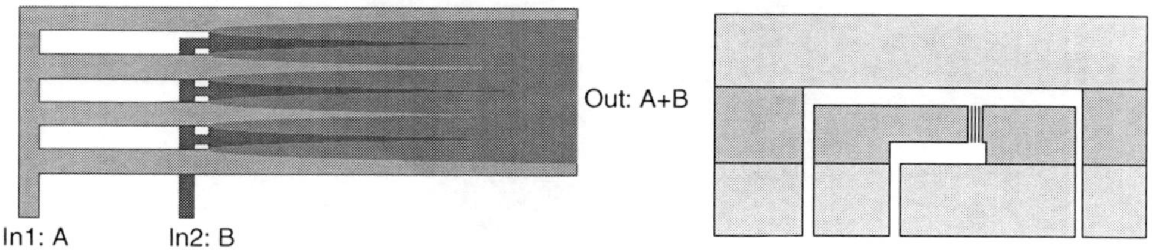

Figure 6.4 Left: Top view of Lamination Mixing principle: fluid B is injected at multiple inlets into a stream of fluid A. Right: crossectional view of chip.

The way this is realized is by creating an array of 39 shielded inlets for fluid B in the stream of fluid A. See also the SEM picture in Figure 6.5. The shielding of the individual inlets is necessary as otherwise the incoming fluid would stay at the bottom of the channel, resulting in a distribution of fluid B at the bottom of the channel with fluid A above that. This would result in a diffusion-distance of over 50 microns which would take a relatively long time – especially for large molecules as DNA.

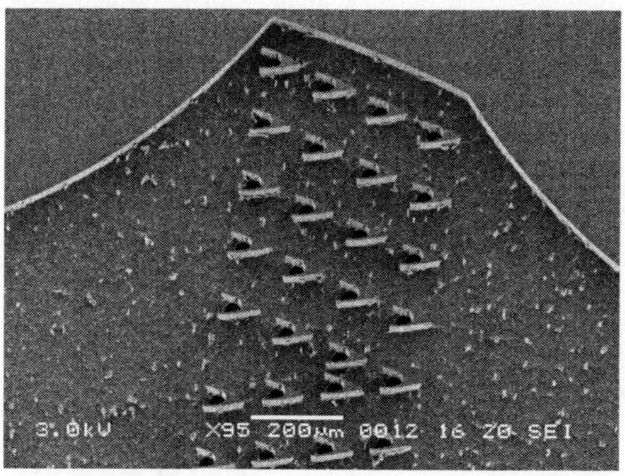

Figure 6.5 SEM picture of the mixing section.

In the mixing section, 39 streams of fluid B are intertwined in 40 streams of fluid A. When this combined stream has arrived at the channel section of 100 microns wide, the average laminate-width is only 1,25 microns, ensuring almost instantaneous mixing. From simulations it was learnt that the distribution of the flow is nearly homogeneous over the inlets (see Figure 6.6). The center of the channel shows a slightly higher flow-rate, locally resulting in a slightly wider section. This does not significantly interfere with the mixers' performance.

With the two most basic fluidic functions of valving and mixing established, the advanced functions will be implemented. First the DNA isolation will be discussed.

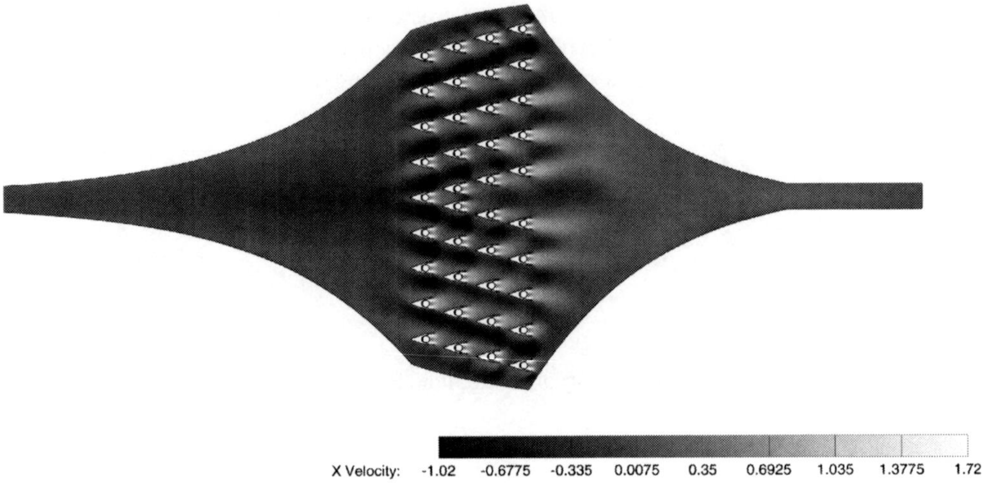

Figure 6.6 Simulation on the mixer's performance.

6.3.3 The DNA isolation component

For the extraction of DNA from a sample liquid, commercial kits are available. These kits require a number of pipetting, agitation and flushing steps to obtain a sample with the desired DNA extracted from the cells in the original sample solution. The working principle of these kits is that the DNA from the disrupted cells binds to a glass surface (mostly in the form of small glass beads). A flushing step makes sure that all other cell-material is disposed of. Using an appropriate buffer solution, the bound DNA is re-eluted from the glass beads into the buffer solution.

The procedure from a standard commercial kit (Roche MagnaPure LC total nucleic acid-kit) is taken as example upon which our chip-system will be based (Figure 6.7).

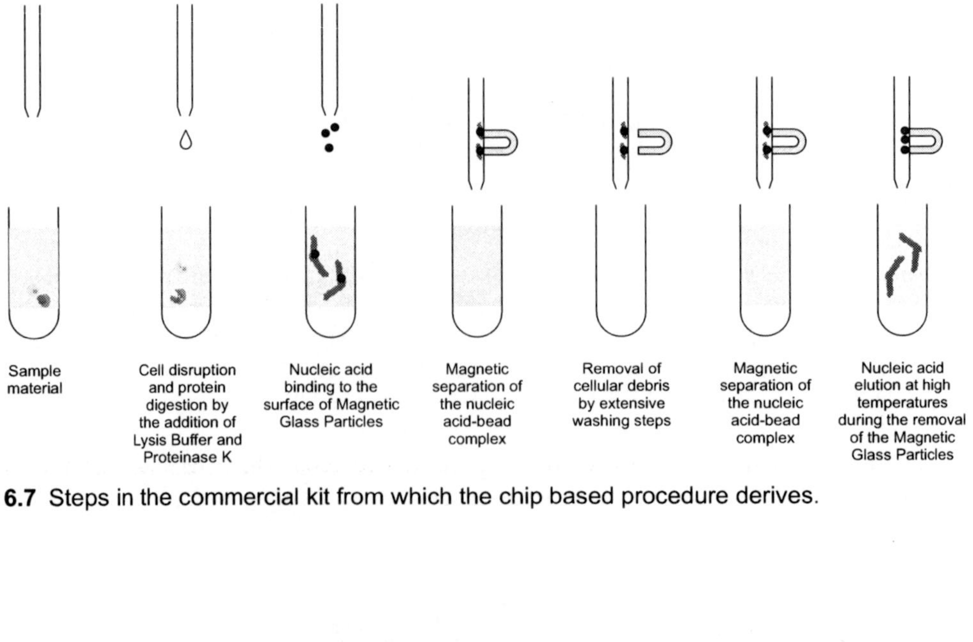

Figure 6.7 Steps in the commercial kit from which the chip based procedure derives.

To translate the "open fluidics" commercial kit into a "closed fluidics" system, some means of trapping the glass material within the chip has to be realized. To be able to do this, a switch is made from glass beads to a glass fibre material. The glass fibre material is placed within a cavity in the chip before closing the chip (see also Figure 6.8). The dimensioning of the cavity with respect to the glass fibre pad is critical: if a flow-path is available around the glass fibre pad, the liquid containing the DNA will go around the envisioned binding sites, resulting in a very low DNA-extraction efficiency of the chip. A donut-shaped cavity is proposed. With the inlet on the outside of the donut and the outlet in the centre, no short-circuiting flow-path exists around the perimeter of the fibre-pad, as long as the height of the chamber is controlled correctly during the fabrication: all sample liquid must pass through the fibre pad.

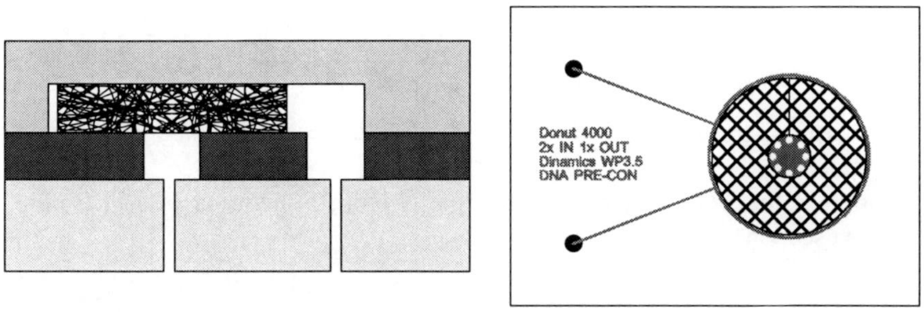

Figure 6.8 Left: schematic of the proposed geometry of the DNA isolation chip, showing how the fiber-pad is placed in the chip. Right: Top view of the chip. Two inlet channels connect to a glass-fiber pad (hatched area). The outlet of the chip is in the center of the glass-fiber pad.

The chip has two fluidic inlets; one is used for supply of sample liquid, the other is used for the re-elution buffer. The flow of the re-elution buffer is controlled through a syringe pump with a small syringe, such that a high degree of control over the disposed/displaced volume is available.

As seen in Figure 6.8, the channel is closed with a Viton (or PDMS) seal, ensuring a leak-tight assembly. This way it is also possible to exchange the fibre pad for example the investigation of different binding materials. Figure 6.9 shows how the chip in its holder looks before and after assembly.

Figure 6.9 Photographs of the chip and its holder. The white patch seen on the right is the glass fiber pad.

Preliminary tests have been performed with the DNA-isolation chip. It was shown that the DNA-capturing efficiency was in the same range as for the commercial kit (see Figure 6.10).

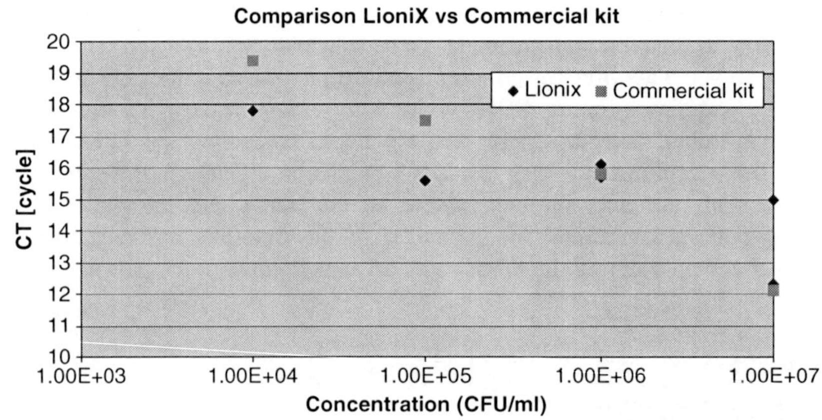

Figure 6.10 Comparison of the efficiency of the reference kit to the implemented chip-procedure.

After the isolation of the DNA-material, the sample is processed further in a PCR-chip. This chip is described in the next section.

6.3.4 The PCR reaction component

The chip described in this section has as task the multiplication of the amount of (targeted) DNA present in the sample-solution. This multiplication is done by the classic Polymerase Chain Reaction (PCR) procedure (see vast amounts of literature for details on "PCR"). For this reaction, the sample needs to be mixed with a "PCR-mix", after which a number of temperature steps need to be performed. The temperature steps are indicated in Figure 6.11. The temperature is cycled a number of times between 95, 55 and 75°C, effectively doubling the amount of targeted DNA with each cycle.

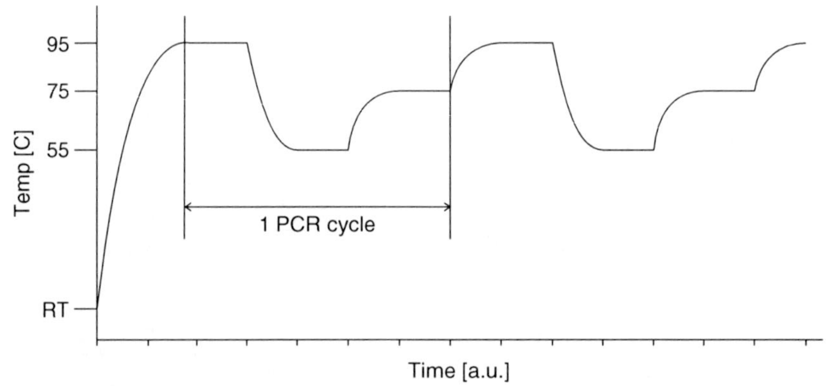

Figure 6.11 The temperatures of the PCR cycle in time.

The chip that has to perform a PCR reaction needs to fulfil quite a number of essential requirements. First, the sample liquid needs to be mixed with the 'PCR buffer'. For this, the mixer described earlier can be applied. Next the sample liquid has to be run through a number of the temperature cycles of the PCR-cycle. As the temperature will get as high as 95°C, the pressure within the chip will reach approximately 1 Bar. Without any counter measures, the PCR chip would boil empty. To prevent this, all the inlets and outlets of the chip must be closed tightly to prevent any liquid leaving the chip during the temperature cycling. For the first chip-series, it was chosen to use external commercial zero-dead-volume valves to fulfil this function (Upchurch rotary valves). Finally, some form of temperature management has to be implemented. To this end, a temperature dependant resistor was implemented on the chip with which it is possible to measure the temperature directly on the chip. Heating of the chip was achieved with an integrated heater-resistor, whereas the cooling-function was implemented with a Peltier cooling element in the chip-holder. See also Figure 6.12 and Figure 6.13.

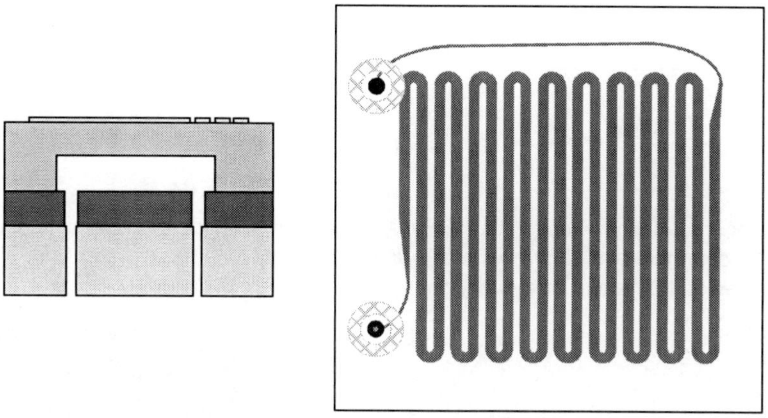

Figure 6.12 Schematic drawings of the geometry of the PCR-chip. Left: cross-sectional view. Right: Top view. The PCR-chamber/channel is strongly elongated to avoid no-flow regions.

Photograph of chip (Front) Photograph of chip (Backside) Photograph of (chip in) holder

Figure 6.13 Photographs of the realized PCR chips and its holder.

Last but not least, the inner surface had to be coated in order to be compatible with the compounds within the PCR-mix. This coating is realized by flushing the chip (when mounted in its holder) with SigmaCote™. Following the prescribed procedure for the sigma-cote, the chips interior is fully covered with a monolayer of water-repellent protective molecules.

Testing the PCR-chips, it was found that the performance is just below the theoretical level; the multiplication factor of 2 per cycle was not fully met. Adding a few extra multiplication cycles solved this issue easily.

The sample coming from the PCR-chip should contain enough DNA-material for detection. This sensing-part is covered in the following section.

6.3.5 The hybridization chamber component

After the DNA has been multiplied in the PCR-unit, it should be possible to detect the DNA using an appropriate sensor. The sensor shown here is employed by the University of Bologna. The working principle of the sensor is a change in the capacitance between two electrodes as DNA selectively binds to the electrodes' surface. The change in electric capacitance or of other label-free or label-dependent electrochemical properties of the electrode interface then can be translated to a DNA-concentration in the sample liquid. Details on the working principle can be found in the chapter by Gazzola and co-workers in this book (Chapter 7). The event of DNA binding to the surface is called "hybridization", hence the chip is named the hybridization chip.

The pictures and photographs in Figures 6.14 and 6.15 show the realized hybridization chip. Four parallel channels with each a private electrode-set are employed on the chip.

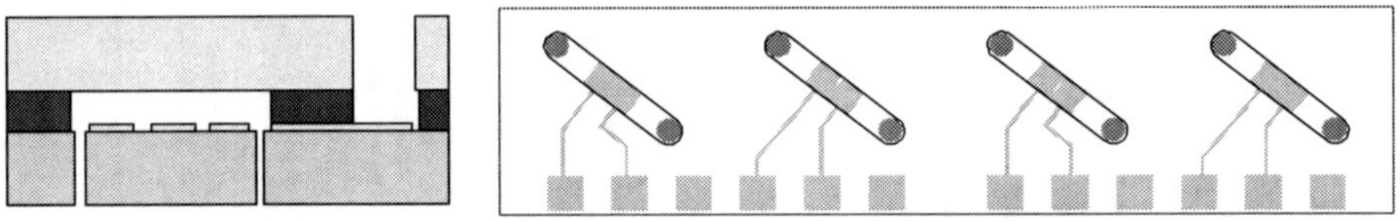

Figure 6.14 Left: schematic cross-section of the Hybridization chip. Right: top view of design of the chip. The oval outlines of the channels are shown with the fluidic in- and out-lets at each end. The measurement electrodes are connected to the two arrays of connector pads.

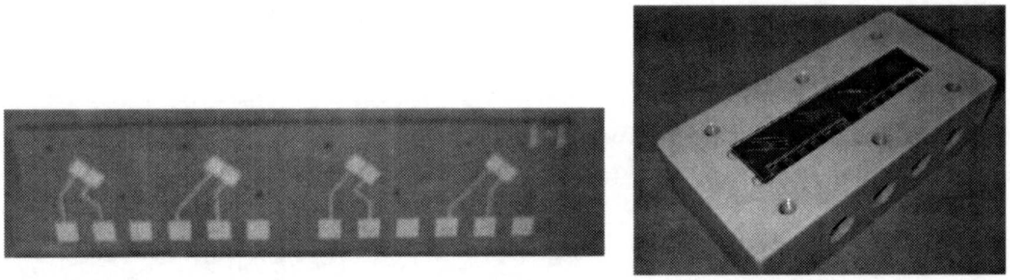

Figure 6.15 Photographs of the chip-half with the electrodes for the Hybridization-chip (left) and the hybridization chip mounted in its holder (right).

6.3.6 Fabrication – Technology platform

All fluidic components from the previous sections were fabricated in silicon and glass technology (MEMS) using different fabrication schemes. These fabrication schemes are summarized in Figure 6.16 and Table 6.1. Some of the dimensions are dictated by the function of the component, whereas others could freely be chosen.

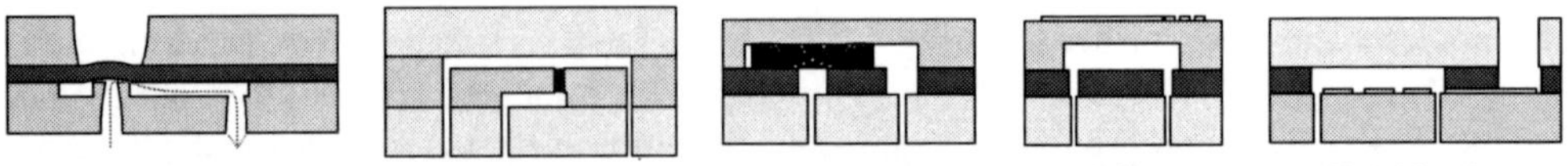

Figure 6.16 Different fabrication schemes of the different components.

As the ultimate goal for the DINAMICS project was the integration of all microfluidic functions into one single chip, all fabrication schemes of the standalone components have to be aligned to fit within one fabrication technology.

Table 6.1 Used fabrication techniques for the different parts of the microfluidic chips.

	Valve	Mixer	DNA isolation	PCR	Hybridization
Channels	DRIE etching 150 μm	DRIE etching 100 μm	DRIE etching 240 μm	DRIE etching 200 μm	Cut in Viton (250 μm)
Closing of channel	Viton	Viton/PDMS	Viton	PDMS	Clamped glass
Through holes	Powder Blasting	DRIE etching 350 μm	Powder blasting	DRIE etching 350 μm	Powder blasting
Channel2	–	DRIE etching 100 μm	–	DRIE etching 100 μm	–
Closing of channel2	–	Glass plate Anodic bonding	–	Glass plate Anodic bonding	–
Through holes2	–	Powderblasting	–	Powder blasting	–
Electrodes	–	–	–	Sputtered Ta/Pt on thin SiO_2 layer	Sputtered Cr/Au on thick SiO_2 layer

The used fabrication schemes for the single components were mostly selected for their convenience at the single component level. Now some compromises have to be made in order to be able to fit all components into the single production scheme. For example this might mean that the depth for a certain section of the chip will be leading, forcing another component to a non-optimal depth. This reconfiguration process is schematically shown in Figure 6.17.

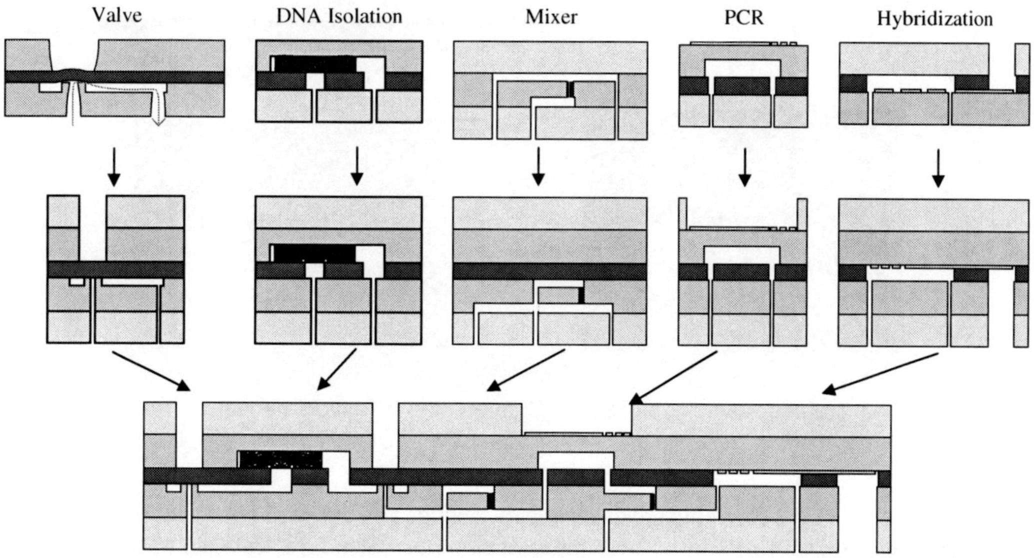

Figure 6.17 Overview of the technology shifts in the different componenents in order to align the fabrication technologies into the fully integrated chip.

From the last schematic in Figure 6.17 it can be seen that the depth of the PCR chamber is dictated by the depth of the Isolation chamber, due to the thickness of the fiber-glass pad. All other etched channels are used only for transportation purposes, for which purpose the exact depth is not too important. Here a trade-of between internal volume [low volume gives lower sample broadening (*dispersion*)] and pressure-drop over the channel is made. A depth of around 100 microns works out fine for these channels.

6.4 CURRENT STATUS

At the moment of writing this section, all components have been fabricated and shown to be able to perform their function. Two functional integration steps have been taken. Firstly, the mixing unit has been integrated with the PCR-unit and secondly, the valve units have been combined with the hybridization chamber. The latter will be described here in somewhat more detail.

As a single component, all fluid switching actions for the hybridization chamber were performed manually. Here the same (geometrically) hybridization layout is shown with valves integrated for the control of three inlets (sample, hybridization buffer, flush buffer) and control over the outlet (directly to outlet or to outlet through hybridization chamber). The layout of this chip is shown in Figure 6.18.

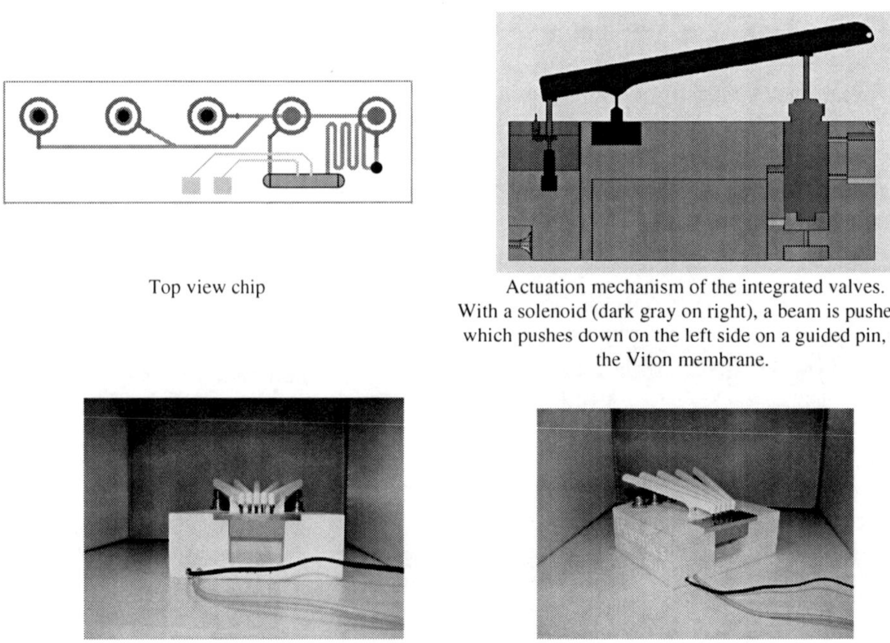

Top view chip

Actuation mechanism of the integrated valves. With a solenoid (dark gray on right), a beam is pushed up, which pushes down on the left side on a guided pin, onto the Viton membrane.

Photos of assembly

Figure 6.18 Schematic of the valved hybridization chamber, the actuation mechanism and a photograph of the chip mounted in its holder.

The valves in this hybridization chip are mechanically actuated by a guided pin which is pushed down by a solenoid. In order to have fine control over the actuation, the solenoid is coupled to the pin/valves by means of a long lever. A high travel at the solenoid side corresponds to a small travel-length at the valve's side.

It has been shown that liquids from three sources could be directed to either the hybridization chamber or to waste. This enables the transport of sample and buffers with very low losses, since no liquid is wasted during washing steps due to dead volume in the connecting tubing.

6.5 CONCLUSION

In this chapter, the current status of the Microfluidic parts of the DINAMICS system has been shown. The microfluidic part of the DINAMICS system begins directly after lysis of the captured cells. The DNA is isolated form the solution, after which the amount of available DNA is multiplied in a PCR reactor. The resulting sample is brought to a detection cell in which the DNA is bound to a gold electrode, causing a change in the electrochemical properties of the electrodes interface of the electrode. This change eventually leads to a signal which can be interpreted in terms of either a safe or a dangerous pathogen-count in the original sample.

It was also discussed how all microfluidic components could be aligned in terms of fabrication technology. Though the system was not realized as a fully integrated system, the functional integration was shown to be possible in the PCR-chamber as well as the hybridization chamber (detection). The integration of valves and mixers with the functional blocks eliminates dead volume and the associated loss of sample/sensitivity since the sample is not smeared out in long connecting tubing. The on-chip valving also leads to lower liquid buffer usage as less volume has to be used when refreshing the critical volumes in the chip.

REFERENCES

DINAMICS, 6th European Framework Programme Project Diagnostic Nanotech and Microtech Sensors IP 026804-2, information at http://www.dinamics-project.eu/ (accessed 1 March 2012)

Hessel V., Lowe H. and Schonfeld F. (2005). Micromixers – a review on passive and active mixing principles. *Chemical Engineering Science*, **60**(8–9), 2479–2501.

Chapter 7

Electrochemical biosensor strategies for pathogen detection in water security

Daniele Gazzola, Simone Bonetti and Giampaolo Zuccheri

7.1 INTRODUCTION

This chapter addresses the limitations of conventional laboratory analytical techniques, and presents an introduction to electrochemical biosensing for the detection of pathogens in water supply systems. The detection principle of biosensors is based on the biological recognition of a molecule that is specific to the organism of interest. Of the many strategies available, the ones based on antibodies and on DNA recognition are evaluated and compared to conclude that DNA sensors are the optimal available choice mainly because of the short assay response time, and of the capability to detect organisms that are present in small concentrations and in small fractions with respect to other organisms present in the same sample. The second part of the chapter introduces the technologies used for the assays, in particular some electrochemical analytical techniques are introduced, and the technology of choice developed in the framework of the DINAMICS project is described.

7.2 BIOSENSOR STRATEGIES FOR PATHOGEN DETECTION IN WATER SECURITY

As described in Chapter 1, there is an increasing demand for testing the quality of public water systems, because of both natural contaminations, and biological terroristic attacks. Being able to perform accurate and fast tests to reveal the presence of hazardous microorganisms can be critical in certain situations. Unfortunately, currently available tests are not as fast as desirable, and the results are usually available after one or more days, which makes them not very effective. In this scenario, a number of alternative technologies are under development to solve the issue of accurately and timely measuring very small amounts of microbes in large water supply systems.

Among the many technologies that can be used to detect microorganisms, the ones based on biosensors are certainly the most promising, also because such type of sensors are already well established in the market of portable medical devices. There are research projects aiming at the translation of the required knowledge from the medical field to security, and the DINAMICS project was one of them. As an example, the most successful biosensor application in the medical market is the glucometer, which measures glucose in diabetic patients. The device is portable and the measurement is fast (less than a minute), as required for security specifications. Furthermore, the instrument can be used by untrained personnel, generally by the patient itself, and the active element that is used for the measurement, the biochip, is cheap and disposable, to avoid cross contamination between samples and to guarantee testing quality.

The choice of the specific biosensor technologies to be used depends strongly on the application, and in the following sections, after a short description of conventional laboratory analytical techniques, we will discuss the biosensor solutions that have highest potential for water security, in particular the technologies based on affinity recognition through antibodies and on DNA hybridization.

7.2.1 Conventional laboratory analysis

Most of the currently used microbiological procedures consist of two fundamental steps: preconcentration/enrichment and detection/quantification. The first step commonly uses filtration or elution to increase the number of microbes to be tested, so that their concentration becomes measurable. In fact, due to the variability of the infectious dose in the population, even a small number of viruses, bacteria or protozoa can be considered as potentially able to cause a disease if ingested. For example, the median hazardous dose for Salmonella is ten thousand cells, while this number lowers to just a few cells for *Cryptosporidium parvum*. The detection of such small quantities is very challenging, and a preconcentration step of those microbes is required with all available technologies. However, the amplification yield of microorganisms preconcentration must be proportional to the limit of detection of the measurement technique: if a lower concentration of microbes can be recognized by the instrument, the amplification can be more modest. In the case of conventional laboratory techniques, the limit of detection is often not small enough, so it is required to cultivate the microbes for one or two days in optimal temperature and nutrient conditions before they can reach a detectable concentration.

Following the preconcentration steps, the organisms are usually cultured for further amplification until plaques, colonies, or cysts are formed and visible. Finally a counting procedure is used for the calculation of the initial concentration of pathogens in water. All the cultivation procedures are optimized so that the organisms of interest are selectively stimulated to grow faster than the others, resulting in an improvement of assay selectivity. Also, the cultivation steps exclude dead organisms from the assay response, as they do not replicate.

This schematization of the general architecture of water pathogens recognition is performed through a number of technologies, each optimized for a specific microbial targets, as summarized in Table 7.1. (Koster *et al.*, 2003)

Table 7.1 Schematization of the steps involved in the analysis of different microorganisms in conventional laboratory analysis.

	Microbial groups		
	Viruses	**Bacteria**	**Parasitic protozoa**
Preconcentration/ enrichment	Adsorption-elution	Membrane filtration	Cartridge filtration, immunomagnetic separation
Detection/enumeration	Cell culture, cytopathic effect, count of plaque forming units	Selective growth on agar, count of colony forming units	Immunological staining, count of fluorescent cysts

A consequence of the described multi-step approach is that the currently used protocols have a very long response time, mainly due to the cultivation step, which can take more than one day. In this way, the results are obtained after one or two days of laboratory analysis since the collection of the sample. This scenario is clearly not optimal for current security requirements, and motivates the search for novel, faster technologies.

7.2.2 Biosensor analysis

Biosensors are analytical devices for the detection of a target analyte. The working principle is based on two main elements: a biological recognition unit, called the receptor, which can interact specifically with the target, and a physicochemical detection unit, the transducer, which converts a change in property (optical, electrochemical, mechanical, etc.) of the sample surface or solution, into an electric output signal. The signal is then analyzed and elaborated by a computing unit to output the response of the measurement.

The ability of such tests to be highly accurate depends on two factors: on one side, their use of a biological receptor element enables the biochip to bind molecules of the desired target with low interference from other organisms and molecules present in the sample. On the other side, the transducer unit transforms the biochemical recognition of the target into an electric signal through technologies that can sense small differences in the presence of microbes or microbe markers close to the surface of the biochip.

In the medical field, the principal advantage of biosensors on conventional assays is that the detection of molecular interactions happens as they take place, at the point of care, and without requiring auxiliary procedures thanks to their high sensitivity and low limit of detection. As explained above, in the case of water security systems, a preconcentration step is anyways required, as there is the need to recognize tiny quantities of highly toxic pathogens. However, the limit of detection of biosensors allows to be less stringent on the amplification steps, as described herein.

Of the many existent types of biosensors, developed primarily for medical applications, we herein describe in detail the two that can be translated towards the detection of pathogens in drinking waters.

Immunosensors

Immunosensors use antibodies as receptors for the identification or quantification of the target material in a sample. Antibodies are an efficient recognition tool because, if properly developed, they rely on their unique ability to bind their respective antigen with much higher affinity than other molecules. Furthermore, immunogenic diversity allows to monitor potentially any compound that can produce a response of an immune system. In fact, antibodies are produced by B-cells of the immune system to identify and neutralize non-self antigens such as bacteria and viruses.

Antibodies recognize a specific part of a foreign molecule, called the antigen, and bind it. Through this molecular interaction, an antibody can fasten to the surface of a microbe, hence providing a tag for the target selection and target recognition. A further advantage of this technology is that the general structure of all antibodies is very similar, with just an extremely variable small region at the tip of the protein, known as the hypervariable region (Figure 7.1), which is responsible for the recognition of the antigens. This allows the development of generally functioning transduction techniques, relatively independently on the specific target specie out of the wide variety of pathogenic organisms.

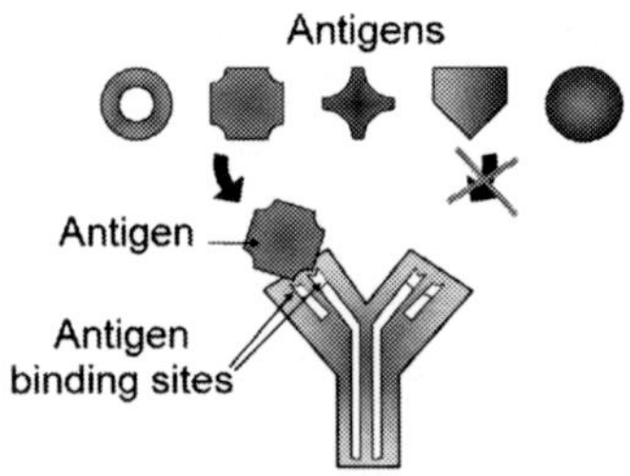

Figure 7.1 Structure of an antibody. The general structure of all antibodies is very similar, but different antibodies can specifically bind a certain antigen thanks to a short part of the sequence which is extremely variable, called the hypervariable region (indicated with the arrow).

A wide range of immunological methods that take advantage of the interaction antibody-antigen is available. As an example, the target organism can be caught by capture antibodies that are stably deposited on the surface of the biochip transducer. The persistence of the target organism on the surface can be directly detected by the transducer, or the detection can be aided by the addition of a label, normally carried by a secondary reporter antibody in a so called sandwich assay, as pictured in Figure 7.2. While direct label-free detection uses an easier biochemical protocol, its limit of detection is generally not sufficient for most applications, therefore the sandwich assay is a more frequent configuration.

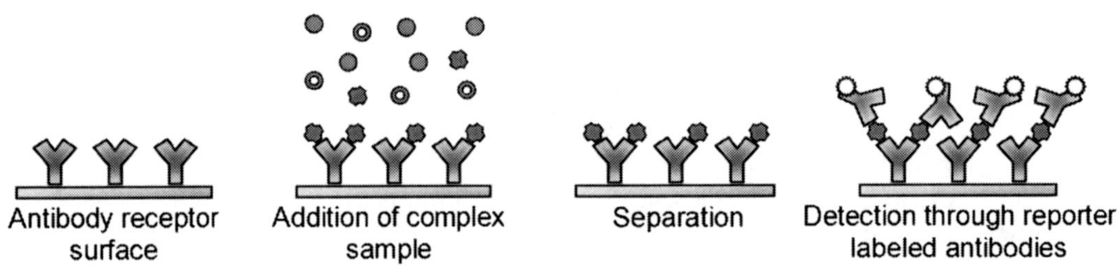

Figure 7.2 Schematization of an immunosensor sandwich assay. Generally a layer of probe antibodies deposited on the biosensor surface is used to capture the target analyte and separate it from the whole sample. Secondary reporter antibodies then bind to the captured target antigens and are used for the signal transduction and detection.

In another type of immunosensor, magnetic beads coated with capture antibodies specific for a target organism are used (Gijs, 2004). After incubation and efficient mixing of the particles in the cell suspension, the target and the magnetic beads are bound through affinity recognition, so they can be separated from the rest of the suspension with the help of a magnetic field. As in the previous example, a set of reporter antibodies can be used for the signal transduction.

To operate an immunosensor, a basic lab infrastructure could be an advantage, but it is not necessary. The assays are developed to be easy to perform and fast, with a typical response time inferior to one hour. Immunosensors have a potential for applications in water security, but the limit of detection needs to be lowered for direct application of such devices in the field, in order to decrease the requirements on the preconcentration procedure, and hence the response time.

A further use of immunosensors stems from them being also sensitive to non-viable microorganisms. Poorly perpetrated deliberate contamination attempts might result in unusual but dead microbial agents flowing in the water system: it might still be desirable to detect the breach in the microbiological security system even if it does not directly lead to an increased health risk.

DNA sensors

The use of nucleic acid recognition layers is a relatively new and exciting area in sensors technology. Hybridization biosensors are in fact considerably promising for obtaining sequence-specific information in a simple, fast and cheap manner compared to conventional assays. Their advantages are very similar to those offered by immunosensors, with an additional opportunity coming from the existence of established and fast DNA amplification procedures, such as the polymerase chain reaction (PCR) or other newer strategies (see Chapter 8 of this book), which are significantly faster than the time consuming amplification of whole organisms through selective cultivation.

The working principle of these devices is the Watson-Crick DNA base pairing. The biosensors rely on the immobilization of a single-stranded DNA sequence, defined as probe, on the transducer surface. In turn the transducer gives an electrical signal upon hybridization with the desired complementary region of the target nucleic acid (Figure 7.3) (Wang, 2000).

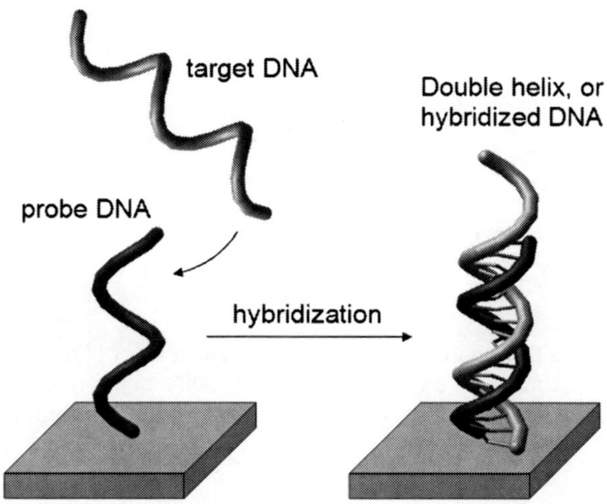

Figure 7.3 Concept of the formation of a DNA double helix through hybridization.

The required DNA sample extraction and preparation steps, including amplification processes such as PCR, have been demonstrated at the microscale (Mir *et al.*, 2004), with examples reported also in this book, in Chapters 5 and 6. This makes such devices very suitable for complete on-chip DNA analysis.

PCR exponentially multiplies the number of target DNA strands by using the DNA target as a template, and replicating it through the DNA polymerase enzyme, which is natively used by eukaryotic and prokaryotic cells, and by some viruses for replication and repair of DNA. The oligonucleotide sequence to be amplified is selected by the primers, which are short oligonucleotides (usually less than 30 bases) that bind to complementary strands flanking the sequence of interest of the target DNA. A thermocycler allows iteration of the three necessary steps for the reaction: denaturation, annealing and extension of the primers. On each cycle, the amount of the DNA sequence specifically selected by the primers is roughly doubled, allowing for a selective amplification of only the nucleic acid of organisms of interest. The

DNA amplification is a particularly desirable step when very small amounts of microorganisms are available, since it can specifically amplify the desired DNA sequence by many orders of magnitude in a timescale of about one-two hours.

The PCR output can then be injected on the biochip, where probe oligonucleotides complementary to a part of the amplified target sequence capture the target molecules by base-sequence recognition. The double helix can be directly recognized, or it can be labelled for example with a reporter DNA sequence that binds to a part of the target that is still free (in a sandwich assay configuration). A readily detectable signal can be produced (Figure 7.4).

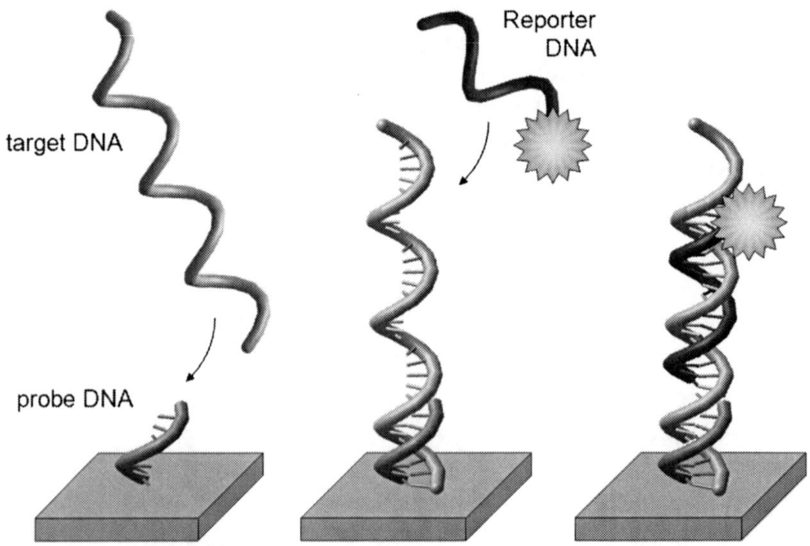

Figure 7.4 Schematization of a DNA sandwich assay. Generally a layer of probe oligonucleotides deposited on the biosensor surface is used to capture the target DNA sequence only. A secondary reporter oligonucleotide then forms a double-helix to an unbound segment of the target sequence and is used for the signal transduction and detection.

The DNA biosensor technology has been developed at the end of the 20th century for biomedical applications, and it led to the well known products called DNA microarrays, where the probe can be produced by conventional or in situ synthesis (20 to 30 bases probe) (Lemieux *et al.*, 1998; Lipshutz *et al.*, 1999), or it can be synthesized through the enzymes reverse transcriptase and DNA polymerase from a biological source (500 to 5000 bases cDNA probe) (Ekins *et al.*, 1999).

As in the case of immunosensors, DNA devices do not require a lab infrastructure for the measurement, and the assays are sensitive, easy and fast to perform, with a typical time of 2–4 hours including the target amplification step. DNA biosensors are very promising for applications in water security, with the only downside that they need to be integrated with a sample preconcentration module and a DNA extraction module. Of particular relevance is the great improvement in response time with respect to conventional laboratory assays. Furthermore, like immunosensors, DNA biosensors maintain the ability to recognize both viable and non-viable microbes.

In Table 7.2, the main advantages and disadvantages of the classes of biosensors so far mentioned are summarized.

Table 7.2 Summary of the advantages and disadvantages of conventional laboratory analysis, immunosensors, and DNA biosensors.

Method	Characteristics & advantages	Limitations & disadvantages
Conventional laboratory analysis	Well established Precise	Long response time, mainly due to the detection step, which generally requires the cultivation of the microbe. The analysis must be performed in a biological laboratory. Usually unable to detect dead organisms.
Immunosensors	Compatible with portable devices, and the assay can be performed by untrained personnel Typical response time inferior to one hour, without considering sample pretreatment. Micromanufacturing techniques allow testing of several organisms in a single assay, on a single biochip.	Sample selective concentration is not sufficient for very low concentrations of pathogenic organisms, and in some cases an amplification step through cultivation is necessary.
DNA biosensors	Compatible with portable devices, and the assay can be performed by untrained personnel. Possibility to integrate with DNA amplification established techniques. Sensitive, selective and specific enough for water pathogens. Typical response time of 2 to 4 hours, including sample amplification. Micromanufacturing techniques allow testing of several sequences in a single assay, on a single biochip.	The sample needs to be pretreated for DNA extraction. It is difficult to achieve absolute quantification.

7.2.3 Conclusions

In this section, we have analyzed the characteristics of the most promising technologies that can be translated from the biomedical field for water security, with a clear competitive advantage for the fast, sensitive and selective DNA biosensors. In the following section, the most common methods for the electrochemical conversion of the hybridization event into an electrical signal are presented.

7.3 COMMON ELECTROCHEMICAL DETECTION SYSTEMS FOR DNA BIOSENSORS

Each biosensor is composed of a biological receptor, as described in the former section, and of a transducer. The receptor, which is generally located on the surface of the biochip, captures the target molecule, inducing a consequent modification in a property of the solution close to the sensor surface (e.g. dielectric constant, or availability of electrochemically active molecules). The transducer is the unit that detects such property change and converts it into a readable electrical signal. A number of technologies are used for the transduction, each using a particular physicochemical property change that can be electrochemical, optical, and mechanical (D'Orazio, 2003; Rodriguez-Mozaz *et al.*, 2004). Each technique has specific advantages, for example, the use of surface plasmons in Surface Plasmon Resonance (SPR) devices allows for real-time label free kinetic measurements, colorimetric sensors give a response within minutes that can be read by the naked eye, and electrochemical sensors are the most miniaturizable and integrable.

In this section, we present the basic principles of DNA biosensors based on electrochemical transduction. This class of sensors offers several advantages over other technologies, mainly related to its intrinsic capabilities of miniaturization and integration with electronic circuits. The strong connection with electronics both at the level of the control circuitry and of the biochip fabrication technology, allows to take advantage and incorporate knowledge from the very well established field of information technology. Not only the response signal is already electrical and can be directly processed by conventional electronics in a cheap and fast manner, but also the biochip itself can be considered part of the electrical circuit, so that electronic architectures typical of sensing devices, such as the Wheatstone bridge, can be replicated with the inclusion of the biochip as one of the sensing impedances (Luong *et al.*, 2008). The high sensitivity, simplicity, and cost competitiveness gave a wide popularity to this class of technologies, so that a large fraction of the market and of the research publications on biosensors are based on electrochemical transducers (Meadows *et al.*, 1996).

In summary, electrochemical detectors have fast response, high sensitivity, small dimensions, low cost, easy signal integration, and are compatible with microfabrication technology. All this makes such class of transducer technologies the best candidate for the integration in microchip devices.

In the next sections, the most used electrochemical detection techniques are presented: amperometry and voltammetry.

7.3.1 Amperometry

Amperometric biosensors are based on the measurement of an oxidation or a reduction process occuring at the measuring electrode (Mir *et al.*, 2007). Such redox processes take place at a rate that depends on the voltage – called the excitation voltage – which is applied and kept constant by the measuring system. The signal measured is the oxidation or the reduction current as a function of time. The detection process generally involves the use of suitable enzymes which convert substrate molecules to measurable electroactive species. The majority of amperometric biosensors use receptor molecules immobilized on the surface of the measuring electrode, as in (Liu *et al.*, 2004), but this is not necessary, as the electroactive product can be produced in the bulk by the enzyme, and then diffuse to the electrode, where it is detected.

The first amperometric biosensor developed for a commercial application is the glucose biosensor (1962 by L. C. Clark), which still has huge medical implications in the treatment and follow-up of diabetic patients. (Harvey, 2000) The selectivity to glucose is guaranteed by the presence of a set of isolating membranes: a polycarbonate membrane, which is permeable to glucose and dissolved oxygen, a second membrane bound to glucose oxidase molecules, which catalyzes glucose oxidation to gluconolactone and hydrogen peroxide, and finally a cellulose acetate membrane, through which H_2O_2 diffuses and reaches the platinum or gold electrode. This system makes use of an amperometric chemical sensor of hydrogen peroxide that Clark developed few years earlier.

The hydrogen peroxide reaches a system of two electrodes positioned behind the cellulose acetate membrane (Figure 7.5a). The two electrodes and the measurement setup are designed so that the system applies a constant voltage at the interface between a platinum electrode (called the working electrode) and the sample. In this case, the second electrode (called the counter electrode) is a silver ring treated with a surface coating of silver chloride.

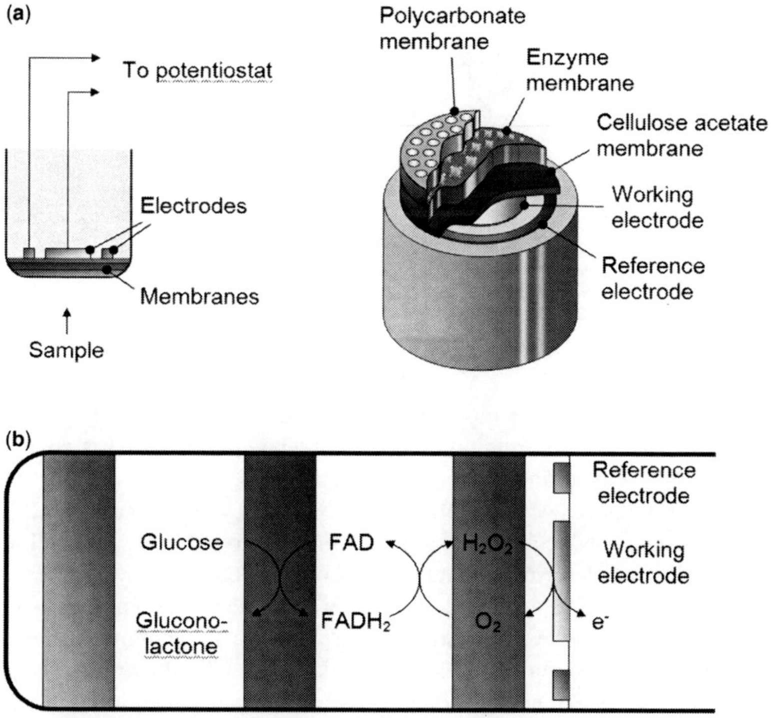

Figure 7.5 The glucose biosensor. In (a) the structure of the biosensor, which consists of a set of membranes deposited on the two measuring electrodes, is represented. In (b) the conceptual use of the membranes to pre-filter the sample, to host glucose oxidation, and to further filter the solution that reaches the electrodes for the measurement, is shown.

At the electrode the hydrogen peroxide is oxidized by an excitation potential, following the reaction:

$$H_2O_2(aq) + 2OH^-(aq) \rightarrow O_2(g) + 2H_2O(l) + 2e^-$$

The working scheme of the biosensor is represented in Figure 7.5b.

Recently, electrochemical mediators have been introduced to substitute hydrogen peroxide as a charge carrier. They operate at lower electrode potential, and give the advantage of decreasing the interferences by other electrochemically active species found in complex matrices (Farré *et al.*, 2005).

In another configuration, amperometry can be used in DNA biosensors by employing electroactive indicators to monitor the hybridization event (Mikkelsen *et al.*, 1996). Several biochemical strategies can be used: for example the electrochemical indicator can be a molecule that has higher affinity towards DNA in the double helix form, compared to the affinity for single strands. After exposing the biosensor to the sample, a certain amount of probe strands on the surface of the biosensor hybridizes with the target, with a consequent increase in the amount of detectable indicator bound to DNA on the electrode. Another strategy uses a sandwich configuration (Figure 7.6), where a secondary reporter probe DNA hybridizes a free sequence of the target. If the reporter probe is tagged with a redox enzyme, such as horseradish peroxidase, a mediator can be continuously oxidized by the enzyme, and then reduced at the electrode, with a resulting reduction current measured by the amperometric electronics (Wang, 2000).

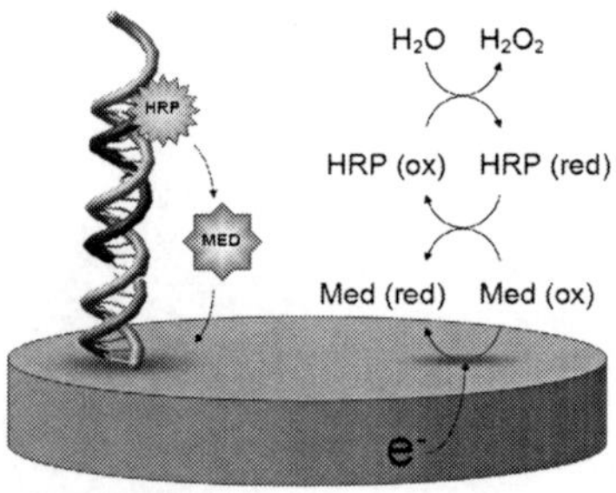

Figure 7.6 Example of a DNA sandwich configuration for an enzymatic detection of a target sequence.

7.3.2 Voltammetry

Voltammetry reveals the amount of redox molecules reduced or oxidized when the electrochemical cell is excited with a controlled potential. In voltammetric experiments, the potential is changed with time in a predetermined way, while the resulting current is measured. Herein, the working principles and the measurement interpretation of linear sweep voltammetry (LSV) and cyclic voltammetry (CV) are introduced.

LSV and CV are well established techniques, as they were developed in the first half of the 20th century (Matheson & Nichols, 1938), and described theoretically about ten years later by Randles and Sevcik (1948). In LSV measurements, the excitation voltage is ramped at a constant rate while the current is monitored (Christensen & Hamnett, 1994). The resulting signal is displayed in a voltammogram (Figure 7.7a), where the current is represented as a function of the applied voltage. In CV, the voltage is ramped linearly as well, but the potential is swept back and forth one or several times between a minimum and a maximum value. The response is displayed in a voltammogram (Figure 7.7b), like for LSV signals.

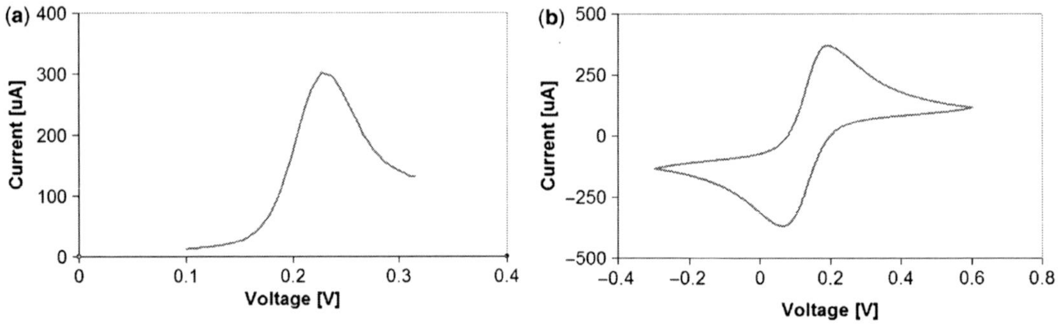

Figure 7.7 Typical results of LSV (A) and CV (B) measurements. The voltammograms represent the current generated at varying excitation potential.

In typical voltammograms of LSV and of CV, a peak-shaped current stands on top of a background current that increases with the voltage. The background current can be associated to the movement and collection of charges on the surface of the *working electrode* (charging of the double layer capacitance), and in normal conditions it does not contain useful information. On the other hand, the informative parts of the curves are the peaks, whose shape depends on the concentration of the electroactive molecule, as well as on other features, such as the reversibility of the redox reaction.

In the special case of electroactive molecules bound to the surface, the rise of the redox peaks can be easily interpreted as the oxidation (or reduction) which starts at a sufficiently high potential, as defined by the Nernst equation. After some time, at a higher potential, most of the molecules initially available have reacted, thus the reaction drops off because the molecules already oxidized (or reduced) cannot oxidize (or reduce) again. When there are no more reduced (or oxidized) molecules available to react, the reaction stops, and the current goes back to the background value, completing the peak-shaped curve in Figure 7.7.

In the case of electroactive species floating in the bulk, the number of available redox molecules is not limited, but still the shape of the voltammetric response curve is very similar to the former case. In fact, even if the bulk contains a large quantity of redox molecules, only the ones that are close enough to the electrode so that they can diffuse to it in the timescale of the experiment are able to react with the electrode. In other words, distant molecules cannot reach the electrode. Also in this case, the availability of redox species close to the surface decreases as the reaction goes along. The shape of the oxidation and reduction peaks depends not only on the concentration of redox agents in solution, but also on the rate at which the voltage scan is performed. Also, if more than one redox molecules are used, the many signals can be identified and separated because they happen at different potentials.

Both for LSV and for CV setups, the most relevant operative feature used for DNA detection is the quantification of the redox agent that has reacted at the electrode. This amount is correlated with the total area of the peak. Both LSV and CV are very precise in this measurement with respect to other electrochemical techniques including amperometry, also because the measuring instrument, called the potentiostat, does not simply apply a voltage across two electrodes, but it uses a three electrodes setup to precisely control the potential drop across the interface between the one measuring electrode (called the working electrode) and the bulk (Figure 7.8). The electronics and the other two electrodes are specifically designed so that the result depends only on the molecules reacting at the working electrode, while it is independent of the reactions occurring elsewhere. The other electrodes are the *reference electrode*, which maintains a constant potential with respect to the bulk, and sets the reference for the control of voltages, and the *counter electrode*, which closes the electrical circuit to allow the electrons exchanged between the working electrode and the sample to travel through the potentiostat to the sample again, and so it keeps electroneutrality.

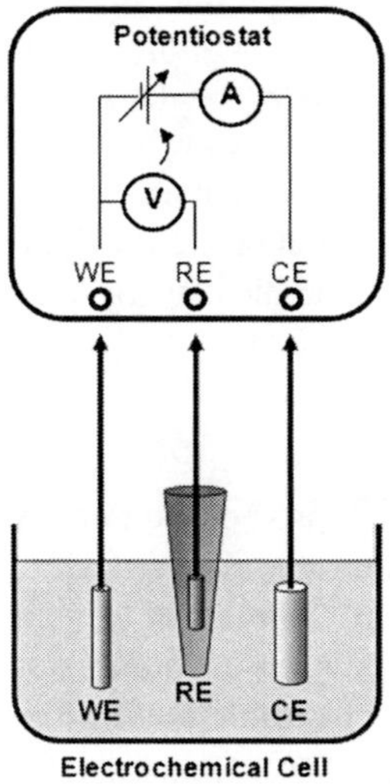

Figure 7.8 Conventional electrochemical cell and schematization of the electrical circuit used for the measurement.

7.4 THE ON-CHIP SIMPLIFIED ELECTROCHEMICAL TECHNIQUE DEVELOPED IN THE DINAMICS PROJECT

The electrochemical biosensor team of the DINAMICS EU collaborative project aimed at developing a device able to detect the presence of pathogens in short times and at the point of care. As expressed before, a number of DNA detection technologies have a sufficient sensitivity for the recognition of a target after amplification for example with PCR, so that a low limit of detection is not a primary issue. Hence, the most useful advancements are related to other aspects, such as the ease and speed of the assay. To this aim, our team at the University of Bologna worked on a project to design and develop a method for the electrochemical DNA detection of pathogens, aimed at a large simplification of the sensing protocol, of the measurement hardware, and of the biochip structure. The method is based on the translation of an existent measurement strategy to a setup that uses only 2 electrodes, instead of the conventional 3 electrodes used in electrochemical measurements.

The undertaking of the project was the development of a biosensor strategy that uses pairs of equal gold electrodes, in order to reduce the difficulties related to biochip fabrication, validation, and pretreatment, while maintaining the same sensitivity as its 3-electrode counterpart. In fact, it is known that some electrode characteristics such as construction material, dimensions, and surface properties are relevant for the performance of electrochemical biosensors. This is valid not only for the working electrode, where the sensing takes place, but also for the counter electrode and for the reference electrode that should maintain a constant potential independently on the target recognition and on the electrochemical measurement. As a consequence, the electrodes on a conventional biochip are made of at least two different materials, which results in a strong limitation towards miniaturization, and in a problem for production costs.

Moreover, a secondary objective of our project was to develop a procedure that could function properly using a biochip where all the electrodes are pretreated in the same way (before the injection of the sample), hence making the two measuring electrodes identical, with a consequent simplification of the biochip development process, as well as lower fabrication and procedure costs. In perspective, such biochip technology opens the road for dense miniaturization and integration in more complex devices, where a large number of sensing spots on the same biochip can give the possibility to independently test a large number of species in parallel, or the possibility to take measurements for a large number of trials before replacing a disposable used biochip, hence incrementing the working lifetime of each disposable biochip.

A first demonstrator has been designed and fabricated in collaboration with the project partner LioniX BV (Enschede, The Netherlands) which includes the electric biochip, a microfluidic motherboard and the electric connections for the measurement (for further information, see Chapter 6). The voltage excitation and signal readout unit was designed and produced in collaboration with the project partner, IDEA s.r.l., Italy.

In the following sections the electrochemical biosensing strategy developed within the framework of the DINAMICS project is outlined.

7.4.1 On-chip voltammetry

The biosensor developed for the DINAMICS project is based on the simplification of an already functional DNA detection technique, which is based on the measurement of the concentration of a redox-active dye with a conventional electrochemical setup (Ahmed *et al.*, 2009; Ahmed *et al.*, 2010). The proposed simplification is expected to provide not only an economic advantage, but also a competitive advantage for the development of a simpler portable electronic measuring unit. The ensuing reduction of the size of the biosensor, which can be relevant for the integration with a miniaturized microfluidic injection system, or when a highly parallel detection of multiple target nucleic acids is required, could be useful in the case of pathogen detection for water security.

7.4.2 Electrochemical reporter

The technique is based on the oxidation of Hoechst 33258 (2′-[4-Hydroxyphenyl]-5-[4-methyl-1-piperazinyl]-2,5′-bi-1H-benzimidazole trihydrochloride pentahydrate), which binds more efficiently double-stranded DNA than single stranded DNA (Figure 7.9). The application of an oxidizing potential generates a faradaic current which is dependent on the quantity of Hoechst 33258 on the surface. Hoechst 33258 is generally regarded as a minor groove binder, with some preference for AT-rich regions of the DNA double-helix (Guan *et al.*, 2007). Other binding modes have also been reported, such as binding to the phosphodiester backbone (Guan *et al.*, 2007), and also GC-rich stretches (Saito *et al.*, 2004). Hoechst 33258 oxidizes irreversibly at about 470 mV *vs.* Ag/AgCl, via oxidation at the benzimidazole nitrogen (Sufen *et al.*, 2002) providing with an anodic current directly proportional to the amount of dye close to the electrode, which is bound to the immobilized double helix and with lower efficiency to the probe DNA. Previous exploitation of Hoechst 33258 in DNA analysis and sensing with classical electrochemical techniques showed that it can be used to screen for the presence of DNA single-nucleotide polymorphisms, or to detect DNA oligonucleotides with a concentration down to 1–10 nM (Choi *et al.*, 2005).

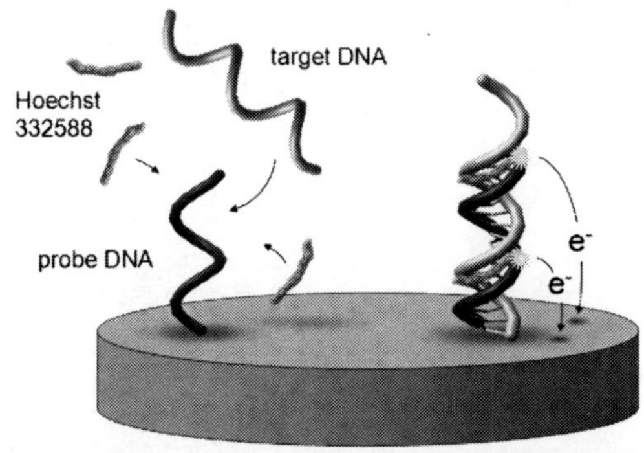

Figure 7.9 Working principle of the electrochemical measurement strategy chosen for the DINAMICS project, based on the oxidation of Hoechst 33258.

7.4.3 Measurement principle

A standard electrochemical measurement makes use of a 3-electrodes setup (the working principle is depicted in Figure 7.8, Section 7.3.2), which has the advantage of being sensitive only to the reactions that occur at the surface of the working electrode, independently of the reactions occurring at the counter or reference electrode. Nevertheless, the conventional setup presents two main downsides: 1) the three electrodes must be carefully designed to function properly, in particular the reference electrode, which in most of miniaturized biochips is made of silver coated with a thin layer of silver chloride. Commercial three-electrode biochips are available, but the fabrication cost is much larger than an equivalent biochip with electrodes made of a single material, and most importantly the miniaturization capabilities are reduced. 2) The electrochemical control is actuated by a complex feedback-based electronics, which makes miniaturization and portability costly, energy expensive, and difficult to implement.

As described in the previous section, the potential is commonly applied to a working electrode with respect to a reference electrode in a conventional voltammetric system. In our simplified measurement, one of the two identical oligonucleotide-derivatised gold electrodes is used as a working electrode, and its potential is set with respect to its twin gold electrode, which is also functionalized in the same manner. The electrical current flowing between the two is measured as a function of the applied potential, as in a regular linear voltammetry scan. Such current can be identified as an oxidation peak laying on top of a background current, similarly to a standard three electrode measurement. The shape of the resulting voltammogram can be traced down to two main factors: a peak shaped faradaic current due to the oxidation of Hoechst 33258, and an exponential background current due to capacitive charging at the electrode interfaces.

7.4.4 Interpretation of the measurements

While in a standard measurement system, the current response to the excitation voltage scan is solely due to the reactions on the working electrode, in our simplified system, the potential applied to the cell is partitioned between the two electrodes, without a direct control over the potential at a single electrode, as shown in the equivalent model of Figure 7.10a.

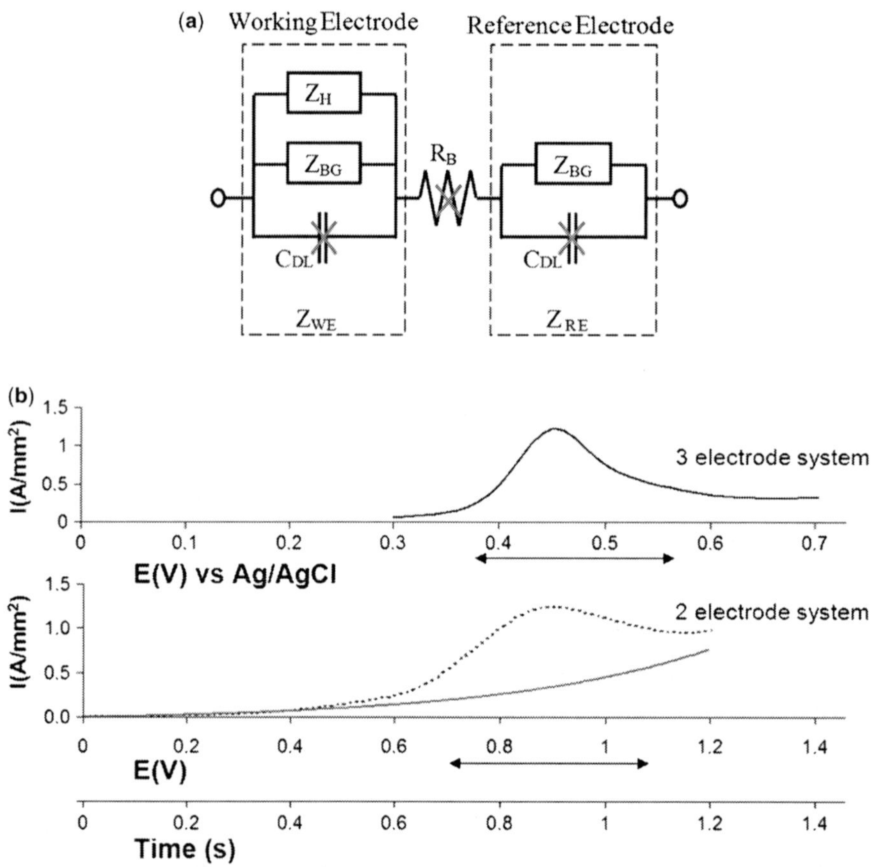

Figure 7.10 Equivalent model of the simplified two-electrodes electrochemical system (a), and comparison between the voltammograms generated in a three-electrodes and a two-electrodes systems (b).

Such model aids the interpretation of the experimental response signal, in particular, in relation to the standard three electrodes measurement that has already been discussed above. At potentials where the dye oxidation does not occur, the impedance representing Hoechst 33258 oxidation Z_H is negligible, and the system approximates to a series of two equal background impedances Z_{BG}. In this case, the equivalent model of the two electrodes is roughly symmetric, and the applied potential is nearly equally partitioned at the interfaces of the two electrodes with the sample. Hence, assuming that the voltage drop at each electrode is approximately half of the excitation voltage, a comparison between 2-electrode and 3-electrode voltammograms is meaningful only if the voltage dependent parameters are suitably adjusted. In particular, the voltage scan rate of the 2-electrode system must be double than the scan rate of the classical system. Also, for the sake of a graphical comparison, the voltammogram of the 2-electrode LV should be readjusted on a 2X-shrunk voltage axis, with respect to the 3-electrode LV (see Figure 7.10b). Under these conditions, the shape of the background component of the voltammograms of the two systems should be similar, and the two oxidation peaks should have comparable maximum currents and widths. The experimental results confirm these conclusions, as shown in Figure 7.10b. Here are the results of a 3-electrode and a 2-electrode LV performed on identical working-electrodes (functionalized with oligonucleotides and exposed to the same concentration of target DNA analyte). The peak current intensity and the (scaled) peak width agree quite well (statistical agreement over a number of measurements has also been reached).

7.4.5 Results

In the previous section, we have shown the validation of the measurement principle against a more established electrochemical technique. Here, we summarize some of the features of the simplified 2-electrodes biosensor, such as inter- and intra-chip reproducibility, limit of detection, and the sensing range.

Reproducibility

The reproducibility of the measurements is assessed by considering the distribution of all the measurements from several biochips on 10 nM target DNA. The experimental relative standard deviation of all the measurements is comparable to state-of-the-art results (~2%). The measurements were taken on five biochips, where each single chip hosts four independent electrode pairs, and hence it hosts four independent measurement sites.

Due to fabrication heterogeneity, in some cases, the outputs from different biochips could be incompatible, while the outputs of the measuring sites on the same single biochip could still be reproducible. In such unfavourable scenario, the inter-chip reproducibility would be worse than the intra-chip reproducibility. We compared the distributions of single chips with each other, and with the distribution of the whole set of 20 measurements, and concluded that the inter- and intra-chip reproducibility is the same: about 2% in both cases. This result suggests that the 2-electrodes setup is in fact as reliable as its 3-electrodes counterpart, probably, thanks to the attention paid in the choice of the fabrication parameters and of the measurement strategy.

Limit of detection and sensing range

A preliminary response curve of the biosensor is shown in Figure 7.11, where the response signal is the oxidation peak current after background subtraction, measured at different bulk concentrations of target DNA. The biosensor response depends almost logarithmically on target concentration, at least in the range 1 nM – 1 μM. If the limit of detection is defined as the concentration that induces a response in the sensor equal to the background signal plus three standard deviations, then in our case it turns out to be approximately 1 nM. Control experiments performed with 100 nM non specific DNA (marked as '100 ctrl' in Figure 7.11) behave well and yield signals comparable with the background signal ('0 prb' in Figure 7.11).

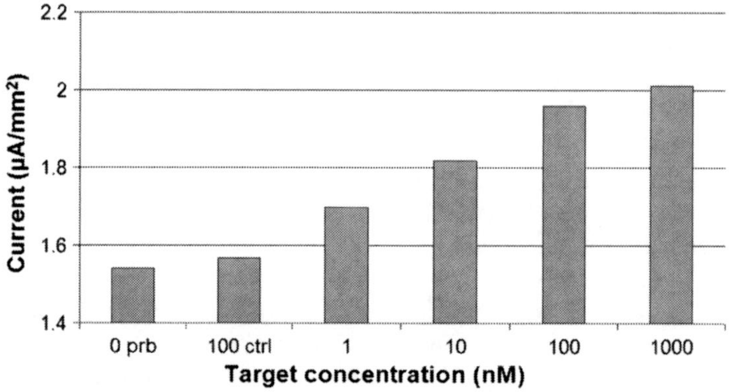

Figure 7.11 Preliminary response curve of the biosensor based on the two-electrodes electrochemical simplified system.

Scan inversion

In addition, after a first measurement on the two gold electrodes, the Hoechst 33258 dye on the electrode used as a working electrode for the measurement is in its oxidized form. On the contrary, the Hoechst 33258 is still in its reduced form on the other electrode. In fact, during pre-treatment both the electrodes are exposed to the identical functionalization procedure. This condition can be exploited by performing a second oxidizing measurement on the second electrode, simply by inverting the voltage scan. The experimental curves for direct (black trace in Figure 7.12) and inverse measurements (grey trace in Figure 7.12) of 10 nM target DNA have comparable peak currents, which suggests that it could be possible to use both measurements as independent test sites, hence maximizing the number of assays to one per electrode, which can be a critical improvement for certain applications. This could turn out as a plus in the medical field, where miniaturization of implanted devices is essential. Another potential advantage in the field of water security can be the increase of the working lifespan of the single biochips, as explained earlier.

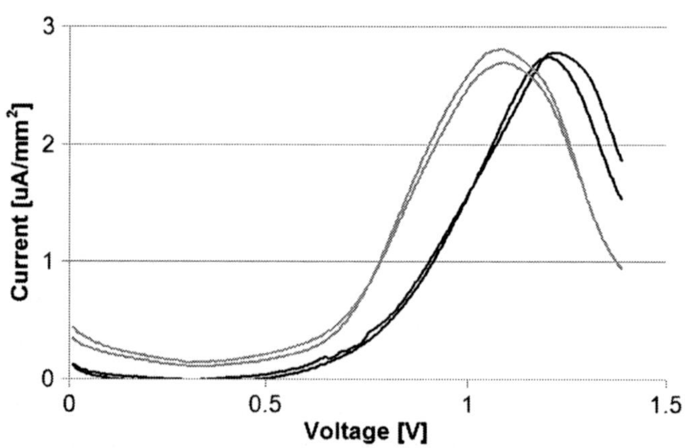

Figure 7.12 Typical direct (black) and inverse (grey) response curves of the biosensor, obtained by a voltage inversion.

7.4.6 Conclusions

The described system is a demonstration that under certain measurement conditions, it is possible to adapt the electrochemical transduction principle of a biosensor from a standard 3-electrodes setup to an on-chip 2-electrodes setup. In this communication, we present an example of such principle in the case of a DNA biosensor based on linear sweep voltammetry measurements of an electrochemically active groove binder.

The simplified electrochemical detection method has a sensitivity of 1 nM, which is similar or better than what previously reported in the literature for a 3-electrode detection of DNA with Hoechst 33258. The demonstrated technique exhibits two main advantages, while being significantly cheaper, simpler to implement and more keen to miniaturization and parallelization:

(1) Our proposed method makes it extremely straightforward to fabricate the biochips using standard ICT manufacturing procedures, with no adaptation of a standard protocol. This is a major benefit in the market competition of cheap disposable biosensors fabricated in large numbers, and it could even facilitate the development of biochips where microsized electrodes are integrated with active microelectronic components to reduce the noise and enhance the readout automation.

(2) The number of measurements per biochip is maximized. In this way, it is possible to perform one independent measurement for each electrode present in the biochip, instead of using three electrodes for each single measurement. This advantage can be exploited either to maximize the number of target nucleic acid analytes measured in a single biochip, to improve the sensitivity of the biosensor to one target analyte by averaging several measurements, or to maximize the working lifespan of the biochip.

Furthermore and most importantly, such improvement did not degrade the limit of detection of this transducing technique (1 nM), which is the same value as the best one present in the literature with a standard setup.

Acknowledgements

We kindly acknowledge support from the DINAMICS "Diagnostic Nanotech and Microtech Sensors" FP6-EU project ((IP 026804-2) and thank Dr. Theo Veenstra (LioniX B.V., The Netherlands) and Dr. Manuele Onofri (University of Bologna, Italy) for helpful discussion.

REFERENCES

Ahmed M. U., Hasan Q., Hossain M. M., Saito M. and Tamiya E. (2010). Meat species identification based on the loop mediated isothermal amplification and electrochemical DNA sensor. *Food Control*, **21**(5), 599–605.

Ahmed M. U., Saito M., Hossain M. M., Rao S. R., Furui S., Hino A., Takamura Y., Takagi M. and Tamiya E. (2009). Electrochemical genosensor for the rapid detection of GMO using loop-mediated isothermal amplification. *Analyst*, **134**(5), 966–972.

Choi Y. S., Lee K. S. and Park D. H. (2005). Single nucleotide polymorphism (SNP) detection using microelectrode biochip array. *Journal of Micromechanics and Microengineering*, **15**(10), 1938–1946.

Christensen P. A. and Hamnett A. (1994). Techniques and Mechanisms in Electrochemistry. Kluwer Academic Publishers, Blackie Academic & Professional, Glasgow.

D'Orazio P. (2003). Biosensors in clinical chemistry. *Clinica Chimica Acta*, **334**(1–2), 41–69.

Ekins R. and Chu F. W. (1999). Microarrays: their origins and applications. *Trends in Biotechnology*, **17**(6), 217–218.

Farré M., Brix R. and Barcelò D. (2005). Screening water for pollutants using biological techniques under European Union funding during the last 10 years. *Trends in Analytical Chemistry*, **24**(6), 532–545.

Gijs M. A. M. (2004). Magnetic bead handling on-chip: new opportunities for analytical applications. *Microfluidics Nanofluidics*, **1**(1), 22–40.

Guan Y., Shi R., Li X. M., Zhao M. P. and Li Y. Z. (2007). Multiple binding modes for dicationic Hoechst 33258 to DNA. *Journal of Physical Chemistry B*, **111**(25), 7336–7344.

Harvey D. (2000). Modern Analytical Chemistry. McGraw–Hill, USA.

Köster W., Egli T., Ashbolt N., Botzenhart K., Burlion N., Endo T., Grimont P., Guillot E., Mabilat C., Newport L., Niemi M., Payment P., Prescott A., Renaud P. and Rust A. (2003). Chapter 8: Analytical methods for microbiological water quality testing. In: Assessing Microbial Safety of Drinking Water: Improving Approaches and Methods, Book published on behalf of WHO and OECD: Water Sanitation and Health Programme, WHO, and Organisation for Economic Co-operation and Development, OECD. IWA Publishing, London, UK.

Lemieux B., Aharoni A. and Schena M. (1998). Overview of DNA chip technology. *Molecular Breeding*, **4**, 277–289.

Lipshutz R. J., Fodor S. P. A., Gingeras T. R. and Lockhart D. J. (1999). High density synthetic oligonucleotide arrays. *Nature Genetics*, **21**(Suppl S), 20–24.

Liu D. J., Perdue R. K., Sun L. and Crooks R. M. (2004). Immobilization of DNA onto poly(dimethylsiloxane) surfaces and application to a microelectrochemical enzyme-amplified DNA hybridization assay. *Langmuir*, **20**, 5905–5910.

Luong J., Male K. and Glennon J. (2008). Biosensor technology: technology push versus market pull. *Biotechnology Advances*, **26**(5), 492–500.

Matheson L. A. and Nichols N. (1938). The cathode ray oscillograph applied to the dropping mercury electrode. *Transactions of the Electrochemical Society*, **73**, 193–210.

Meadows D. (1996). Recent developments with biosensing technology and applications in the pharmaceutical industry. *Advanced Drug Delivery Reviews*, **21**(3), 179–189.

Mikkelsen S. R. (1996). Electrochemical biosensors for DNA sequence detection. *Electroanalysis*, **8**(1), 15–19.

Mir M., Dondapati S., Vreeke M. and Katakis I. (2007). Enzymatic self-wiring. *Electrochemistry Communications*, **9**(7), 1715–1718.

Mir M., Homs A. and Samitier J. (2004). Integrated electrochemical DNA biosensors for lab-on-a-chip devices. *Electrophoresis*, **30**, 3386–3397.

Randles J. E. B. (1948). A cathode ray polarograph. *Transactions of the Faraday Society*, **44**, 322–327.

Rodriguez-Mozaz S., Marco M. P., Lopez de Alda M. J. and Barcelo D. (2004). Biosensors for environmental monitoring of endocrine disruptors: a review article. *Analytical and Bioanalytical Chemistry*, **378**(3), 588–598.

Saito M., Kobayashi M., Iwabuchi S. I., Morita Y., Takamura Y. and Tamiya E. (2004). DNA condensation monitoring after interaction with Hoechst 33258 by atomic force microscopy and fluorescence spectroscopy. *Journal of Biochemistry*, **136**(6), 813–823.

Sufen W., Tuzhi P. and Yang C. F. (2002). Electrochemical studies for the interaction of DNA with an irreversible redox compound – Hoechst 33258. *Electroanalysis*, **14**(23), 1648–1653.

Wang J. (2000). Analytical Electrochemistry, Chapter 6 Electrochemical Sensors. 2nd edn, Wiley–VCH, New York, NY, U.S.A.

Chapter 8

Biochemical and nanotechnological strategies for signal enhancement in the detection of nucleic acids with biosensors

Alessandra Vinelli, Manuele Onofri and Giampaolo Zuccheri

8.1 INTRODUCTION

In the context of a research laboratory, it is possible to detect and characterize individual molecules one at a time. If properly confined, for example by adsorption on a surface, technologies such as atomic force microscopy or fluorescence microscopy can have the sensitivity to detect one molecule in a defined volume (Joo *et al.* 2008; Walter *et al.* 2008). Notably, single molecule fluorescence is nowadays used frequently in research to detect and study single molecules of nucleic acids, after labelling them with high quantum yield, highly photostable fluorophores that can be excited by high intensity laser light. The emitted fluorescence light can then be picked up by devices such as avalanche photodetectors or intensified cameras. While possible in the lab, single-molecule techniques, or highly sensitive detection in general, are difficult to implement and require expensive instrumentation even when performed on the scale of the research laboratory; additionally, they require a great deal of professional skill and time for each measurement.

For applications such as point-of-need biosensing, it is fundamental to develop sensors that should be automatic, inexpensive, reliable and require a professional skill that should be as low as possible, to ensure market penetration and a positive impact on public health or safety even in poor countries. The need of producing a high number of preferably disposable measuring chips, to be read by a relatively inexpensive electronic apparatus, implicitly points toward the direction of not employing sophisticated laboratory techniques. This practical need pushes back the limit for the detection of analyte molecules by attainable biosensors by many orders of magnitude with respect to research-level sensitivities. In the case of the detection of nucleic acids in a solution (being for diagnostic or environmental testing purposes), this problem has been classically tackled by exposing the biosensors or other sensing elements to the result of analyte amplification, usually by means of techniques such as the polymerase chain reaction (PCR). This can be implemented in solution (in a laboratory) prior to introduction of the specimen in the sensing device or directly in the biosensor through a number of reported on-chip implementations of the PCR protocol.

It was one of the proposals of the DINAMICS European collaborative research project to develop signal enhancement strategies that could take place after target analyte recognition by the biosensor probes, thus in an independent stage from a target nucleic-acid amplification step (such as PCR). In this report, we will review a number of presented methods that try (or could be amenable) to attain post-recognition signal enhancement of the nucleic-acid specific binding inherent in sensing. This commonly means that a biosensor exposes specific oligonucleotide probes to the analyte flow, capturing the specifically searched for nucleic acids; later some processing would turn such recognition into a more easily detectable structure, to enhance an otherwise weak signal due to the presence of few bound molecules.

The reported methods will be evaluated in terms of their applicability for a surface-bound recognition process (such as that interesting for the goals of the DINAMICS project) and for their applicability towards the realization of a simple-to-use (and implement) sensing-chip. All presented methods are isothermal reactions, so that they can be implemented with relatively low-level technology, often simply at room temperature. Amongst the presented strategies, PCR-based methods have been excluded. PCR is usually employed as a preliminary target amplification method, rather than a signal enhancement strategy. Nevertheless, there exist surface implementations of PCR (also termed solid-phase PCR) (Stamm *et al.* 1991) that could be used towards signal enhancement. Such patented implementations, including technology developed for next-generation DNA sequencing (from Solexa, now Illumina Inc.), albeit interesting, have been excluded from this survey also in view of the expensive and complex technology they require.

Two kinds of considerations, amongst others, are key. First, the enhancement factor is important and it should be as high as possible, while maintaining specificity in the detection (even though false positives are generally less harmful than false negatives in pathogen detection). Secondly, the signal enhancement strategy must be compatible with the detection technology of choice (and this was true also for DINAMICS). In the research project, the development of two types of sensors was pursued: electronic sensors that aimed at the measurement of a chemoelectronic signal as a response to DNA recognition (and signal enhancement) and optoelectronic detection that would look at an increase in optical absorbance or the emergence of chemoluminescence as a result of capturing target nucleic acids on the sensing surface.

While the target of some applications (like in DINAMICS) needs enhancement in order to detect the presence of a small copy number of pathogens (from a small number of their genomes or RNAs), the possibility of a (semi-)quantitative response on a wide dynamic range of concentrations is desirable: this reason would call for a pre-enhancement (post-hybridization) read-out to detect large amounts of nucleic acids and a post-enhancement readout to detect the low concentration nucleic acids. For this to be easily implemented, the detection technology should be the same for both the enhanced and non-enhanced signals. This strategy was taken into consideration during DINAMICS.

8.1.1 Rationale of the classification

In the following sections, a number of the methods of interest found in the scientific literature have been classified according to the chemical strategy employed for the amplification, as this seems to be the basic criterion for the definition of enhancement times and implementation (and also overall yield).

Some of the most widespread (and commercially available) methods for the enhancement of the hybridization of nucleic acids are based on the presence of enzymes. A single recognition event leads to the immobilization of a single enzyme molecule close to the sensing area. In turn, this leads to the enzyme-catalyzed conversion of many substrate molecules into an easily detectable product, thus leading to signal enhancement. The first class of methods in our list makes use of enzymes (or other active molecules) to enhance some measured signal that is dependent on nucleic acids hybridization.

The most common method to read the presence of nucleic acids on a sensor spot or any other system is through a label. Sometimes the target nucleic acid is directly labelled (via PCR, for instance), while more frequently a secondary probe oligonucleotide carries the label and forms a "sandwich" targeting the analyte on a different portion while this is immobilized on a surface through the primary oligonucleotide probe. Usually, the occurrence of many target-binding events leads to the immobilization of many labelled secondary probes and thus to a readable signal. When signal enhancement is needed, “macro-labels” can be used, so that the immobilization of a single labelled oligonucleotide brings in many readable molecules (such as fluorophores) or especially active labels (such as nanoparticles). This type of enhancement strategy will be classified in the second and third class of our list.

As a final class, we enlist such enhancement methods that employ strategies derived from DNA-based technologies. Such processes make use of biochemical reactions peculiar to nucleic acids, or implemented by DNA nanotechnology in order to produce enhancement effects, such as the massive recruitment of labels, as a response to hybridization. Due to the scientific backgrounds of the DINAMICS consortium (such as from our group), this type of enhancement strategy is regarded with particular interest and its implementation was tested as a viable approach towards the requirements of the DINAMICS applications.

8.2 ENHANCEMENT METHODS BASED ON ENZYMATIC (OR CATALYZED) REACTIONS

A number of methods have been proposed in which an enzyme is finally co-immobilized with the target on the sensing spot. The enzyme, often a peroxidase or a phosphatase, will convert a substrate into a to-be-sensed product. The accumulation of this product as a result of a limited number of hybridization events is the signal enhancement strategy (see Scheme 8.1).

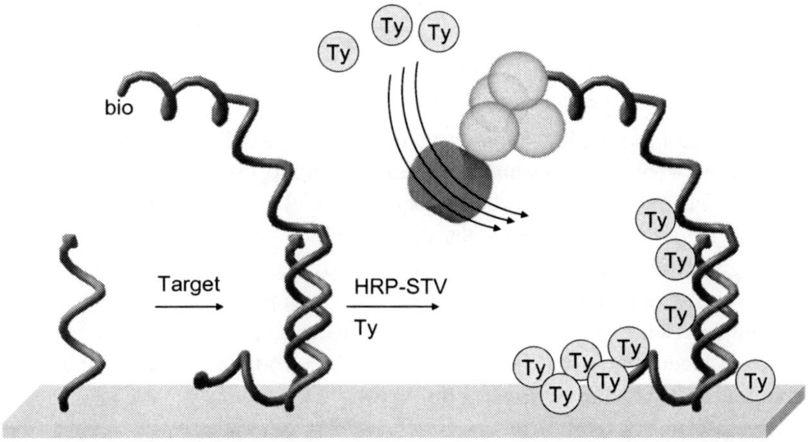

Scheme 8.1 Scheme of the tyramide-peroxidase signal enhancement. Detection of the products (fluorescent tyramine or others, from different substrates of the enzyme) can be done with optical or electrochemical methods. The target DNA can be biotinylated through its amplification using biotinylated primers (as showed in the scheme) or using a secondary complementary oligonucleotide (sandwich approach). Successive binding of peroxidase-streptavidin chimera (HRP-STV) and treatment with tyramide conjugated with a fluorophore (Ty) leads to very quick binding of the many fluorophores to the region of target binding.

Other enzymes have been used as well. Glucose oxidase has been employed several times as the enhancer label in sandwich assays. Xie and coworkers obtain a 1 fmol/l detection limit (working with 1 µl samples) by using an electroactive polymer (containing osmium) as the charge carrier from glucose oxidase to the electrode (Xie *et al.* 2004).

8.2.1 Peroxidase to enhance the signal of nucleic acids detection

Tyramide-mediated fluorescence signal generation

As an example of the basic principle, *tyramide signal amplification* (TSA) is presented. This commercial method (commonly used to enhance the signal due to many different types of molecules, not only nucleic acids) binds a biotinylated target on the sensing spot. Peroxidase-conjugated streptavidin is then bound to the target. Fluorescein-conjugated tyramide is used as an enzyme substrate to produce an insoluble fluorescein derivative, the accumulation of which is then measured. Jin and coworkers applied tyramide signal amplification on DNA microarray for the multiple detection of six waterborne pathogen causing diarrhoea in human (Jin *et al.* 2008). Coupling multiplex PCR and TSA-Cy3 labelling method they could detect 10^3 CFU/ml of each pathogen species, thus the use of tyramide signal amplification represents an effective alternative to the direct labelling with fluorescent dyes.

Peroxidase as an enzyme label in a sandwich assay

Exploiting the same strategy as in the previous example, Alfonta and coworkers directly bonded one peroxidase molecule to a secondary oligonucleotide (Alfonta *et al.* 2001). This is hybridized to the probe-immobilized target DNA on an electrode surface. In the presence of hydrogen peroxide added after rinsing of the unbound peroxidase, 4-chloronaphtol is converted into an insoluble compound by the peroxidases, thus leading to a measurable hindrance to the electron flow through the electrode.

In a similar assay, which exploits the same signal enhancement, Ostroff and coworkers detect the presence of the increasing amount of insoluble product through a change in reflectance of a surface (Ostroff *et al.* 1999). The reported detection limit is 5 pmole/l (150 amol/sample).

Electrogenerated luminescence or electrogenerated precipitate formation as a signal enhancement induced by an electroactive intercalator

Doxorubicin is an intercalator molecule that binds at GC pairs along dsDNA. If a target DNA is immobilized on an electrode surface by a capture oligo in the presence of doxorubicin, then the doxorubicin molecules immobilized close to the electrode can undertake electrochemical reduction (Patolsky *et al.* 2002). Reduced doxorubicin can be cycled by dissolved molecular oxygen which is, in turn, reduced to hydrogen peroxide. According to this strategy, hydrogen peroxide accumulates over time as a result of DNA hybridization. Its presence is revealed by soluble peroxidase, which uses hydrogen peroxide (reducing it to water) to convert luminol to 3-aminophtalate which is chemiluminescent. Alternatively, in the presence of 4-chloronaphtol, an insoluble product is accumulated on the surface of the electrode. The accumulation of this product is detected by the hindrance to the charge transfer through the electrode.

The reported detection limit of this strategy is 10^{-11} M. Even though this type of methods seems to perform well, using intercalators yields to the possibility that portions of double-stranded nucleic acids not related to the recognition event might bind the intercalator, either competing or leading to non-specific positive signals. Several different electroactive intercalators are available and have been used for similar (sometimes unenhanced) detection methods.

8.2.2 One to several instances of alkaline phosphatase for the electrochemical detection of nucleic acids

Detection by redox of an organic mediator

In a sandwich-type detection, the analyte RNA molecule can be immobilized on a sensor spot by an adsorbed oligonucleotide probe. Other regions of known base sequence of the target can be exploited to bind one to several secondary biotinylated oligonucleotides (see Scheme 8.2). These can then bind streptavidin-conjugated alkaline phosphatase molecules. In the presence of alkaline phosphatase, p-aminophenyl phosphate is converted to p-aminophenol, which thus accumulates over time in the case (and in the location) of RNA recognition and binding. p-Aminophenol can be oxidized to the corresponding quinoimide compound. The electric current derived from redox-cycling of this compound can be measured (implementing the chemoelectronic sensor as a series of interdigitated electrodes). The presence of the bound enzyme will make the redox current grow with time due to the progressive accumulation of the electroactive molecule, thus leading to signal enhancement. The method is straightforwardly implemented on a surface of a microfabricated device. Several application of alkaline phosphatase for enhancement of the signal due to 16S rRNA detection have been proposed (Elsholz *et al.* 2006; Elsholz *et al.* 2009; Walter *et al.* 2011). Several pathogens, some of which might be found in drinking water and are possible bio-threat agents (*E. coli*, *P. aeruginosa*, *E. faecalis*, *S. aureus*, *Y. pestis*, *B. anthracis* and *S. epidermidis*) were identified in parallel using nanometric interdigitated gold array electrodes (Elsholz *et al.* 2006; Elsholz *et al.* 2009). Working at constant current and varying the potential of the interdigitated electrodes, the authors monitored the time necessary to chemical conversion of the substrate oxidized by the enzyme, which is directly proportional to the analyte concentration. A limit of detection 0.5 ng/μl was achieved for *E. coli* RNA.

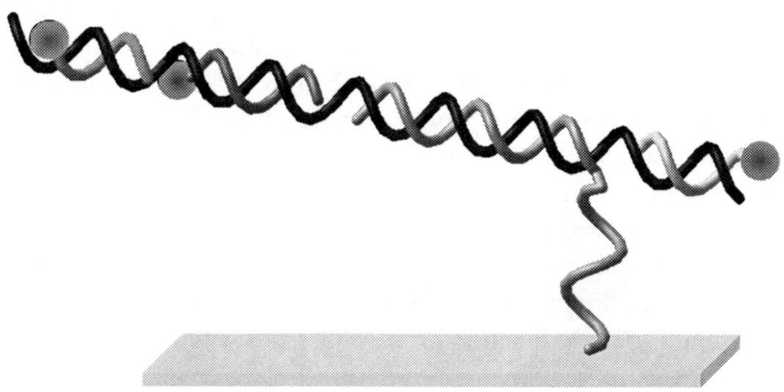

Scheme 8.2 Schematic representation of the binding and labelling of a ss-RNA or DNA (long black strand) by a surface-tethered oligonucleotide, with subsequent (singular or multiple) labelling by oligonucleotides bound to alkaline-phosphatase (represented as a sphere).

The same type of signal enhancement (redox cycling of alkaline phosphatase products) has been proficiently employed for the detection of DNA too, using a quite similar electrochemical set-up and microfrabricated electrodes (Schienle *et al.* 2004).

Detection by redox of a metallic mediator

In another use of alkaline phosphatase (AP) as a signal enhancing mediator, the hybridization event and the subsequent formation of a sandwich compound with an AP-labelled oligonucleotide is measured after the reduction of soluble Ag(I) ions to insoluble silver (Hwang *et al.* 2005). Such reduction is undertaken by aminophenol produced by the reaction of AP on p-aminophenylphosphate (*vide supra*). The presence of even a few copies of the enzyme can lead to the reduction of a considerable amount of silver. Deposited silver is then detected via anodic electrochemical stripping.

Due to the formation of a solid silver deposit, other techniques can be alternatively used to detect the hybridization event, such as a quartz crystal microbalance or a surface plasmon resonance detector, as both techniques are sensitive to the added mass on the surface. It can be foreseen that also optical absorbance properties (or light scattering) will change as well as the reduction proceeds. A detection limit of 10 zmoles (100 aM concentration) is claimed by the authors, as well as the specificity to detect single base mismatches (Hwang *et al.* 2005).

More than one alkaline phosphatase per binding event: macroenzymatic labels

Munge and coworkers prepared carbon nanotubes coated with several protein layers containing alkaline phosphatase molecules together with streptavidin (Munge *et al.* 2005). Such nanotubes can be associated with oligonucleotides to be used in sandwich recognition assays. A functional nanotube is immobilized on the target which has been immobilized by the main capture probe (on a surface of an electrode or of a bead).

The authors claim that such high number of alkaline phosphatase molecules that are recruited by very few DNA-binding events will lead to the detection of as few as 80 copies of a nucleic acid (a concentration of 5 aM in the reported conditions).

8.2.3 Terminal transferase to grow DNA at the recognition site

Terminal deoxynucleotidil transferase (TdT) is a mammalian enzyme that extends the 3′-OH end of a single-stranded DNA in a template-free fashion. The incorporation of dNTP is relatively random, even though preferences can be found in the presence of different cationic cofactors. Homopolymers can be obtained by feeding the enzyme with a single type of nucleotides. It has been shown that TdT can be used to extend surface bound oligonucleotides (Chow *et al.* 2005). The rationale for its innovative use as a possible enhancement technique resides from the consideration that a 3′ bound DNA probe oligonucleotide cannot be extended by TdT, while the target DNA that could bind to it should present a free 3′-OH amenable of TdT mediated extension. Extension would result in the accumulation of nucleic acid in a surface-bound state, and thus to the enhancement of a detection signal if this is proportional to the amount of DNA bound at a specific location on a surface (see Figure 8.1).

At our laboratory, experiments in solution analyzed by gel-electrophoresis showed that the reaction was efficient and relatively fast as to be a good candidate strategy. Higher molecular weight products were obtained starting from short 3′-OH free oligonucleotides. No extension was observed when the oligonucleotide had no free 3′-OH, showing that the specificity of the enhancement reaction could be as high as the specificity of target recognition by the probe, as we desired. Once attempted in a surface-bound format, the reaction resulted as less efficient leading to an enhancement factor of not more than 5, when the amount of DNA was determined by fluorescence measurements (via the non-specific binding of SybrGold to the surface-bound DNA). Alternative labelling strategies or detection techniques might prove more efficient.

8.2.4 Signal enhancement of fluorescence through the use of a nickase

When a fluorescent probe is used for detecting a target (e.g. through a molecular beacon approach), the cleavage of the probe in case of hybridization can release the target in solution for subsequent binding events. In an implementation of this strategy using a nickase, Zheleznaya and coworkers used molecular beacons and a site specific nickase to obtain an enhancement factor of 100 (Zheleznaya *et al.* 2006).

Very interestingly, researchers also noted some interference in the assay in case of the presence of extraneous DNA, that leads to a decrease in the signal (rather than to an expected increase due to non specific binding) probably due to the binding of the nickase to the extraneous DNA.

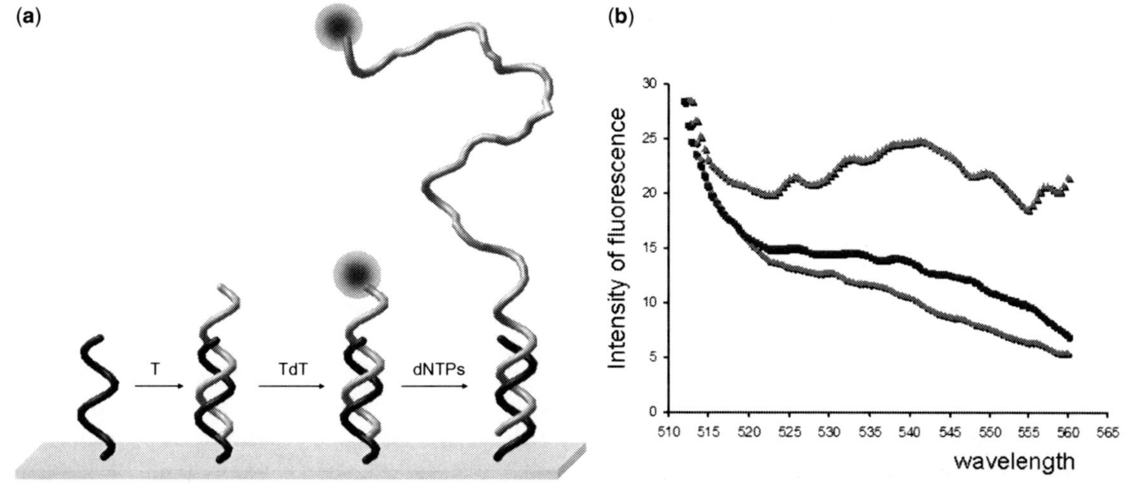

Figure 8.1 a) Scheme of the use of terminal deoxynucleotidyl tranferase (TdT) for signal enhancement in DNA biosensors. TdT catalyzes the addition of nucleotides at the free 3′ terminus of DNA. A target DNA (T) paired at a 3′-blocked, surface-bound probe can then be extended leading to the accumulation of more DNA on the recognition spot. b) background (lower trace, circle markers), target (middle trace, square markers) and enhanced (upper trace, triangle markers) fluorescence emission spectra for the TdT reaction signal enhancement on a biosensor surface. TdT enhancement was performed after target-DNA exposure onto a probe-functionalized gold-electrode. Later, the surface-bound DNA was removed from the surface by mercaptoethanol treatment (Demers *et al.* 2000) and the resulting specimens (no target DNA, target, target + TdT enhancement) were stained with SybrGold dye. The resulting fluorescence spectra (with intensity dependent on the amount of surface-bound DNA) measured an approximately 3-fold signal gain. More specific detection methods could lead to an even larger enhancement factor.

In a recent application, a nicking enzyme sensing assay was coupled with CdSe/ZnS quantum dots amplification for cymbidium mosaic virus detection (Chen *et al.* 2010). A thiolated hairpin DNA probe labelled with biotin was immobilized on gold electrode via S-Au bond. The double strand loop of the hairpin contained the restriction site for the endonuclease BfuCI, the nicking enzyme. In absence of the target, the hairpin bounded on the surface was closed and the restriction enzyme could digest the loop. As a consequence, the oligonucleotide end labelled with biotin was released in solution and the avidin-QD conjugate could not bind. On the contrary, the presence of a target molecule in solution opened the hairpin, blocking the enzymatic digestion and leading to the binding of avidin-quantum dots conjugate to the biotinylated probe. The excess of QD was removed and the electrochemical detection was performed after a treatment with acid solution to dissolve quantum dots. Stripping voltammetric measurements of the Cd^{2+} ions were performed using an *in situ* plated mercury film on a glassy carbon electrode. Using this indirect electrochemical measurements, the author reported the detection of 3.3×10^{-14} target molecules.

8.2.5 RNase H as a target recycling operator for RNA-based sensors

Goodrich and coworkers implemented a method to recycle the (few) target DNA molecules by preparing a sensing surface with RNA oligonucleotide probes (Goodrich *et al.* 2004). After binding with the target (and thus forming a RNA:DNA hybrid) such double strand can be the substrate for RNase H in solution. Its action leads to the hydrolysis of the bound RNA oligonucleotide and to freeing the target in solution again, so that it can bind to another surface-immobilized oligonucleotide. Over time, all the specific RNA oligonucleotide probes are digested by RNase, thus leading to signal enhancement (for instance if read through surface Plasmon resonance or other techniques). The reported detection limit is 10 fM in a 13 μl sample volume. An intrinsic weakness of this method, in our view, is that RNA is a fragile molecule, that can undergo hydrolysis due to a number of agents tha can be present in a complex analyte matrix. Such non specific hydrolysis could probably be detected and distinguished from the specific one, but it will nonetheless reduce the effectiveness and solidity of the proposed method.

8.2.6 Nucleic acid sequence-based amplification (NASBA)

The NASBA, also known as Self Sustained Sequence Replication (3SR) (Guatelli *et al.* 1990) is an enzymatic technique that amplifies an RNA molecule through a complex, though isothermal series of enzymatic reactions. Such technique is often used (also in commercial assay kits) for the pre-detection amplification of RNA target analytes. As depicted in Scheme 8.3 below, the different steps lead to the conversion of one RNA target into a DNA template for various rounds of transcription, leading to amplification. Briefly, the analyte RNA is paired to an oligonucleotide primer (P1) which is elongated to complementary DNA by reverse transcriptase. The RNA analyte paired to such DNA is hydrolyzed with RNaseH, so that an oligonucleotide primer (P2) can bind to the just-synthesized DNA, providing a substrate for reverse transcriptase. Reverse transcriptase produces a full dsDNA molecule. As P1 is made to include a T7-RNA polymerase promoter site, the produced dsDNA can serve as template for the synthesis of many copies of RNA complementary (antisense) to the target RNA sequence by T7-RNA polymerase present in the same mix: this is the true amplification step. In the mix, P2 can now bind to the produced antisense RNA, start the same process again using P2 and reverse transcriptase to start with using the antisense RNA and ending with producing many copies of sense RNA to feed the cycle.

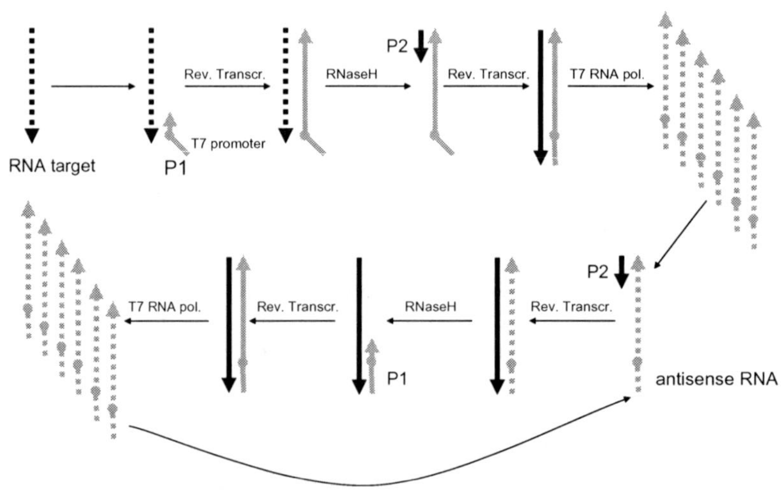

Scheme 8.3 Depiction of the NASBA procedure for nucleic acids amplification. RNA strands are represented as dashed lines, while DNA is solid. Black lines are the 'sense' sequences while gray lines are the 'antisense.' The dot in primer P1 only shows the separation between the target sequence and the added T7 promoter. The NASBA procedure works in a single solution kept in isothermal conditions where all the depicted reactions work in sequence with the result of producing large amounts of antisense RNA in case the sense RNA is present.

In certain implementations, the NASBA has been proficiently employed towards the signal enhancement through its surface implementation (Edman *et al.* 2006). The very high amplification factor of solution implementations of this reaction (detection of 50 copies of nucleic acid) will probably lead to interesting surface-bound applications. In recent applications, NASBA has been successfully applied on microarray systems. Scheler and coworkers optimized NASBA protocol in order to obtain a chemical-modified RNA product (Scheler *et al.* 2009). On it, post-amplification labeling with fluorescent dye was carried out and detection was performed with microarray technology. A lateral-flow based platform for *B. anthracis* amplicon detection was developed by Carter *et al.* (Carter *et al.* 2007). In this application target RNA is not directly labeled but is hybridized in a sandwich formed by an immobilized DNA capture probe and a microsphere-conjugated detection probe. Accumulation of the microsphere in correspondence of the capture feature produce a colorimetric signal which is visible at naked-eye and easily detected at low concentrations using widely available flatbed scanners. The reported sensitivity for the system is sub-femtomolar.

Transcription Mediated Amplification (TMA) is a very similar (and commercially available) amplification procedure that exploits essentially the same mechanism but making use of a different type of reverse-transcriptase and thus avoids the use of RNaseH (Hill, 2001; Chelliserrykattil *et al.* 2009; Bachmann *et al.* 2010; Rao *et al.* 2010).

Even though direct surface implementations of NASBA to lead to post-hybridization signal enhacement are still somewhat lacking, the technique bears some interest for the scope of this review as the reaction can be performed directly in the hybridization/sensing chamber of the biosensor and thus practically work in an integrated fashion with the biosensor, without the need of introducing alternative or additional instrumentation or procedures, other than the addition of the reaction mixture and incubation (usually for 1 hour).

8.2.7 Strand displacement amplification

Along the same line of reasoning of NASBA, strand displacement amplification (SDA) is another strategy towards nucleic acids amplification that can be quite effortlessly integrated in the biosensor detection compartment. It is an isothermal method to amplify DNA that has been proposed in 1992 by Walker and coworkers (Walker *et al.* 1992; Walker *et al.* 1992). It is a rather complex mechanism based on consecutive events of restriction and amplification. The use of four primers is requested (two amplification primers and two restriction primers). Nowadays, kits for TMA are commercially available. Even though the reaction seems to perform well, it quickly amplifies nucleic acids and it could possibly be integrated in biosensors-based detection, all the conditions and rules for the design of the 4 primers and the avoidance of amplification artifacts do not seem as mature as for other techniques (as seen from some attempts performed during the DINAMICS EU research project).

SDA has been used for pathogen detection in solution (Bachmann *et al.* 2010). Recently SDA has been coupled with piezoelectric detection for the real time monitoring of human cytomegalovirus (Chen *et al.* 2009). Quartz crystals have been modified with a specific DNA capture probe: during the strand displacement amplification, products accumulate on the crystal surface. The limit of detection presented is 1 ng/ml.

8.2.8 Loop mediated isothermal amplification (LAMP)

LAMP is a method proposed by Notomi and colleagues (Notomi *et al.* 2000). The LAMP mechanism, like SDA, is based on the strand displacement activity of DNA polymerase. For the reaction, a set of four specific primer is necessary. Primer design is a key point of this complex mechanism: the four primers hybridize to six different specific regions on the template DNA, separated by a defined distance on the sequence. The external primers duplicate the dsDNA and displace the products of the internal primers, which bear loops. The looped product, together with the loop primers allow the progress of multiple polymerizations through self-priming and strand-displacements. Everything occurs in one single step, without interventions either from the operator or any instrumentation. The amplified products have a structure consisting of alternately inverted repeats of the target sequence on the same strand.

LAMP is reported to be highly sensitive and fast: it can detect as few of six copies of hepatitis B virus genome in solution in 45 minutes (Notomi *et al.* 2000). If combined with a reverse transcription step, LAMP can be used for RNA amplification (RT-LAMP) (Notomi *et al.* 2000).

LAMP has been confirmed as a sensitive method for waterborne pathogens in solution as on surface. A specific RT-LAMP assay for *C. parvum* detection has been developed by Inomata and co-workers (Inomata *et al.* 2009). Using 18S RNA as specific target, 6×10^{-3} *Cryptosporidium* oocysts has been detected in 25 μl of reaction mix. Yang *et al.*, have designed specific primers for the detection of Salmonella Enteriditis (Yang *et al.* 2010). The reaction has been performed at 65°C for 20 min and a limit of detection of 4 CFU/μl has been calculated. A first implementation of LAMP on solid support has been proposed by Maruyama and colleagues for the specific amplification of gene stx2 of *E.coli* O157:H7 cells (Maruyama *et al.* 2003).

8.2.9 Metal nanoparticles as active enhancer labels

Besides showing interesting optical properties, metal nanoparticles can also be used as catalysts, and thus permit the implementation of electrochemical signal enhancements when these are bound to reporter oligonucleotide probes. In an example by Polsky and co-workers, a sandwich is obtained on the electrode surface by immobilizing the target and a reporter oligonucleotide that is labelled with a platinum nanoparticle (Polsky *et al.* 2006). Such nanoparticle can catalyze the conversion of hydrogen peroxide to water at a controlled electrochemical potential, enabling the chronoamperometric measurement of the hybridization of 10 pM DNA.

8.3 ENHANCEMENT METHODS BASED ON NANOPARTICLES OR NANOSTRUCTURES

8.3.1 Methods that employ liposomes

Liposomes can be quite versatile vessels for the transportation of many types of molecules. They are vesicles made by a double layer of phospholipids that can bind molecules covalently (on their surface), can include them in the layer or can contain them within the vesicle (Edwards *et al.* 2006). They can be used as labels for the detection of nucleic acids in analytical assays.

Esche *et al.* (Esch *et al.* 2001) used liposomes filled with carboxyfluorescein in a sandwich assay for *C. parvum* detection. The binding is revealed by fluorescence microscopy with a sensitivity of 0.4 fmoles/μl. Baeumner and coworkers published several interesting articles focused on detection of *E. coli, B. anthracis and C. parvum* mRNA transcripts in drinkable water after NASBA amplification (*vide supra*) using liposome-based sandwich assay (Baeumner *et al.* 2003; Baeumner *et al.* 2004; Baeumner *et al.* 2004).

A variation of the liposome labelling includes a lytic stage in which breaking down the liposome frees a large amount of detectable labels in solution (Zaytseva *et al.* 2005). Enhancement strategies that make use of the liposomes can reach a good detection sensitivity but require a complex sequence of operations making use of complex structures, such as the liposomes themselves. Often, target pre-amplification operations are unavoidable.

8.3.2 Fluorescent nanoparticle labels

Tan and co-workers developed fluorescent dye-doped silica nanoparticles functionalized with oligonucleotides as labels for chip-based sandwich DNA assays (Zhao *et al.* 2003). The nanoparticles are composed of a silica matrix that encapsulates a large number of fluorophores. Not only does the nanoparticle increase the fluorescent signal associated with each target recognition event, but it also acts as a protective barrier against fluorophore bleaching. This method results in a detection limit of around 8 fM target and provides ~14:1 differentiation between target DNA and DNA with only one base mismatch. Tan's group used similar particles to detect single bacterial cells.

8.3.3 Colorimetric assays with metal nanoparticles

Metal nanoparticles have been used extensively as labelling agents for the detection of DNA hybridization. Their enhancement properties are due to their stability and easy detectability due to their plasmonic resonance shift (Alivisatos *et al.* 1996; Mirkin *et al.* 1996). Oligonucleotide hybridization on nanoparticles takes place with very high specificity due to increased sharpness in the DNA melting transition.

Storhoff and coworkers (Storhoff *et al.* 2004) could avoid the use of PCR target amplification in their detection of genomic DNA. They measured the plasmonic red-shift due to DNA-induced aggregation of nanoparticles (Storhoff *et al.* 2000). This method was implemented by using two different nanoparticle-conjugated oligonucleotide probes that are complementary to different portions of the target DNA molecule. The binding event can be evidenced with optical methods. 2×10^5 molecules/µl can be detected (333 zmoles)).

Gold nanoparticles were used by Joung an co-workers to enhance 5500-fold the signal originated from 16s rRNA (Joung *et al.* 2008). The target was detected by means of a specific PNA capture probe immobilized on the a surface plasmon resonance (SPR) sensor. As PNA is characterized by a neutral backbone structure, the hybridization with the 16S RNA led to a change in the ionic charge from neutral to negative. Cationic AuNPs were synthesized and used to amplify locally the signal generated by the target binding. This method was applied on *E. coli* total RNA extraction showing a sensitivity of 58.2 ± 1.37 pg/ml RNA. For *S aureus* detection the method was apply without preliminary nucleic acid extraction. In this case the achieved sensitivity was 7×10^5 CFU/ml.

As nanoparticle, alternative nanostructures as nanorods are characterized by interesting physical and chemical properties that make them suitable for biosensing and signal amplification (Yu *et al.* 1997). Recently, Parab and colleagues demonstrated the use gold nanorods (GNR) for the optical detection of a *C. trachomatis* DNA in solution. Monitoring the absorption spectra of GNR-Capture probe they reliable detected target DNA in the range of 250–50 nM in 100 µl sample (Parab *et al.* 2010).

Scanometric detection of DNA

Chad Mirkin and coworkers employed another enhancement step in the nanoparticle-mediated detection of DNA. As their method leads to DNA signals that can be read with a standard flatbed optical scanner, they termed their method "scanometric detection" (Taton *et al.* 2000). In their method, ultra-low levels of DNA labelled with oligonucleotide-functionalized nanoparticles are detected by inducing silver reduction on the surface of the nanoparticles. In a sandwich assay, the surface immobilized nanoparticles, bridged by the target DNA, can then be turned into silver microparticles that are detectable even to the naked eye. The authors claim that nanoparticle labelling leads to signals that are 3–4 orders of magnitude more intense than fluorophores (Storhoff *et al.* 2004). 6×10^6 molecules (200 fM in 50 µl) can be detected.

8.3.4 Signal amplification by conjugate breakdown: the bio-barcode assay

Nam and co-workers use oligonucleotide-nanoparticle conjugates in order to label surface-immobilized nucleic acids (Nam *et al.* 2003; Nam *et al.* 2004), similarly to what done with a secondary binding oligonucleotide in other implementations of the sandwich assays. Such peculiar labelling nanoparticles also carry a high number of characteristic short oligonucleotides. After binding to the target nucleic acid, leading to immobilization of the nanoparticle, the nanoparticle conjugate is disassembled. This leads to the solubilisation of many copies of the previously bound oligonucleotides, easy to detect due to their high number.

Recently, Zhang and coworkers reported the use of a bio-barcode DNA assay for the rapid detection of Salmonella Enteritidis (Zhang *et al.* 2009). The biosensor transducer is composed of two nanoparticles: gold nanoparticles (AuNPs) and magnetic nanoparticles (MNPs). Two specific capture probe were designed: the first one was hybridize to AuNPs, the second one to MNPs. The Au-NPs were conjugated also with fluorescein-labeled barcode DNA in a 1:100 probe-to-barcode ratio. AuNPS, MNPs and sample were mixed and sandwich structure (MNPs-Target DNA-AuNPs-barcode DNA) was formed. A magnetic field was then applied to separate the sandwich from the unreacted materials. Fluorescence intensity of barcode DNA was measured after the oligonucleotide release from the nanoparticles. Using this technique, the detection limit of this bio-barcode DNA assay is as low as 1 ng/mL.

8.3.5 Quantum dots instead of organic dyes

Quantum dots (QD) are progressively gaining their place in applications that were previously making use or organic fluorophores due to their increased photostability and possibility to excite several different quantum dots with the same light source. Even though efficient protocols that use the possible signal enhancement of quantum dots are still under development, examples are emerging of their use in DNA detection. Gerion *et al.* (2003), for example, employed quantum dots to successfully detect SNPs, even though they reported a sensitivity that is still lower than organic dyes.

In the last years quantum dots have found large application in DNA, RNA and protein microarray labeling (Liang *et al.* 2005; Zajac *et al.* 2007; Giraud *et al.* 2009) and more frequent is their use in DNA biosensors. Zhang and colleagues reported a nanosensor based on fluorescence resonance energy transfer (FRET) coupling the use of CdSe–ZnS core–shell nanocrystals as donor and Cy5 dye as acceptor (Zhang *et al.* 2005). In solution, single strand target DNA bound to a biotinylated capture probe and to a reporter probe labeled with Cy5 forming a hybrid sandwich structure. Several hybrids are captured by streptavidin-QD hybrids, accumulating target and Cy5 dyes on it. In this configuration, acceptors dyes were in close proximity to the donor and were able to emit fluorescence by means of FRET after QDs excitation. As a result, fluorescent detection at the emission wavelength of Cy5, indicated the presence of target. The method has reported to be highly sensitive: as the unbound QD produced near-zero background fluorescence while a very clear FRET signal is generated after binding, up to $\sim$50 copies of target DNA can be detected.

8.4 METHOD BASED ON DNA NANOSTRUCTURES

In this section, some methods are included that make use of peculiar nucleic-acids nanostructures (self-assembling during the procedure) in order to create a signal or lead to signal enhancement. The procedures can include the use of enzymes, but are generally simple to perform (with respect to PCR, for example) and isothermal.

8.4.1 Cascade signal amplification by the combined use of rolling circle amplification

Chengde Mao and co-workers proposed a dual amplification strategy towards the amplification of the hybridization signal between complementary DNA molecules (Tian *et al.* 2006). The analyte sequence hybridizes on a circular template, and perfect complementarity at the 3′ end is necessary. The analyte sequence is thus used as a primer for rolling circle amplification by a DNA polymerase so that a long piece of single-stranded DNA is produced as a consequence of recognition. This RCA represents the first amplification strategy of this detection method, as the long DNA fragment is easy to detect.

On a non-probing section of the circular template strand, an additional information content is included, to yield a second amplification step. A sequence is included that will be duplicated many times in case of DNA recognition. This sequence will fold into a G-quadruplex structure with binding capability for hemin, an iron-containing porphyrin that will then work as a peroxidase.

The many copies of such DNA peroxidases can catalyze the oxidation of 2,2′-azino-bis (3-ethylbenzthiazoline)-6-sulfonic acid (ABTS). The reaction product ABTSC+ is blue-green (maximum absorption wavelength, $\lambda_{max} = 415$ nm) and can serve as a convenient, colorimetric output signal. This reaction has multiple turnovers; each enzyme can generate multiple copies of products . This constitutes the second amplification step.

The authors report the signal dependence of the colorimetric detection stating that they can clearly distinguish the presence of the analyte strand down to 1 pM. The output of this method depends on the careful tuning of the two amplification steps, as they cannot be optimized (or pushed) independently: too long a RCA amplification will create a highly entangled DNA molecule that will inhibit the diffusion of reactants for the second amplification step.

Using the same strategy, Cheglakov and co-workers in Itamar Willner's lab take the implementation of the same amplification strategy more towards a biosensor-type application (Cheglakov *et al.* 2007). They reportedly can implement it to detect a long viral DNA down to concentrations like 10^{-14} M (10 fM).

Especially in this last implementation, also exploiting the advantages of the molecular beacons approach for hybridization specificity, it seems plausible to attempt a surface implementation of this amplification strategy, as to make it more amenable of use in a microfabricated sensor, such as those in development within our project.

The most impressive and most effective example of use of rolling-circle amplification is probably still that from Paul Lizardi and co-workers (Lizardi *et al.* 1998), in its proposed surface implementation. In this implementation, a probe is bound to the solid surface via its 3′-end, in order to expose a 5′-phosphate end which cannot serve as primer for RCA. Upon binding of the target, a second oligonucleotide, complementary to the target can also bind and this and the probe oligonucleotide can be ligated in case of perfect match with the target. The second type of oligonucleotide (that can be used to test for sequence variations, such as polymorphisms) is a special oligonucleotide that contains a reversed backbone and thus has two 3′ ends. After ligation and binding, the presence of this second oligonucleotide (and thus the positive binding event) is evidenced by running a rolling circle amplification extension of it, using a soluble circular template. After rolling circle amplification, the presence of the amplified DNA is evidenced by binding several copies of fluorescently labelled oligonucleotides complementary to a tandemly repeated section on the amplified DNA.

Hyper-branched rolling circle amplification is also used by Lizardi and coworkers (even though not in a surface-bound implementation, seemingly). In this variant, an additional oligonucleotide serves as primer of the rolling-circle amplified nucleic acid, so that a new duplication reaction is started on the same molecule for each binding of such second primer to the tandem series of priming sites. Such geometric amplification greatly enhances the amplification factor, as it produce a massive amount of DNA for each positive binding event. It is our opinion that this type of enhancement power, or some of its variation, should be used towards the amplification necessary for our research project.

PNA openers can be used to expose target sequences in double-stranded DNA, thus avoiding the DNA denaturation step. Rolling circle amplification has been showed to function also as a method for the recognition of PNA-exposed sequences. This was implemented, for example, according to an experimental scheme reported by Smolina and co-workers (Smolina *et al.* 2007).

8.4.2 Branched DNA signal amplification

Branching in the probes or in the label-bearing parts will increase the level of signal, sometimes very significantly, sometimes also increasing the level of noise due to non-specific hybridization (Tsongalis, 2006). In a quite interesting method presented at the end of the 90′s by Mark Collins and co-workers, the analyte DNA molecule is held on the surface of a sensor spot by several instances of capture probes, and then labelled by several oligonucleotides bridging the analyte with some amplifier oligonucleotides (Kern *et al.* 1996; Collins *et al.* 1997). These amplifier oligonucleotides bear multiple copied of a capture sequence for alkaline-phosphatase (AP) labelled oligonucleotides (see Scheme 8.4). The result of the successful building of this complex multilayer nanostructure is the recruitment of many AP molecules where only one analyte DNA molecule is bound and thus an impressive signal enhancement. Since the final signal is the result of many events of DNA hybridization, there are various chances for non-specific hybridization that will bring about some false-positive signals. In order to avoid this, the authors use non-natural base pairings in the labelling oligonucleotide probes as to avoid that these bind to the wrong targets.

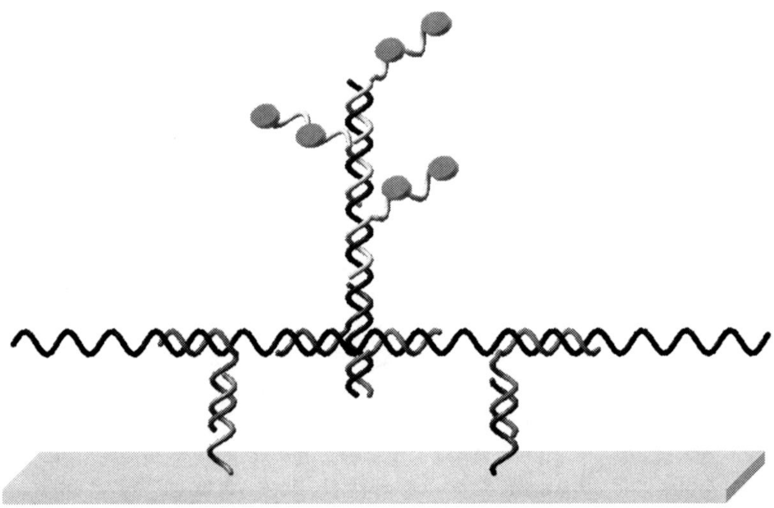

Scheme 8.4 Simplified graphical depiction of the branched DNA approach towards the enhanced detection of nucleic acids targets (such as in (Collins *et al.* 1997). A long ssRNA (or DNA) is immobilized on a surface by multiple capture probes, while labelling oligonucleotides bind a separate portion of the target and recruit onto it multiple instances of signalling oligonucleotides (bearing many copies of alkaline phosphatase molecules, here depicted as disks). Hundreds to thousands of alkaline phosphatase molecules are recruited for each immobilized target molecule, leading to an enzyme-enhanced signal that can be detectable in the presence of as few as a few hundred target molecules.

At the presented optimization stage, the authors claim to be able to reliably detect 60 molecules/ml of RNA, a quite interesting level. Such test has already been employed for the quantitation of the HIV virus in patient serum during pharmacological treatment and could be well amenable of use for the enhanced detection of nucleic acids from waterborne pathogens.

The detection method employs a luminescence meter, a quite complex instrumentation not readily amenable for miniaturization. The protocol timings are quite long (overnight incubations) as this assay was not developed for fast response, rather for therapy follow-up (i.e. determining the level of a known nucleic acids, that is known to be present at some level in the sample). The reactants for this method are quite complex, expensive, numerous and of difficult availability. The branched oligonucleotide concept itself is interesting, and should be considered as a general strategy towards amplification.

8.4.3 Aptamers and DNAzymes

Aptamers are short single-stranded nucleic acids selected (and evolved *in vitro*) for binding specific target molecules. DNAzymes are similarly evolved DNA molecules that can have catalytic properties (such as those in the naturally occurring ribozymes). Such functional nucleic acids nanostructures can be used also for the purpose of assaying the presence of compounds in solutions, and the catalytic properties can be exploited to implement signal enhancement strategies.

Liu and coworkers used a Pb-triggered DNAzyme to progressively break down a DNA-nanoparticle aggregate (Liu *et al.* 2003). Such aggregates are held together by oligonucleotides that can be cleaved by the active DNAzyme (see scheme). The breakdown causes a plasmonic shift that can be visualized simply and also the release of many soluble nanoparticles that can be also detected by other means. One single DNAzyme (even though slowly) can bind and cleave many oligonucleotides, thus explaining the enhancement result. Liu and coworkers implemented the strategy towards the detection of soluble Pb ions, but it is conceivable that the strategy could be extended to the detection of nucleic acids.

A similar strategy towards nucleic acids detections have been presented by Sando *et al.* (Sando *et al.* 2003). A self-cleaving DNAzyme will fold and cleave itself only in the presence of a DNA target. Cleavage of the DNAzyme will lead to the separation of a fluorophore from its quencher (and thus to signal) and to its detachment from the target nucleic acid (recycling the target, which will then be able to bind other DNAzymes, leading to enhancement). Authors report a 4-fold increase of signal with this strategy, and a detection limit of 10 pmol in 100 μl. Biosensor implementation of this strategy should be quite straightforward.

8.4.4 Hybridization chain reaction and surface-initiated DNA polymerization

As DNA is largely used as nanoscale building block for bottom-up assembly of complex nanostructures (Bath *et al.* 2007; Goodman *et al.* 2008), it can also be used for the controlled assembly of DNA nanostructures in response to hybridization, leading to an effective means of signal enhancement.

The hybridization chain reaction (HCR), originally proposed by Dirks and Pierce in 2004 (Dirks *et al.* 2004), is an isothermal, enzyme-free process where an hybridization event triggers the polymerization of oligonucleotides into a long, nicked nucleic acid molecule. Two kinds of co-monomer oligonucleotides are present in solution in the form of folded hairpins. These are opened and added to the growing polymer only after the initial triggering hybridization event (see Scheme 8.5).

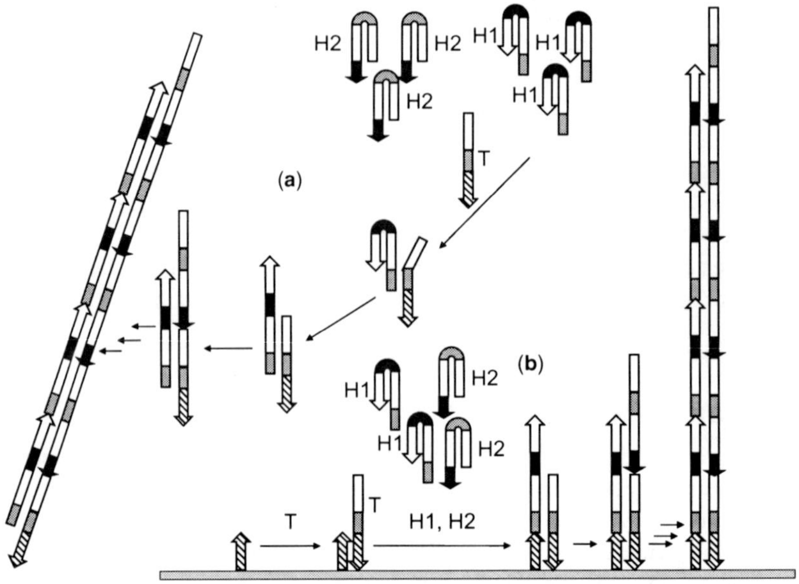

Scheme 8.5 Scheme of the Hybridization Chain Reaction (Dirks *et al.* 2004) or of its implementation on a surface (as implemented in the authors' laboratory). In the structures, strand portions represented with the same colour are complementary. In solution (case A in the figure) the contemporary presence of the hairpins H and H2 and the target/initiator DNA (T) starts the unfolding and polymerization of the hairpins, otherwise stable in solution. On the surface of a biosensor (case B in the figure) polymerization can start on the surface only if the target DNA gets recognized and binds to the surface spot with the complementary oligonucleotide probe. The exposure to H1 and H2 after such recognition starts the growth of the surface tethered dsDNA polymer. In its biosensor implementation, two portions of the target DNA are recognized in the process: one in its binding to the surface probe, another in its binding to the hairpin.

HCR is a conceptually simple mechanism: it involves two species of DNA oligonucleotides which at room temperature are stable in a hairpin conformation with a short loop and a long stem. The key point of the system is that in such conformation, the hairpins are kinetically blocked until an initiator strand is added to the solution. Initiator DNA triggers a cascade of hybridization events during which the hairpin unfold from their conformation. Several initiator strands in solution initiate multiple reactions leading to the formation of a set of nicked double helices of different lengths. In this way, a few hybridization events can use up the available hairpins to make long dsDNA that should be easier to detect.

In our laboratory, in the context of the DINAMICS project, we have implemented the HCR on the surface of an electrochemical DNA biosensor (see Figure 8.2). As the detection methods reads the amount of dsDNA, by labeling it with a groove-binding electroactive dye (HOECHST 33258), the signal should be enhanced by accumulation of DNA on the recognition electrode due to HCR. After some limited optimization, HCR can lead to an about 5-fold increase of signal with respect to the un-enhanced electrochemical readout. This is due to the accumulation of hairpin DNA on the surface which forms a long, nicked dsDNA at the site of recognition. Independent measurements done by using fluorescently labeled hairpins that are displaced from the surface after HCR, it turns out that up to 20 hairpins can bind the surface for each target DNA molecule that is recognized. Regardless of this, and likely due to limitations in the electron exchange from the bound electroactive dye, the effective gain in the signal is lower, even though significant. Preliminary data obtained with the surface plasmon resonance read-out confirm such data.

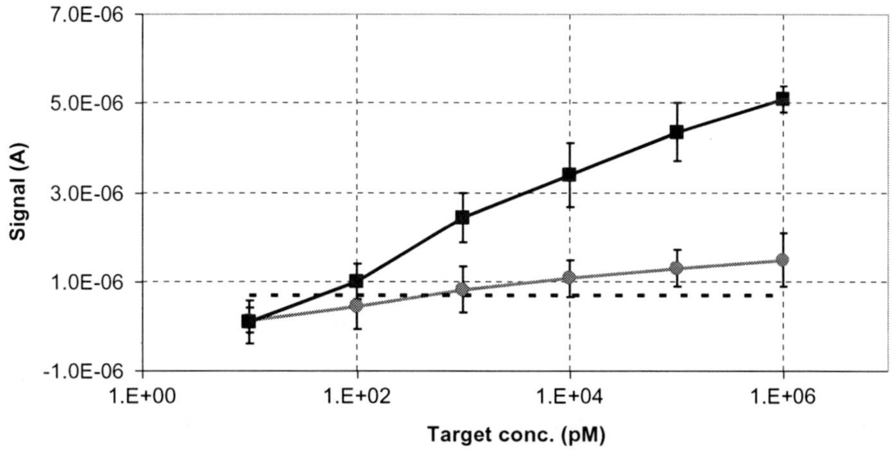

Figure 8.2 Example of the enhancement in the read-out signal of an electrochemical biosensor due to the implementation of surface-bound HCR. The light-colored trace (and circle markers) shows the electrochemical signal due to the recognition of a target DNA oligonucleotide by means of the measurement of the peak current of oxidation of Hoechst 33258 bound to the sensing electrode (Choi *et al.* 2004). The error bars represent ± 1 standard deviation. The black trace in the plot (and square markers) shows the result of the identical detection procedure applied after HCR signal enhancement (1 hr reaction time, 1 μM each hairpin). The dashed horizontal line shows the level of 3 times the standard deviation of the background, as a measure of significance of the read-out: performance of the surface-bound HCR does not lead to a very high signal enhancement factor, still, it significantly lowers the limit of detection and it increases the detectable target concentration range for the biosensor.

We have also showed that HCR can be considered as a general method for the enhancement of the signal due to nucleic acids recognition on the surface. We have designed surface-bound probes and soluble hairpins (see Scheme 8.5) for recognizing nucleic acids from a number of waterborne pathogens, like *Giardia lamblia*, *Cryptosporidium parvum*, Hepatitis E. Even though the requirements for the specific recognition of real pathogens might not always match perfectly with the requirements for the implementation of HCR, we showed that the strategy can be made to work specifically and effectively for the nucleic acids of all the above-mentioned pathogens. It appears that the choice of a target sequence with a limited secondary structure could be enough for the success of surface-bound HCR. This can be achieved by the use of relatively simple bioinformatics tools on the longer sequences normally considered target candidates for such organisms.

Niu and coworkers have combined HCR with enzyme-enhancement system for DNA detection. In this assay, streptavidin-coated magnetic beads are conjugated with a biotinylated hairpin. The hybridization with the DNA target on the particles opens the structure and leads to the consecutive polymerization of complementary biotinylated hairpins present in solution. Biotin moieties exposed on the formed dsDNA is used as anchor group for the binding of HRP-conjugated streptavidin. The target DNA is then quantified through the fluorescent detection of bi-p,p′-4 hydroxyphenylacetic acid (DBDA) generated from the substrate 4-hydroxyphenylacetic acid (p-HPA) under the catalysis of HRP (Niu *et al.* 2010). The reported amplification factor for the system is 30 with a limit of detection of 0.8 fM.

Very recently, Zheng and coworkers have employed another type of surface-initiated DNA polymerization as signal amplification strategy for electrochemical DNA sensors (Zheng *et al.* 2011). They have used the assembly chain reaction developed in Turbefield's Lab (Lubrich *et al.* 2009) to amplify the ssDNA-binding on gold electrodes performing different types of electrochemical measurements. Briefly, the assembly chain reaction is based on the use of double strand monomers and auxiliary "rubbish collector" strands (Lubrich *et al.* 2009).

The methods described above share the characteristics of not requiring enzymes or accurate temperature control. All the driving force is 'stored' in the DNA nanostructures. DNA is normally considered a chemically stable molecule and should have a longer storage life than most enzymes. This should make these methods amenable of use in industrialized biosensors.

8.5 CONCLUSIONS AND PERSPECTIVES

Many are the possible strategies towards signal amplification for nucleic acids biosensors. Some can leverage commercial lab kits for nucleic acids amplification, others exploit technology and solutions coming from ELISA or similar techniques. Full automation and ease of use are still somewhat to come, but it is nowadays clear that PCR is not the only solution at making sensitive nucleic acids detection, even though it will power a number of applications also in the future.

In the arenas of medical diagnostics and environmental testing, the winner strategies will have to put together high-volumes of assays, low-cost, high-reliability, high-sensitivity, high simplicity, at least. Likely, one solution will not satisfy all the needs and all the tasks and many strategies in addition to the ones mentioned in this chapter will emerge and will have a chance to play a role in empowering biosensors even more and participate in the decentralization of bioanalytics.

REFERENCES

Alfonta L., Bardea A., Khersonsky O., Katz E. and Willner I. (2001). Chronopotentiometry and Faradaic impedance spectroscopy as signal transduction methods for the biocatalytic precipitation of an insoluble product on electrode supports: routes for enzyme sensors, immunosensors and DNA sensors. *Biosensors and Bioelectronics*, **16**(9–12), 675–687.

Alivisatos A. P., Johnsson K. P., Peng X., Wilson T. E., Loweth C. J., Bruchez M. P. and Schultz P. G. (1996). Organization of 'nanocrystal molecules' using DNA. *Nature*, **382**(6592), 609–611.

Bachmann L. H., Johnson R. E., Cheng H., Markowitz L., Papp J. R., Palella F. J. and Hook E. W. (2010). Nucleic acid amplification tests for diagnosis of neisseria gonorrhoeae and chlamydia trachomatis rectal infections. *Journal of Clinical Microbiology*, **48**(5), 1827–1832.

Baeumner A. J., Cohen R. N., Miksic V. and Min J. (2003). RNA biosensor for the rapid detection of viable *Escherichia coli* in drinking water. *Biosensors & Bioelectronics*, **18**(4), 405–413.

Baeumner A. J., Leonard B., McElwee J. and Montagna R. A. (2004). A rapid biosensor for viable *B. anthracis* spore. *Analytical and Bioanalytical Chemistry*, **380**(1), 15–23.

Baeumner A. J., Pretz J. and Fang S. (2004). A universal nucleic acid sequence biosensor with nanomolar detection limits. *Analytical Chemistry*, **76**(4), 888–894.

Bath J. and Turberfield A. J. (2007). DNA nanomachines. *Nature Nanotechnology*, **2**(5), 275–284.

Carter D. J. and Cary R. B. (2007). Lateral flow microarrays: a novel platform for rapid nucleic acid detection based on miniaturized lateral flow chromatography. *Nucleic Acid Research*, **35**(10), e74.

Cheglakov Z., Weizmann Y., Basnar B. and Willner I. (2007). Diagnosing viruses by the rolling circle amplified synthesis of DNAzymes. *Organic & Biomolecular Chemistry*, **5**(2), 223–225.

Chelliserrykattil J., Nelson N. C., Lyakhov D., Carlson J., Phelps S. S., Kaminsky M. B., Gordon P., Hashima S., Ngo T., Blazie S. and Brentano S. (2009). Development of a quantitative real-time transcription-mediated amplification assay for simultaneous detection of multiple nucleic acid analytes. *Journal of Molecular Diagnostics*, **11**(6), 680–680.

Chen J., Zhang J., Yang H., Fu F. and Chen G. (2010). A strategy for development of electrochemical DNA biosensor based on site-specific DNA cleavage of restriction endonuclease. *Biosensors & Bioelectronics*, **26**(1), 144–148.

Chen Q. H., Bian Z. H., Chen M., Hua X., Yao C. Y., Xia H., Kuang H., Zhang X., Huang J. F., Cai G. R. and Fu W. L. (2009). Real-time monitoring of the strand displacement amplification (SDA) of human cytomegalovirus by a new SDA-piezoelectric DNA sensor system. *Biosensors & Bioelectronics*, **24**(12), 3412–3418.

Choi Y. S. and Park D. H. (2004). Electrochemical gene detection using multielectrode array DNA chip. *Journal of the Korean Physical Society*, **44**(6), 1556–1559.

Chow D. C., Lee W. K., Zauscher S. and Chilkoti A. (2005). Enzymatic fabrication of DNA nanostructures: extension of a self-assembled oligonucleotide monolayer on gold arrays. *Journal of the American Chemical Society*, **127**(41), 14122–14123.

Collins M. L., Irvine B., Tyner D., Fine E., Zayati C., Chang C. A., Horn T., Ahle D., Detmer J., Shen L. P., Kolberg J., Bushnell S., Urdea M. S. and Ho D. D. (1997). A branched DNA signal amplification assay for quantification of nucleic acid targets below 100 molecules/ml. *Nucleic Acids Research*, **25**(15), 2979–2984.

Demers L. M., Mirkin C. A., Mucic R. C., Reynolds R. A., Letsinger R. L., Elghanian R. and Viswanadham G. (2000). A fluorescence-based method for determining the surface coverage and hybridization efficiency of thiol-capped oligonucleotides bound to gold thin films and nanoparticles. *Analytical Chemistry*, **72**(22), 5535–5541.

Dirks R. M. and Pierce N. A. (2004). Triggered amplification by hybridization chain reaction. *Proceedings of the National Academy of Sciences of the United States of America*, **101**(43), 15275–15278.

Edman C. and Nerember M. I. (2006). Electronically Mediated Nucleic Acid Amplification in NASBA. Nanogen/Becton Dickinson Partnership, USA. US7070961.

Edwards K. A. and Baeumner A. J. (2006). Liposomes in analyses. *Talanta*, **68**(5), 1421–1431.

Elsholz B., Nitsche A., Achenbach J., Ellerbrok H., Blohm L., Albers J., Pauli G., Hintsche R. and Woerl R. (2009). Electrical microarrays for highly sensitive detection of multiplex PCR products from biological agents. *Biosensors & Bioelectronics*, **24**(6), 1737–1743.

Elsholz B., Wörl R., Blohm L., Albers J., Feucht H., Grunwald T., Jürgen B., Schweder T. and Hintsche R. (2006). Automated detection and quantitation of bacterial RNA by using electrical microarrays. *Analytical Chemistry*, **78**(14), 4794–4802.

Esch M. B., Baeumner A. J. and Durst R. A. (2001). Detection of cryptosporidium parvum using oligonucleotide-tagged liposomes in a competitive assay format. *Analytical Chemistry*, **73**(13), 3162–3167.

Gerion D., Chen F., Kannan B., Fu A., Parak W. J., Chen D. J., Majumdar A. and Alivisatos A. P. (2003). Room-temperature single-nucleotide polymorphism and multiallele DNA detection using fluorescent nanocrystals and microarrays. *Analytical Chemistry*, **75**(18), 4766–4772.

Giraud G., Schulze H., Bachmann T., Campbell C., Mount A., Ghazal P., Khondoker M., Ross A., Ember S., Ciani I., Tlili C., Walton A., Terry J. and Crain J. (2009). Fluorescence lifetime imaging of quantum dot labeled DNA microarrays. *Internation Journal of Molecular Science*, **10**(4), 1930–1941.

Goodman R. P., Heilemann M., Doose S., Erben C. M., Kapanidis A. N. and Turberfield A. J. (2008). Reconfigurable, braced, three-dimensional DNA nanostructures. *Nature Nanotechnology*, **3**(2), 93–96.

Goodrich T. T., Lee H. J. and Corn R. M. (2004). Enzymatically amplified surface plasmon resonance imaging method using RNase H and RNA microarrays for the ultrasensitive detection of nucleic acids. *Analytical Chemistry*, **76**(21), 6173–6178.

Guatelli J. C., Whitfield K. M., Kwoh D. Y., Barringer K. J., Richman D. D. and Gingeras T. R. (1990). Isothermal, *in vitro* amplification of nucleic acids by a multienzyme reaction modeled after retroviral replication. *Proceedings of the National Academy of Sciences of the United States of America*, **87**(5), 1874–1878.

Hill C. (2001). Molecular diagnostic testing for infectious diseases using TMA technology. *Expert Review of Molecular Diagnostic*, **1**(4), 445–455.

Hwang S., Kim E. and Kwak J. (2005). Electrochemical detection of DNA hybridization using biometallization. *Analytical Chemistry*, **77**(2), 579–584.

Inomata A., Kishida N., Momoda T., Akiba M., Izumiyama S., Yagita K. and Endo T. (2009). Development and evaluation of a reverse transcription-loop-mediated isothermal amplification assay for rapid and high-sensitive detection of Cryptosporidium in water samples. *Water Science and Technology*, **60**(8), 2167–2172.

Jin D., Qi H., Chen S., Zeng T., Liu Q. and Wang S. (2008). Simultaneous detection of six human diarrheal pathogens by using DNA microarray combined with tyramide signal amplification. *Journal of Microbiological Methods*, **75**(2), 365–368.

Joo C., Balci H., Ishitsuka Y., Buranachai C. and Ha T. (2008). Advances in single-molecule fluorescence methods for molecular biology. *Annual Review of Biochemistry*, **77**, 51–76.

Joung H.-A., Lee N.-R., Lee S. K., Ahn J., Shin Y. B., Choi H.-S., Lee C.-S., Kim S. and Kim M.-G. (2008). High sensitivity detection of 16s rRNA using peptide nucleic acid probes and a surface plasmon resonance biosensor. *Analytica Chimica Acta*, **630**(2), 168–173.

Kern D., Collins M., Fultz T., Detmer J., Hamren S., Peterkin J. J., Sheridan P., Urdea M., White R., Yeghiazarian T. and Todd J. (1996). An enhanced-sensitivity branched-DNA assay for quantification of human immunodeficiency virus type 1 RNA in plasma. *Journal of Clinical Microbiology*, **34**(12), 3196–3202.

Liang R.-Q., Li W., Li Y., Tan C.-Y., Li J.-X., Jin Y.-X. and Ruan K.-C. (2005). An oligonucleotide microarray for microRNA expression analysis based on labeling RNA with quantum dot and nanogold probe. *Nucleic Acid Research*, **33**(2), e17.

Liu J. and Lu Y. (2003). A colorimetric lead biosensor using DNAzyme-directed assembly of gold nanoparticles. *Journal of the American Chemical Society*, **125**(22), 6642–6643.

Lizardi P. M., Huang X. H., Zhu Z. R., Bray-Ward P., Thomas D. C. and Ward D. C. (1998). Mutation detection and single-molecule counting using isothermal rolling-circle amplification. *Nature Genetics*, **19**(3), 225–232.

Lubrich D., Green S. J. and Turberfield A. J. (2009). Kinetically controlled self-assembly of DNA oligomers. *Journal of the American Chemical Society*, **131**(7), 2422–2423.

Maruyama F., Kenzaka T., Yamaguchi N., Tani K. and Nasu M. (2003). Detection of bacteria carrying the stx2 gene by in situ loop-mediated isothermal amplification. *Applied and Environmental Microbiology*, **69**(8), 5023–5028.

Mirkin C. A., Letsinger R. L., Mucic R. C. and Storhoff J. J. (1996). A DNA-based method for rationally assembling nanoparticles into macroscopic materials. *Nature*, **382**(6592), 607–609.

Munge B., Liu G., Collins G. and Wang J. (2005). Multiple enzyme layers on carbon nanotubes for electrochemical detection down to 80 DNA copies. *Analytical Chemistry*, **77**(14), 4662–4666.

Nam J.-M., Stoeva S. I. and Mirkin C. A. (2004). Bio-bar-code-based DNA detection with PCR-like sensitivity. *Journal of the American Chemical Society*, **126**(19), 5932–5933.

Nam J.-M., Thaxton C. S. and Mirkin C. A. (2003). Nanoparticle-based bio-bar codes for the ultrasensitive detection of proteins. *Science*, **301**(5641), 1884–1886.

Niu S. Y., Jiang Y. and Zhang S. S. (2010). Fluorescence detection for DNA using hybridization chain reaction with enzyme-amplification. *Chemical Communications*, **46**(18), 3089–3091.

Notomi T., Okayama H., Masubuchi H., Yonekawa T., Watanabe K., Amino N. and Hase T. (2000). Loop-mediated isothermal amplification of DNA. *Nucleic Acid Research*, **28**(12), e63.

Ostroff R. M., Hopkins D., Haeberli A. B., Baouchi W. and Polisky B. (1999). Thin film biosensor for rapid visual detection of nucleic acid targets. *Clinical Chemistry*, **45**, 1659–1664.

Parab H. J., Jung C., Lee J.-H. and Park H. G. (2010). A gold nanorod-based optical DNA biosensor for the diagnosis of pathogens. *Biosensors and Bioelectronics*, **26**(2), 667–673.

Patolsky F., Katz E. and Willner I. (2002). Amplified DNA detection by electrogenerated biochemiluminescence and by the catalyzed precipitation of an insoluble product on electrodes in the presence of the doxorubicin intercalator. *Angewandte Chemie*, **114**(18), 3548–3552.

Polsky R., Gill R., Kaganovsky L. and Willner I. (2006). Nucleic acid-functionalized Pt nanoparticles: Catalytic labels for the amplified electrochemical detection of biomolecules. *Analytical Chemistry*, **78**(7), 2268–2271.

Rao V., Fabrizi F., Pennell P., Schiff E., de Medina M., Lane J. R., Martin P. and Ivor L. (2010). Improved detection of hepatitis C virus infection by transcription-mediated amplification technology in dialysis population. *Renal Failure*, **32**(6), 721–726.

Sando S., Sasaki T., Kanatani K. and Aoyama Y. (2003). Amplified nucleic acid sensing using programmed self-cleaving DNAzyme. *Journal of the American Chemical Society*, **125**(51), 15720–15721.

Scheler O., Glynn B., Parkel S., Palta P., Toome K., Kaplinski L., Remm M., Maher M. and Kurg A. (2009). Fluorescent labeling of NASBA amplified tmRNA molecules for microarray applications. *BMC Biotechnology*, **9**(1), 45.

Schienle M., Frey A., Hofmann F., Holzapfl B., Paulus C., Schindler-Bauer P. and Thewes R. (2004). A fully electronic DNA sensor with 128 positions and in-pixel A/D conversion. Solid-State Circuits Conference, 2004. Digest of Technical Papers. ISSCC. 2004 IEEE International, San Francisco, CA, U.S.A.

Smolina I., Lee C. and Frank-Kamenetskii M. (2007). Detection of low-copy-number genomic DNA sequences in individual bacterial cells by using peptide nucleic acid-assisted rolling-circle amplification and fluorescence in situ hybridization. *Applied and Environmental Microbiology*, **73**(7), 2324–2328.

Stamm S. and Brosius J. (1991). Sanchored PCR – PCR with cDNA coupled to a solid-phase. *Nucleic Acids Research*, **19**(6), 1350–1350.

Storhoff J. J., Lazarides A. A., Mucic R. C., Mirkin C. A., Letsinger R. L. and Schatz G. C. (2000). What controls the optical properties of DNA-linked gold nanoparticle assemblies? *Journal of the American Chemical Society*, **122**(19), 4640–4650.

Storhoff J. J., Lucas A. D., Garimella V., Bao Y. P. and Muller U. R. (2004). Homogeneous detection of unamplified genomic DNA sequences based on colorimetric scatter of gold nanoparticle probes. *Nature Biotechnology*, **22**(7), 883–887.

Taton T. A., Mirkin C. A. and Letsinger R. L. (2000). Scanometric DNA Array detection with nanoparticle probes. *Science*, **289**(5485), 1757–1760.

Tian Y., He Y. and Mao C. (2006). Cascade signal amplification for DNA detection. *ChemBioChem*, **7**(12), 1862–1864.

Tsongalis G. J. (2006). Branched DNA technology in molecular diagnostics. *American Journal of Clinical Pathology*, **126**(3), 448–453.

Walker G. T., Fraiser M. S., Schram J. L., Little M. C., Nadeau J. G. and Malinowski D. P. (1992). Strand displacement amplification: an isothermal, in vitro DNA amplification technique. *Nucleic Acid Research*, **20**(7), 1691–1696.

Walker G. T., Little M. C., Nadeau J. G. and Shank D. D. (1992). Isothermal *in vitro* amplification of DNA by a restriction enzyme/DNA polymerase system. *Proceedings of the National Academy of Sciences of the United States of America*, **89**(1), 392–396.

Walter A., Wu J., Flechsig G. U., Haake D. A. and Wang J. (2011). Redox cycling amplified electrochemical detection of DNA hybridization: application to pathogen *E. coil* bacterial RNA. *Analytica Chimica Acta*, **689**(1), 29–33.

Walter N. G., Huang C.-Y., Manzo A. J. and Sobhy M. A. (2008). Do-it-yourself guide: how to use the modern single-molecule toolkit. *Nature Methods*, **5**(6), 475–489.

Xie H., Zhang C. and Gao Z. (2004). Amperometric detection of nucleic acid at femtomolar levels with a nucleic acid/electrochemical activator bilayer on gold electrode. *Analytical Chemistry*, **76**(6), 1611–1617.

Yang J. L., Ma G. P., Yang R., Yang S. Q., Fu L. Z., Cheng A. C., Wang M. S., Zhang S. H., Shen K. F., Jia R. Y., Deng S. X. and Xu Z. Y. (2010). Simple and rapid detection of Salmonella serovar Enteritidis under field conditions by loop-mediated isothermal amplification. *Journal of Applied Microbiology*, **109**(5), 1715–1723.

Yu Chang S.-S., Lee C.-L. and Wang C. R. C. (1997). Gold nanorods: electrochemical synthesis and optical properties. *The Journal of Physical Chemistry B*, **101**(34), 6661–6664.

Zajac A., Song D., Qian W. and Zhukov T. (2007). Protein microarrays and quantum dot probes for early cancer detection. *Colloids and Surfaces B: Biointerfaces*, **58**(2), 309–314.

Zaytseva N. V., Goral V. N., Montagna R. A. and Baeumner A. J. (2005). Development of a microfluidic biosensor module for pathogen detection. *Lab on a Chip*, **5**(8), 805–811.

Zhang C.-Y., Yeh H.-C., Kuroki M. T. and Wang T.-H. (2005). Single-quantum-dot-based DNA nanosensor. *Nature Materials*, **4**(11), 826–831.

Zhang D., Carr D. J. and Alocilja E. C. (2009). Fluorescent bio-barcode DNA assay for the detection of *Salmonella enterica* serovar *Enteritidis*. *Biosensors and Bioelectronics*, **24**(5), 1377–1381.

Zhao X., Tapec-Dytioco R. and Tan W. (2003). Ultrasensitive DNA detection using highly fluorescent bioconjugated nanoparticles. *Journal of the American Chemical Society*, **125**(38), 11474–11475.

Zheleznaya L. A., Kopein D. S., Rogulin E. A., Gubanov S. I. and Matvienko N. I. (2006). Significant enhancement of fluorescence on hybridization of a molecular beacon to a target DNA in the presence of a site-specific DNA nickase. *Analytical Biochemistry*, **348**(1), 123–126.

Zheng Y., Li Y., Lu N. and Deng Z. (2011). Surface-initiated DNA self-assembly as an enzyme-free and nanoparticle-free strategy towards signal amplification of an electrochemical DNA sensor. *Analyst*, **136**(3), 459–462.

Chapter 9

Computational modelling of aqueous environments in micro and nanochannels

D. Mantzalis, K. Karantonis, N. Asproulis, L. Könözsy and D. Drikakis

9.1 INTRODUCTION

Recently, flows in micro and nanoscales have triggered the scientific and industrial interest. The development of micro and nanofluidic devices is continuously increasing along with their applicability that spans from material and environmental sciences to bioengineering and medicine. To further develop their performance and the range of applicability a systematic investigation is requires concerning their physics. As the scales are downsized to nanoscales, the surface-to-volume ratio increases and the interfacial interactions start to dominate the flow properties (Asproulis & Drikakis, 2010a). Fluids that are nanoconfined, exhibit different behaviour compared with fluids in macro- and micro-scale environments. This type of flows, are mainly characterized by the weakened volume forces with the simultaneous boost of surface forces and therefore both their static and the dynamic behaviour is strongly affected by their interfacial characteristics.

The majority of the problems related with the dynamics of Newtonian flows are related with solving the Navier-Stokes equations in a desired formation according to the investigated problem. One of the key topics in the investigation related with the flows near solid surfaces is the no-slip assumption. Macroscopically, the effect of surface interactions between a fluid and a wall are modelled by the traditional no-slip boundary condition that implies immobilization of fluid in the proximity of the solid surface. This hypothesis suggests that the fluid, in the solid-fluid interface, flows with velocity equal to the one that the wall moves with. In large scale systems the well known no-slip condition is widely accepted having been confirmed by a great number of numerical and experimental works (Priezjev, 2005; Karniadakis, 2005). However, in micro- and nano-fluidic systems, the no-slip assumption is usually violated and this type of flows exhibits considerable amount of slippage in the interfacial regions. The main reason for the break of the continuum hypothesis in nanoscale systems can be located in the fact that the molecular free path is analogous to the characteristic size of the fluidic channels. Aiming to quantify the slip, the slip length is utilized defined by the extrapolated distance from the wall to the point where the component velocity tangent to the wall equals to zero. In general, based in both experimental and computational modelling, a number of parameters that influence the slip length have been suggested such as the surface energy, wettability and shear rate.

Molecular theories can foresee molecular slip lengths with typical dimensions up to hundreds of Angstroms in case of hydrophobic systems (Li *et al.* 2010). If calculations suggest larger values, then the flow cannot be considered only as dynamical. A great part of the studied cases focus in hydrophobic surfaces for the slip investigation since this kind of surfaces exhibit slip. Several works have been devoted in studying the relation between hydrophobic surfaces and the slip, since it has been assumed that this kind of surfaces exhibit a slip condition (Li *et al.* 2010; Voronov & Papavasiliou, 2008; Pratt & Pohorille, 2002). Ho *et al.* (2011) recently showed that hydrophilic surfaces display attributes directly related with hydrophobicity such as the slip of liquid water.

Computational techniques such as molecular dynamics (MD) can be employed for simulating processes and phenomena taking place at micro and nanoscales, where the continuum models fail to provide accurate physical insights. In MD a combination of Newtonian's equation of motion and intermolecular potential functions are used for projecting the system of particles' trajectories. Through proper interpretation of this information the molecular systems can be utilised for investigating transport and physico-chemical phenomena. In the literature MD simulations have been performed to investigate the slip mechanisms and the slippage effects across the liquid-solid region (Tropea *et al.* 2007). A decrease in solid-liquid interactions leads to an increase of the molecular slip. Similarly the molecular slip performs the same trend as the liquid density or the density of the wall decreases simultaneously. On the contrary, slip decreases with pressure increase. Numerous works have described in more detail the physics on slip conditions.

MD simulations produce information corresponding in microscopic level while the conversion of these data into macroscopic terms is required characterized as a very computationally demanding task. The computational capabilities limit the size of the simulated systems up to several thousands of particles keeping the spatial and temporal s scales in low levels. The basic input parameters to control a MD simulation are the molecular mass m, the interaction energy ε, and the molecular size σ. The majority of the studies in the literature for this type of problems utilises reduced units corresponding to spatial scales of the order of σ and temporal scales of the order of $\tau \sim \sqrt{m\sigma^2/\varepsilon}$ limiting therefore the system to some hundreds of Angstroms and nanometers. MD simulations can enlighten the slip mechanisms in confined flows; Thompson and Robbins (1989) presented a detailed study of LJ liquids flow between two walls. The degree of slip in a boundary condition is strongly dependent upon the wall-fluid interaction potential that determines the strength of intra and inter molecular forces between the wall and fluid particles. For strong interactions a significant epitaxial distribution is observed with strong ordering of single or double fluid layers being observed in the vicinity of the solid wall introducing no-slip conditions. As the interactions are weakened, the ordering near the walls is smaller and the slip becomes more apparent.

It has to be underlined that MD studies, except equilibrium studies (Bocquet & Barrat, 1993; Bocquet & Barrat, 1994), catechize the system with shear rates with much higher values compared to experimental measurements (Tropea *et al.* 2007). An eloquent issue occurring in the framework of interpretation MD results in the continuum limit has been focused in the diffusion of particles in the proximity of a solid wall. Brenner and Ganesan render the interfacial diffusion as part of their study in the frame of their work for the scale separation between the continuum and molecular phenomena (Brenner & Ganesan, 2000). They employed a perturbation analysis to appeal that the boundary condition in continuum scale is represented by matching the asymptotic limit of the transport processes' physical description (more exact macroscopically) with the outer limit of the inner molecular-scale conditions. They showed that slip lengths should not be computed at molecular scales; they should be measured by an extrapolation of the far field results (at the boundaries).

9.2 EFFECTS OF PHYSICAL CHARACTERISTICS

9.2.1 Surface roughness

In nanoscale flows, surface roughness along with its geometrical attributes has been considered as pivotal factor that triggers different slip behaviours. Asproulis and Drikakis (2010b) studied the effects of rectangular surface roughness with variable height on the induced slip using MD simulations. The presence of a rectangular corrugation suggested a mechanism for fluid particles layering propagation towards the centre of a nanochannel. As the corrugation height increases the density fluctuations are displaced closer to the centre of the channel. Investigating the effects of attraction energy combined with surface roughness they showed that an increase of the latter the fluid's velocity tends to zero making the effects of the former unnoticeable. In cases where surface roughness has not been introduced, variations in the attraction energy suggest a linear impact on the slip velocity. On the contrary, as a roughness is introduced in the solid wall, non-linear phenomena arise considering the flow characteristics and the slip velocities. They proposed an exponential decrease of the slip length as the height of the roughness increases for different various surface properties and it has been demonstrated that the hypothesis of no-slip boundary condition is valid exceeds a height of 2σ.

9.2.2 Surface stiffness

A recently investigated factor, influencing the slip process is the surface stiffness of the wall. Asproulis and Drikakis (2010a) performed MD studies to investigate the dependency of slip length on the wall stiffness for a LJ fluid. To perform a minimization on the influence of the shear rate a comprehensive investigation of the effects of the wall stiffness on slip phenomena has been realized for a wide range shear rates values. In Asproulis and Drikakis, (2010a) the particles' wall are attached to their equilibrium lattice site $\boldsymbol{r}_0$ via an elastic spring force $F = -k(r_i - r_0)$ with k to be the wall stiffness. The introduction of stiffness is a critical parameter providing a coupling between the solid model and the properties of the materials defining the physical behaviour at the wall. Lower values in κ imply that wall particles vibrate around their equilibrium positions with higher amplitude and lower frequency thus the fluid molecules are allowed to move closer to the solid wall theoretically. The solid walls have been modelled by employing two (111) fcc lattice walls planes with the density of the particles to be equal with the fluid ones. The interaction between fluid and solid simulated with a simple LJ potential function. Also of great importance is the sensitivity of slippage phenomena to the solid-fluid interaction parameters. Specifically, there is a decrease of the momentum that is being transferred across the interface as surface energy decreases which leads to higher slip values (Thompson & Troian, 1997). The authors calculated the averaged density profiles in the proximity of the solid walls for various values of surface stiffness. Near the solid walls the density distribution exhibits oscillations independently of the surface stiffness. They showed that the variations of the stiffness do not alter the topology of the density profile as κ varies affecting only the absolute maxima and minima resting to the bulk values after a certain length ranging in $(5\text{–}7)\sigma$. Using LJ parameters with values $\varepsilon_{wf} = 0.4\varepsilon$, $\sigma_{wf} = 0.75\sigma$ and an external driving force $f_x = 0.01\varepsilon\sigma^{-1}$, they investigated the behaviour of the slip length as a function of surface stiffness. Smaller values of κ suggest an increase to the surface roughness due to the larger displacements of the solid wall particles. As κ increases, the wall evolves into a smoother surface, resulting higher slips. It has been demonstrated that no monotonic increase in the slip length with the wall stiffness is observed. On the contrary it obtains a maximum value followed by a decreasing part. In case of stiffer walls the vibration amplitude of the solid surface decreases, however the oscillating frequency increases and becomes the dominant factor that slightly increases the momentum transfer across the interface leading to smaller values of slip. In general, they showed that the function of slip length as surface stiffness varies can be estimated by a master curve.

9.2.3 Wetting – surface energy – contact angle

Since early studies (Goldstein, 1938) it is widely recognized that slip at the liquid-solid interface is influenced on various physicochemical properties of both the solid and the liquid. In detail, the wetting properties of the solid have a vital role, as has been suggested in many experimental studies, and their effects have also been studies computationally. De Gennes (1985) and De Gennes *et al.* (2004) examined the wettability of solids by liquids, attempting to quantify their effect by introducing a spreading coefficient S as follows:

$$S = \gamma_S - (\gamma_L + \gamma_{LS}) \tag{9.1}$$

where γ_S, γ_L and γ_{LS} are the solid, liquid and liquid-sold interfacial energies. The spreading coefficient is defined as the difference in surface energy observed between a dry solid surface and the same surface after being wetted by a liquid layer. The degree of wetting can be characterized according to the spreading coefficient's values. When $S > 0$, the solid is completely wet and negative values of S ($S < 0$), corresponds to partially wetting. To investigate the hydrophilic or hydrophobic behaviour of a surface the introduction of contact angle, θ_C is needed, where θ_C is defined as the angle on the droplet side of the contact line tangent to the liquid-solid interface. Equation (9.1) combined with Young's law suggests a relation between S and θ_C

$$S = \gamma_L(\cos\theta_C - 1) \tag{9.2}$$

A fluid exhibiting a wetting behaviour has a contact angle $0° \leq \vartheta_C \leq 180°$, whereas a fluid displays a non-wetting behaviour has a contact angle $90° \leq \vartheta_C \leq 180°$ corresponding to hydrophilic and hydrophobic surfaces respectively. Realistic surfaces, displaying roughness and various other non-idealistic properties, produce the phenomenon of contact angle hysteresis which is the difference between the maximum (advancing) and the minimum (receding) contact angles which depend on the relative speed between the solid surface and the contact line. In practice, hysteresis estimates the deviation of the contact line which depicts a change in the size of the droplet. Slip has been measured in cases of complete (Bonaccurso *et al.* 2003; Bonaccurso *et al.* 2002; Pit *et al.* 2000) and partial wetting (Baudry *et al.* 2001; Boehnke *et al.* 1999; Cheikh & Koper, 2003; Zhu & Granick, 2001; Churaev, 1984; Cho *et al.* 2004), indicating that slip increase as a function of contact angle in a systematic mode (Zhu & Granick, 2001) or only for nonpolar liquids (Cho *et al.* 2004). Tropea *et al.* (2007) summarizes several results about the connection between slip and contact angle suggesting a poor correlation.

9.2.4 Shear rate – pressure

Priezjev (2007) studied the effect of molecular scale surface roughness on rate-dependent slip in simple fluids. He showed that in case of smooth, in molecular level, rigid surfaces accompanied with weak solid-fluid interactions, the slip length possess a linear increasing behavior with the shear rate. Thermal, random and periodic roughness has been considered as the variations of surface roughness. Periodically and randomly corrugated rigid walls, characterized by amplitudes smaller that the molecular diameters exhibited a high reduction at the slip length along with the dependence on shear rate. He investigated thermal surface roughness with finite values in solid-surface stiffness of the wall atoms illustrating that it alters the slip behavior. In case of soft walls weak rate dependence has been observed originating from the large infiltration of the vibrated wall atoms into the fluid. However as the surface stiffness increases, the walls tend to smooth providing a linear rate dependence of the slip length. Tretheway *et al.* (2004) performed experimental studies, investigating the effect of pressure in slip length. An increase in the absolute pressure leads to a decrease in the slip length. In case of water, he showed that the validity of the no-slip condition arises when the pressure reaches 6 atm. Ruckenstein and Rajora (1983) studied the possibility of surface slip caused by pressure gradients, employing elements from equilibrium thermodynamic theory. They based their idea that the presence of a pressure gradient induces a gradient in the chemical potential.

The chapter is structured as follows: In the present chapter, a review effort is taking part for the behaviour of liquid nanoflows in confined geometries. The water behaviour flows through nanostructures is also discussed along with their molecular modelling.

9.3 COMPUTATIONAL APPROACHES

9.3.1 Atomistic modelling

The core of the MD simulations is based in the integration of the Newton's equation of motion for single particles. Applying an integration method among the proposed in the literature, the basic dynamics quantities such as position, velocity and the interaction force for each particle, can be determined. Statistical mechanics computations can be employed afterward for the computation of thermophysical properties such as pressure, mean velocity, temperature number density and so on. In the MD framework, the second law of Newton is written for each atom i as a point mass as follows

$$m_i \ddot{\mathbf{r}}_i = -\frac{\partial V_i}{\partial \mathbf{r}_i} \tag{9.3}$$

where m_i is the atomic mass, $\mathbf{r}_i$ the acceleration of the atom i and V_{ij} the potential energy computed as the sum of semi-empirical analytical functions that typify the real interatomic forces. The most basic potential function is the Lennard Jones (LJ) pair-wise potential modelling the Van der Waals attraction along with the repulsive forces gives by:

$$V_{ij} = 4\varepsilon\left[\left(\frac{\sigma}{r_{ij}}\right)^{12} - \left(\frac{\sigma}{r_{ij}}\right)^{6}\right] \tag{9.4}$$

where r_{ij} is defined as the distance between i and j particles, ε is an interaction strength parameter, σ is the molecular length scale defining the location of the zero potential energy. In case of multi-atomic fluids such as gaseous mixtures, the Lorentz-Berthelot mixing rules are used:

$$\sigma_{ij} = \frac{\sigma_i + \sigma_j}{2}, \quad \varepsilon_{ij} = \sqrt{\varepsilon_i \varepsilon_j} \tag{9.5}$$

To calculate the total potential energy a summation of all the individual pairs is employed:

$$V = \sum_{i}^{N} \sum_{j>i}^{N} V(r_{ij}) \tag{9.6}$$

where N is the total number of atoms in the simulation box. To calculate the potential of a single atom V_i a summation of all the potential interactions where atom i is involved as follows:

$$V_i = \sum_{i \neq j}^{N} V(r_{ij}) \tag{9.7}$$

In case a two body interaction potential is applied to the simulation, the interaction force between a pair of particles is computed from the following equation:

$$f_i = \nabla_{r_i} V_i \tag{9.8}$$

The trajectories of each particle i are calculated by the time integration of the Equation (9.4). To perform the time integration a finite difference scheme is used such as the Verlet method or the predictor-corrector method (Verlet, 1967; Allen & Tildesley, 1987). The application of the above schemes may be characterized by a mathematical simplicity; however they are extremely computational demanding especially in cases of simulated systems with large number of particles. The most common time integrator is the Verlet algorithm, suggesting a direct solution of the Newton's equation of motion, based in a second-order approximation for the ith particle.

$$\mathbf{r}_i(t + \delta t) = 2\mathbf{r}_i(t) - \mathbf{r}_i(t - \delta t) + \delta t^2 \cdot \boldsymbol{\alpha}_\iota(t) + O(\delta t)^4 \tag{9.9}$$

with δt to be the time step of the simulation and $\alpha_i(t)$ is the acceleration of each particle. To calculate the position o the ith particle on the next time step $t + \delta t$, it is not required the knowledge of the current velocity $\boldsymbol{u}(t)$. However in case of more complex molecular systems to obtain more accurate results additional terms are included for the calculation of the total potential energy of the system. The basic idea is to interpret the molecular system in an equation containing bonded and non-bonded interactions. Atomistic vibrations are observed, resulting deviations from the equilibrium values of bond and angle values that need to be modelled as follows:

$$V = \sum_{\text{bonds}} \frac{k}{2}(l_i - l_{\text{eq}})^2 + \sum_{\text{angles}} \frac{k}{2}(\theta_i - \theta_{\text{eq}})^2 + \sum_{\text{torsion}} \frac{V_n}{2}(1 + \cos(n\omega - \gamma))$$
$$+ \sum_{i}^{N} \sum_{j>i}^{N} 4\varepsilon \left[\left(\frac{\sigma_{ij}}{r_{ij}} \right)^{12} - \left(\frac{\sigma_{ij}}{r_{ij}} \right)^{6} \right] + \frac{q_i q_j}{4\pi\varepsilon_0 r_{ij}} \tag{9.10}$$

The first three terms in Equation (9.10) model any intra-molecular interactions observed in the system; the first one calculates the bonded interactions with a harmonic functional form, where the second is a summation over all angles in the molecule (valence angles) with the same functional form. Advancing to the third term, it depicts the change of energy as a bond rotates, with k being the Hooke's law constant and l_{eq} and θ_{eq} the reference bond and angle lengths respectively. The fourth term is the non-bonded term with its first part being modelled through LJ potential as analysed previously. Concerning the second part, it describes the long-range electrostatic interactions through the Coulomb law, where q_i refers to the electric charge of the ith atom and ε_0 is the vacuum permittivity.

Modelling heterogeneous nanoflows, by employing MD calculations, usually requires computations for the interactions that include a huge number of atoms. Among the calculations made during a simulation, the one that has the highest computational cost is the estimation of the long-range interactions summarized in the electrostatic and dispersion interactions. Given that there is not a certain cut-off the computation cost of a system with N particles equals to $O(N^2)$. Various algorithms have been implemented to reduce the time consuming evaluation, have been proposed spanning from the aforementioned Verlet algorithm with the trivial sorting scheme, to the particle-particle particle-mesh (PPPM) technique (Hockney & Eastwood, 1973; Hockney & Eastwood, 1988), the Ewald summation (Ewald, 1921; Petersen, 1995) and the particle-mesh Ewald method (PME) (Darden *et al.* 1993) with the requiring operations to be $O(N\log N)$, $O(N^{1.5})$, $O(N\log N)$ respectively.

9.3.2 Continuum modelling

Most of the problems encountered in micro- and nanoscale flows cannot be confronted solely by the continuum theory. As the operational dimensions are downsized to smaller scales, the continuity assumption the continuity assumption and the conventional computational fluid dynamics (CFD) models fail to fully capture the flow physics. In that case, molecular or atomistic modelling techniques, such as those presented in the preceding section have to be employed.

Despite the fact that continuity does not hold in certain regions, the rest of the domain can still be described through the traditional continuum based approaches. Continuum level modelling has been widely applied to an overwhelming number of practical engineering problems ranging from fluid mechanics, solid mechanics and structural mechanics to bio-engineering mechanics systems. Several studies have been also undertaken for engineering problems related to microflows, after making use of the continuum hypothesis. Besides, continuum models play significant role in multiscale approaches, where the solution of fluid flow and mass transport problems (in the macro-, micro- and nano-scale level) presupposes the coupling of continuum with atomistic models.

The scope of this section is twofold. Firstly, to briefly address the fundamental concepts of continuum modelling with focus on the equations that govern flow motion, and then to describe the discretisation process of the equations which is necessary for the system to be solved numerically. The main purpose here is to emphasise the physical meaning of the governing equations, as well as to discuss the most common discretisation approaches used in the framework of finite-volume applications. Comprehensive description of numerical tools and methods can be found in a number of text books available in the field of CFD (Anderson, 1995; Laney, 1998; Drikakis & Rider, 2005; Karniadakis *et al.* 2005; Toro, 2009).

The derivation of the fundamental equations of fluid motion is based on three basic physical principles.

- The mass of fluid is conserved
- The rate of change of momentum equals the sum of the forces that are imposed on a fluid particle
- The rate of energy change equals the sum of the rate of work done on a fluid particle, and the rate of heat addition to the fluid particle, according to the First law of Thermodynamics.

The aforementioned physical principles are applied to a suitable flow model, which can be either a finite control volume V or an infinitesimally small fluid element with a differential fluid dV. The corresponding fluid-flow equations can then be expressed either in integral form, or in partial differential form. Furthermore, the governing equations analogous to the form of the finite control volume (CV) or the infinitesimal fluid element (i.e., if they are fixed in space and the flow passing through them, or if they are moving with the fluid) can be in conservative or non-conservative form. As previously mentioned, the governing equations of the fluid motion will be outlined here, by means of a finite CV fixed in space. The conservation laws of mass, momentum and energy will be presented, followed by a general description of the complete system of the Navier-Stokes Equations (NSE) for a compressible, time-dependent, viscous flow. Note, however, that an overwhelming number of different submodels and simplified versions of the governing equations can be found in the literature, whose description is beyond the scope of the present section.

The continuity equation is based on the physical principle that the mass is conserved, which means that the net mass flow out of a CV through its surface S, is equal to the time rate of decrease of mass inside the CV. Consider the flow model shown in Figure 9.1. The net mass flow of a moving fluid across any fixed space is equal to the product of (density) × (velocity component normal to the surface S) × (surface area dS). Hence, the elemental mass flow across the area dS is given by:

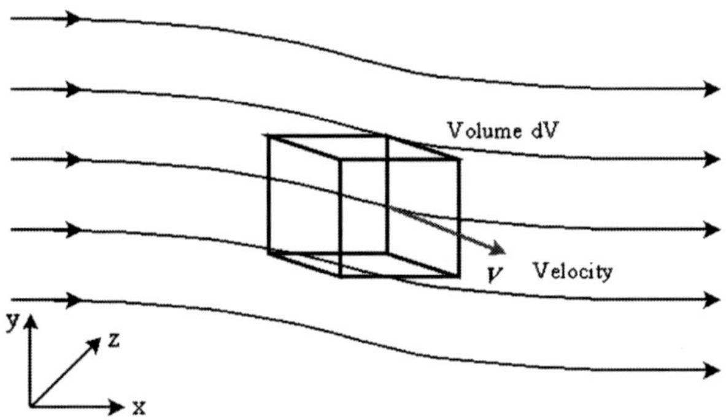

Figure 9.1 Infinitesimal control volume fixed in space with the fluid moving through it.

$$\rho V_n dS = \rho \mathbf{V} \cdot \mathbf{dS} \tag{9.11}$$

where $\mathbf{V} = ui = vj = wk$ is the velocity vector.

The net mass flow over the entire CV through surfaces S, is the summation over S of the elemental mass flow, expressed in Equation (9.11). In the limit, this becomes a surface integral given by:

$$\rho \mathbf{V} \cdot \mathbf{dS} \rightarrow \iint_S \rho \mathbf{V} \cdot \mathbf{dS} \tag{9.12}$$

Note also that when **V** points out of the CV, the product $\rho \mathbf{V} \cdot \mathbf{dS}$ is positive and the flow state can be considered as *outflow*. On the other hand, when **V** points into the CV the product $\rho \mathbf{V} \cdot \mathbf{dS}$ is negative and the flow state can be considered as *inflow*.

Accordingly, the time rate of decrease or increase of mass inside the CV is given by the following expression:

$$(-)\frac{\partial}{\partial t}\iiint_{\mathrm{V}} \rho d\mathrm{V} \tag{9.13}$$

Thus, the final formulation of the integral form of the continuity equation in conservative form can be obtained, by adding equation (9.13) to equation (9.12) as follows:

$$\frac{\partial}{\partial t}\iiint_{\mathrm{V}} \rho d\mathrm{V} + \iint_S \rho \mathbf{V} \cdot \mathbf{dS} = 0 \tag{9.14}$$

The momentum equation is based on the physical principle of Newton's second law of motion, which states that "the force exerted on a body is equal to the time rate of change of momentum".

The time rate of change of momentum as the body moves through a fixed CV is the sum of two terms.

The variation in time of momentum inside a fixed CV can be expressed in integral form as:

$$\frac{\partial}{\partial t}\iiint_{V} \rho \mathbf{V} dV \tag{9.15}$$

where $\rho\mathbf{V} = [\rho u, \rho v, \rho w]^T$. The second term completing the left hand side of the momentum equation comprises the convective flux tensor, which describes the momentum transfer across the boundary of the CV, given by the following formula:

$$\iint_{S} \rho \mathbf{V}(\mathbf{V} \cdot \mathbf{dS}) \tag{9.16}$$

Note that in the Cartesian coordinate system, the three components of the convective flux tensor in equation (9.16) can be analysed as follows:

$$\begin{Bmatrix} x - component : \rho u\mathbf{V} \\ y - component : \rho v\mathbf{V} \\ z - component : \rho w\mathbf{V} \end{Bmatrix} \tag{9.17}$$

The forces that the two elements are exposed to, can be separated into two categories. The first category includes the body forces, that is, gravitational forces, inertial forces, electromagnetic forces to name but a few. These forces act directly on the volumetric mass of the fluid element, causing the element to accelerate. The contribution of the body forces per unit volume, ρf_b, to the momentum equation is given by:

$$\iiint_{V} \rho f_b dV \tag{9.18}$$

The second category includes the surface forces, which act directly on the surface S of a CV and are, in general, due to two sources: a) the pressure distribution, which also acts on the surface of the fluid element and is imposed by the outside fluid surrounding the volume, and b) the normal and shear stresses originating from a force vector parallel and perpendicular to the surface normal vector, respectively. The contribution of the pressure distribution along with the normal and shear stresses to the momentum conservation equation is as follows:

$$-\iint_{S} p\mathbf{dS} \iint_{S} \bar{\bar{\tau}}\mathbf{dS} \tag{9.19}$$

where p is the pressure, and $\bar{\bar{\tau}}$ represents the shear stress tensor. Adding all the above contributions to the general conservation law, the momentum conservation inside an arbitrary control volume V that is fixed in space can be expressed as:

$$\frac{\partial}{\partial t}\iiint_{V} \rho \mathbf{V} dV + \iint_{S} \rho \mathbf{V}(\mathbf{V} \cdot \mathbf{dS}) = \iiint_{V} \rho f_b dV - \iint_{S} p\mathbf{dS} + \iint_{S} \bar{\bar{\tau}}\mathbf{dS} \tag{9.20}$$

Finally, the last equation based on the principle of the first law of thermodynamics (i.e., energy can be neither destroyed nor created, but only can change forms), is the energy equation. The rate of change of energy inside a fluid element is equal to the heat flux q through the CV surface, plus the rate of work done on element thanks to body and surface forces. The net heat flux consists of two parts, specifically the heat flux due to the volumetric rate of heat addition per unit mass or due to chemical reactions, $\dot{q}_h$, and the net heat flux due to temperature gradients. According to Fourier's law of heat conduction, the local heat flux is equal to the product of the thermal conductivity coefficient k and the negative local temperature gradient $-\nabla T$.

$$q = -k\nabla T \tag{9.21}$$

Moreover, the surface forces, f_s, related to the time rate of work done by pressure, as well as by normal and shear stresses on a fluid element are given by the following formula:

$$f_s = -p\bar{\bar{I}} + \bar{\bar{\tau}} \tag{9.22}$$

where $\bar{\bar{I}}$ is the unit vector.

Taking into account the above properties encountered in the energy equation, the final formula takes the following form:

$$\begin{aligned} \frac{\partial}{\partial t}\iiint_{V} \rho E d\mathrm{V} + \iint_{S} \rho E \mathbf{V} d\mathbf{S} &= \iiint_{V} \dot{q}_h\, d\mathrm{V} + \iint_{S} k\nabla \mathbf{T} d\mathbf{S} + \\ &+ \iiint_{V} (\rho f_b \cdot \mathbf{V}) d\mathrm{V} - \iint_{S} p\mathbf{V} d\mathbf{S} + \iint_{S} (\bar{\bar{\tau}} \cdot \mathbf{V}) d\mathbf{S} \end{aligned} \tag{9.23}$$

In the above equation E is the total energy per unit volume, consisting of the internal energy and the kinetic energy component as:

$$E = \rho(e + \frac{1}{2}V^2) \tag{9.24}$$

The normal and shear stresses emanate from the friction between the surface of an element and the fluid. The time rate of change of the shearing deformation is related to the shear stress, τ_{xy}, whereas the time rate of volume change of the fluid element is related to the normal stress, τ_{xx}, as illustrated in Figure 9.2. In Cartesian coordinates, the general form of the stress tensor $\bar{\bar{\tau}}$ is given by:

$$\bar{\bar{\tau}} = \begin{bmatrix} \tau_{xx} & \tau_{xy} & \tau_{xz} \\ \tau_{yx} & \tau_{yy} & \tau_{yz} \\ \tau_{zx} & \tau_{zy} & \tau_{zz} \end{bmatrix} \tag{9.25}$$

where τ_{xx}, τ_{yy} and τ_{zz} are the normal stresses in the x, y and z direction, respectively, while the remaining six components represent the shear stresses of the fluid

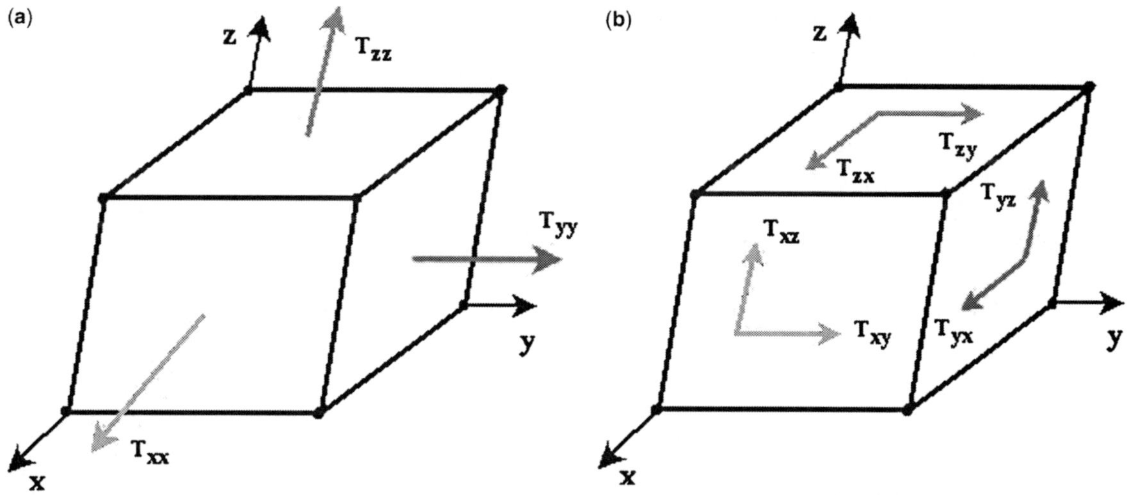

Figure 9.2 Illustration of normal (left) and shear (right) stresses exerted on a fluid element in the Cartesian coordinate system.

By convention, the notation τ_{ij} denotes a stress component that works on a plane that is perpendicular to the i-axis, in the direction of the j-axis, as shown in Figure 9.2. For a Newtonian fluid, where the shear stress is proportional to the velocity gradient, the viscous stresses can be expressed as follows:

$$\begin{aligned}\tau_{xx} &= \lambda\left(\frac{\partial u}{\partial x}+\frac{\partial v}{\partial y}+\frac{\partial w}{\partial z}\right)+2\mu\frac{\partial u}{\partial x}\\ \tau_{yy} &= \lambda\left(\frac{\partial u}{\partial x}+\frac{\partial v}{\partial y}+\frac{\partial w}{\partial z}\right)+2\mu\frac{\partial v}{\partial y}\\ \tau_{zz} &= \lambda\left(\frac{\partial u}{\partial x}+\frac{\partial v}{\partial y}+\frac{\partial w}{\partial z}\right)+2\mu\frac{\partial w}{\partial z}\end{aligned} \tag{9.26}$$

$$\begin{aligned}\tau_{xy} &= \tau_{yx} = \mu\left(\frac{\partial u}{\partial y}+\frac{\partial v}{\partial x}\right)\\ \tau_{xz} &= \tau_{zx} = \mu\left(\frac{\partial u}{\partial z}+\frac{\partial w}{\partial x}\right)\\ \tau_{yz} &= \tau_{zy} = \mu\left(\frac{\partial v}{\partial z}+\frac{\partial w}{\partial y}\right)\end{aligned} \tag{9.27}$$

According to Stokes hypothesis, the second viscosity coefficient $\lambda = -\frac{2}{3}\mu$, is particularly responsible for the energy dissipation in a fluid of uniform temperature, once a change in volume at finite rate takes place. However, note that for flows such as blood, paint, starch suspensions, and so on, where the shear stress is not proportional to the fluid velocity gradient, different forms of stress tensor might be more appropriate.

It is important to underline that the various terms appeared in the conservation laws of mass, momentum and energy [see Equations (9.14), (9.20) and (9.23)] can be all collected in one system of equations. However, it can be seen that the number of unknown flow-field variables is always larger than the number of equations, thus the system cannot be considered as closed. Specifically, the number of unknowns is six (u, v, w, ρ, e, p), while the total number of equations is five. To overcome this difficulty, we generally assume that the gas is a perfect gas (i.e. the intermolecular forces are negligible) and the thermal equation of state can then be added to the system of equations, in order to close it.

$$p = \rho RT \tag{9.28}$$

where R is the specific gas constant.

Despite the fact that the number of equations is increased to six, one new unknown variable – the temperature T – is added to the system of equations, increasing, hence, the number of unknowns to seven. A thermodynamic relation between the state variables is then used as a seventh equation to close the system. For a calorically perfect gas with constant specific heats, this equation is written as:

$$e = \frac{p}{(\gamma - 1)\rho} \tag{9.29}$$

where γ is the ratio of specific heats, known also as the adiabatic exponent.

The discretisation of the governing equations (i.e., Navier-Stokes, Euler, Stokes equations, to name but a few) that describe the fluid motion can be achieved by using one of the following methods: a) finite difference (FDM), b) finite element (FEM), and c) finite volume methods (FVM). Each of the aforementioned methods can potentially provide accurate approximations under certain circumstances, which are mainly related to the nature of the problem to be solved.

Continuum modelling of liquid and gas nanoflows is the way of mathematical flow field modelling, where the unknown physical quantities are assumed to be continuously differentiable functions in the governing transport equations. For modelling nanoscale flow phenomena, the validity of continuum hypothesis is limited, because of the physics of small scales. According to the literature (Heller, 2005), the characteristic length scale of the fabricated system has to be a bigger value than the average interatomic space

$$\lambda_A = \left(\frac{M_A}{\rho N_A}\right)^{\frac{1}{3}},$$

where M_A is the molar mass, ρ is the fluid density, and $N_A = 6.022 \cdot 10^{23}$ mol^{-1} is the Avogadro's number. After satisfying the afore-mentioned criterion, the mass, momentum, energy, and species transport equations of continuum physics are considered to be valid for mathematical modelling of nanoflows.

The continuity equation is derived from the mass conservation principle (Versteeg & Malalasekera, 1995), which can be written as follows

$$\frac{\partial \rho}{\partial t} + \nabla \cdot (\rho \mathbf{u}) = 0,$$

where $\mathbf{u}$ is the velocity field. The conservation of momentum is derived from Newton's II law for fluid motion, which can be written in a general form as

$$\rho \frac{D\mathbf{u}}{Dt} = \rho \mathbf{g} + \nabla \cdot \underline{\underline{\sigma}} + \rho \mathbf{f},$$

where $\mathbf{g}$ is the force field of gravity, $\underline{\underline{\sigma}}$ is the stress tensor, and $\mathbf{f}$ is any other external (additional) force. The stress tensor $\underline{\underline{\sigma}}$ consists of two main parts such as

$$\underline{\underline{\sigma}} = -p\mathbf{I} + \underline{\underline{\tau}},$$

where the first part represents the surface forces with the hydrodynamic pressure p, $\mathbf{I}$ is the unit tensor, and the second part is the viscous stress tensor $\underline{\underline{\tau}}$. This viscous part of the stress tensor $\underline{\underline{\tau}}$ plays an important role to derive momentum equations for different physical fluid flow situations. The viscous stresses for ideal fluid are vanished $\left(\underline{\underline{\tau}} = 0\right)$, and the Euler momentum equation is gained in a vector form as

$$\rho \frac{D\mathbf{u}}{Dt} = \rho \mathbf{g} - \nabla p + \rho \mathbf{f}.$$

For Newtonian fluid and gas flows, the viscous stress tensor analogously to the Hooke law is derived by Navier and Stokes as

$$\underline{\underline{\tau}} = \mu(\mathbf{u} \otimes \nabla + \nabla \otimes \mathbf{u}) - \frac{2}{3}\mu(\nabla \cdot \mathbf{u}) \cdot \mathbf{I},$$

where μ is the dynamic viscosity of fluid. Consequently, the Navier-Stokes equations can also be written in a vector form as

$$\rho \frac{D\mathbf{u}}{Dt} = \rho \mathbf{g} - \nabla p + \mu \nabla^2 \mathbf{u} + \frac{\mu}{3} \nabla(\nabla \cdot \mathbf{u}) + \rho \mathbf{f}.$$

For incompressible flows, the divergence of the velocity field is vanished ($\nabla \cdot \mathbf{u} = 0$), therefore, the Navier-Stokes momentum equation becomes

$$\rho \frac{D\mathbf{u}}{Dt} = \rho \mathbf{g} - \nabla p + \mu \nabla^2 \mathbf{u} + \rho \mathbf{f}.$$

The principle of the conservation of energy can be written in differential form as follows

$$\rho \frac{DE}{Dt} = -\nabla \cdot \left(p\mathbf{u}\right) + \nabla \cdot \left(\underline{\underline{\tau}} \cdot \mathbf{u}\right) + \nabla \cdot (\lambda \nabla T) + \Phi_E + \Phi_D,$$

where the specific energy E is the sum of internal energy i, kinetic energy k and potential energy U. Furthermore, T is the temperature, Φ_E is the source of internal heat, and $\Phi_D = \underline{\underline{\tau}} \cdot (\nabla \otimes \mathbf{u})$ is the function of dissipation. In order to obtain an equation for internal energy and temperature, it is necessary to use the changes of mechanical kinetic energy equation (Versteeg & Malalasekera, 1995). The kinetic energy transport equation can be derived from the general form of momentum equation multiplied by the velocity field $\mathbf{u}$ obtaining the following

$$\rho \frac{Dk}{Dt} = \rho \mathbf{u} \cdot \mathbf{g} - \mathbf{u} \cdot \nabla p + \mathbf{u} \cdot \left(\nabla \cdot \underline{\underline{\tau}}\right) + \rho \mathbf{u} \cdot \mathbf{f}$$

differential equation. Subtracting the mechanical kinetic energy equation from the equation of conservation of energy, a differential equation is obtained for the changes of internal energy as

$$\rho \frac{Di}{Dt} = -p(\nabla \cdot \mathbf{u}) + \nabla \cdot (\lambda \nabla T) + \Phi_E + \underline{\underline{\tau}} \cdot \cdot (\nabla \otimes \mathbf{u}) - \rho \mathbf{u} \cdot \mathbf{f},$$

where $i = c_p T$ for incompressible flows, and taking into account the divergence-free constraint ($\nabla \cdot \mathbf{u} = 0$), a differential equation is gained for the temperature as

$$\rho c_p \frac{DT}{Dt} = \nabla \cdot (\lambda \nabla T) + \Phi_E + \underline{\underline{\tau}} \cdot \cdot (\nabla \otimes \mathbf{u}) - \rho \mathbf{u} \cdot \mathbf{f},$$

where c_p is the specific heat at constant pressure. For modelling nanoflows, the species transport equation for multicomponent fluid may also be part of the governing equation as

$$\frac{\partial c_i}{\partial t} + \nabla \cdot \mathbf{J}_i = 0,$$

where c_i is the concentration of ith species, and $\mathbf{J}_i$ is the flux of ith species.

Modelling of physical phenomena in aqueous environment is often fallen into the modelling of multiphysics problems including diverse fields of mathematical, physical, engineering, and computational sciences. Under the continuum hypothesis, any other external force field in the momentum equation, such as electric and magnetohydrodynamic effects, can be taken into account, and, for instance, single-stranded DNA or double-stranded DNA Eulerian-Lagrangian coupled flow modelling. Another important issue is that the relevant dimensionless numbers such as Reynolds-, Mach-, and Knudsen numbers appear in the non-dimensionalised system of governing transport equations.

The Reynolds number is expressed as the ratio of inertial to viscous forces as:

$$\mathrm{Re} = \frac{\rho u_0 L}{\mu} \tag{9.30}$$

where u_0 is the characteristic reference velocity, L is the characteristic length of the physical problem, and μ is the dynamic viscosity of the fluid. The Mach number is the ratio of characteristic velocity of flow to the speed of sound as:

$$Ma = \frac{u_0}{a} \tag{9.31}$$

where α is the speed of sound in the medium. The Mach number is also a measure of compression that can relate the speed of the fluid to the speed of sound. Note that for flows with low Mach number, where the characteristic fluid-flow velocity is small compared to the speed of sound, the incompressibility assumption can be used. The flow density in that particular state is considered as constant, and the mass conservation Equation (see Equation 9.14) becomes:

$$\nabla \cdot \mathbf{u} = 0 \tag{9.32}$$

Furthermore, by making use of the incompressibility condition, the simplified momentum equation is given by:

$$\frac{\partial \mathbf{u}}{\partial t} + \mathbf{u} \cdot \nabla \mathbf{u} = -\frac{1}{\rho} \nabla p + \nu \nabla^2 \mathbf{u} \tag{9.33}$$

where v is the kinematic viscosity of the fluid and p is the incompressible pressure.

The Knudsen number can be expressed as the ratio of the free mean path λ over a characteristic geometric length L. The Knudsen number is also related to the Mach number and Reynolds number as follows:

$$Kn = \frac{\lambda}{L} = \sqrt{\frac{\pi \gamma}{2}} \frac{M}{\mathrm{Re}} \tag{9.34}$$

For flows with $Kn > 1$, the definitions of instantaneous macroscopic values in complex microgeometries, where three-dimensional spatial gradients are expected, become problematic. The continuum hypothesis is no longer valid, as the macroscopic property distribution breaks down. As the Knudsen number increases further, the flow cannot be considered as a continuum and rarefaction effects become dominant. In this flow regime, the common flow and heat transfer models based on the continuum hypothesis are not adequate of predicting the heat flux, the shear stress and the corresponding mass flow rate. Schaaf and Chambre (1961) demonstrated an empirical classification of different flow regimes based on the Knudsen number.

For $Kn \leq 10^{-2}$, the fluid is considered to be in the continuum regime. Thus the governing equations of the fluid flows (for example, the Navier-Stokes and Euler equations) can be treated with the conventional continuum models. In that particular fluid flow regime, no-slip boundary conditions close to solid surfaces can be also employed.

For $10^{-2} < Kn < 10^{-1}$, the flow lies in the slip flow regime, where the rarefied gas flow can be considered neither as an absolutely continuum nor as a free-molecular flow. Navier-Stokes equations or Burnett equations can be used under certain circumstances analogous to the problem geometry.

For $10^{-1} < Kn < 1$, the flow regime is considered as transitional. The Navier-Stokes equations, as well as Burnett equations can be used in conjunction with high-order slip conditions.

For $Kn > 1$, continuum models are no longer valid.

In terms of boundary conditions for gas flows, the no-slip boundary condition for the velocity is valid at the fluid-solid interface up to $Kn < 0.001$ [see in Gad-el-Hak (2001)]. Therefore, different slip models, such as Maxwell's first-order slip condition and higher-order slip models are introduced to overcome the no-slip boundary condition break down in the field of micro- and nanofluidics [see more detail in Gad-el-Hak (2002) and in Karniadakis, Beskok, and Aluru (2005)]. Gad-el-Hak (2002) studied the molecular and continuum modelling approaches together. Karniadakis, Beskok, and Aluru (2005) overviewed the numerical methods for both liquid and gas micro-, and nanoflows.

Bahukudumbi and Beskok (2003) developed a semi-analytical slip-based model to predict the velocity and shear stress distribution, load capacity and pressure profile in slider bearings using a wide range of Knudsel numbers ($Kn < 12$). The developed model was derived from the Navier-Stokes equations in conjunction with generalised slip models. Closure of the system was achieved by using the Poiseuille flowrate database obtained from the two-dimensional Boltzmann equation solution. The method demonstrated in their study found to give reasonable results, with the level of physical information being similar to direct simulation Monte Carlo (DSMC) computations. However, the new self-consistent slip-based model was not capable of capturing the flow properties in the free molecular flow regime.

A high-order velocity-slip boundary condition was developed by Beskok (2001), validated for gas microflows. Model predictions were compared against the first-order slip condition and the DSMC results. Besides, continuum-based slip models were employed to analyse and test their robustness and validity at $Kn < 0.1$. For that purpose, the backward-facing step geometry was used to investigate the behaviour of the gas microflows in channels with strong adverse pressure gradients, separation and reattachment. Modelling separated rarefied gas flows in such complex geometries as the BWFS, is of great interest in microfluidics. A good agreement between the proposed method and the DSMC technique was achieved, for both attached and separated flows.

The continuum-scale, single-stranded and double-stranded DNA flow modelling play also important role in the field of nanofluidics in order to detect biological agents, such as bacteria and viruses. The numerical modelling can result in a better understanding of physical processes in detection systems, and can improve further the optimisation methods for the process parameters. The detection methods may rely on a nucleic acidic reaction, which is known as hybridization.

Das, Das, and Chakraborty (2006a) proposed analytical solutions for the rate of DNA hybridization in electroosmotic flows in a microchannel. Their results can be used in designing of DNA microchips and microarrays. Das, Das, and Chakraborty (2006b) studied the momentum, heat and solute transport during DNA hybridization considering electroosmotic effects. They observed that significantly higher DNA can be achieved at the capture probes, by increasing the accumulation time. These results can be used to investigate the rate of DNA hybridization (Das, Das, & Chakraborty, 2006b). Benke, Shapiro, and Drikakis (2008) proposed an efficient multiscale macromolecule modelling approach for single-stranded DNA motion in fluid flow, introducing a new fast linear corrector (FALCO) algorithm. A multiscale modelling strategy was provided by the metamodelling approach based on the coupled solution of governing equations. They considered a simplified mechanical macromolecule model in a lid-driven microcavity. This new algorithm improves computational efficiency and numerical stability compared to the widely used SHAKE algorithm. This method can be also used for simulating flow of macromolecules in detection systems.

9.4 LIQUID FLOW IN CONFINED GEOMETRIES

9.4.1 Flow behaviour in nanochannels

The liquid confinement in nanoscaled environments may promote phase transitions that cannot be observed in bulk phases. In case the liquid is based in water, the drying transition is induced, driven by the strong hydrogen bonding developed between water molecules. Advances in nanometer scaled structures fabrication techniques, along with numerical and experimental studies based in fluid flows through confined geometries have brought about the scientific interest around nanofluidic devices with various applications. A great number of works has been devoted for water-based liquid flows through carbon based nanostructures such as carbon nanotubes (CNTs). The focussing point in the aforementioned studies is located in the thorough interpretation of the interaction between the liquid and the nanostructure. In case of CNTs which have a main role in the literature, the cylindrical hollow geometry that they posses, can be presented as nanovessels. These nanotubes interact weakly with the majority of a flow in adverse with the fraction of the liquid that flows in their proximity and mostly in their interior; the latter exhibits various phase transitions in the confined space. Several studies showed that the properties of the liquids experiencing confinement change in case a sequestration occurs when it is related with solid walls in atomistic scales. MD and Monte Carlo (MC) simulations are considered among the basic computational tools employed for the simulation of nanoscale systems including various molecules and structures. Bitsanis *et al.* (1990) investigated the dynamics of flow in molecularly narrow pores, illustrating that there is a deviation from the continuum theory. The latter was explained by the variations observed in the density in the studied pore changing the relationship between the local viscosity and density. In pores with diameters greater that four molecular diameters can be characterized by redefining the viscosity locally. However as the diameter decrease, the effective viscosity radically increases because the fluid cannot be subjected smoothly and quickly to the planar flow. Asproulis (2010a) studied a Poiseulle flow, observing velocities parallel to wall that deviate from the form suggested by the assumptions made by the continuum theory. Tuzun *et al.* (1996) studied the dynamics of various LJ fluids in carbon nanotubes, showing that the assumption of rigidity that in some cases is used in order to reduce the computational cost affect the fluid behaviour in a quantitative way and qualitative. In case of a tube that is modelled with including dynamic interactions among the atoms that is made of, exhibits a faster reduction in the velocity of the molecules that in case of a rigid one. Kalra *et al.* (2003) showed that water molecules can flow through membranes of open-ended CNTs experiencing an osmotic gradient. Such a gradient results water to flow in a single-file mode almost frictionless governed by microscopic deviations. Striolo *et al.* (2006) demonstrated that the transport of water molecules through an infinitely long narrow CNTs, that can be achieved by the application of periodic boundary conditions, occurs via an aligned motion independently on the degree of filling of the CNTs. In nanotubes, the mechanism of diffusion through them, is time dependent. The presence of long-lasting Hydrogen bonds between adsorbed water molecules, the smooth carbon surfaces (C*) in combination with the weak C*-H_2O attractive interactions trigger a ballistic motion for an exceptionally longer period of time (~500ps) that expected. However, simulating for more timesteps, a change in the diffusion mechanism is observed, tending to Fickian diffusion.

The adsorption isotherms of water in SWNTs, at room temperature, proved to be type V (Striolo *et al.* 2005), indicating a mesoporous adsorbent with weak affinities with an adsorption/desorption hysteresis (Figure 9.3). The width of the aforementioned hysteresis is related with the diameter of the investigated tube. Studying narrower pores like a (6,6) or a (8,8) SWNT, showed that there is no exhibition of an adsorption/desorption hysteresis. Moreover, in case of tubes corresponding to (8,8)-(20,20) the relative pressure at which the filling of the pores takes part, decreases as the pore width decreases. This trend is different in a (6,6) nanotube exhibiting a pore sieving effect with a higher pressure for the filling process than in case of wider CNTs. The arrangement of the water molecules' confinement strongly depends on the tubes width. For diameters larger that a (8,8) possesses, the radial density profiles suggested a structured layer formation compared with a (6,6) CNTs where an one-dimensional configuration for the water molecules is presumed. The main attribute for the later liquid formation in based in a hydrogen-bond network that defines the structured of the formulated water molecules in confined geometries in general.

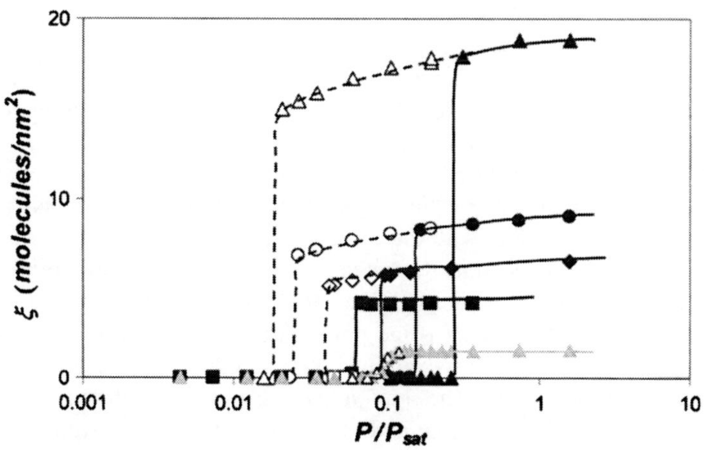

Figure 9.3 Simulated adsorption isotherms computed at 298 K. The coverage of the porous surface, ξ (number of adsorbed water molecules per square nanometer of porous surface) is reported as a function of the bulk relative pressure. Triangles are for water adsorption isotherm in (20:20) SWCNs; circles are for water in (12:12) SWCNs; diamonds are for water in (10:10) SWCNs; squares are for water in (8:8) SWCNs; gray triangles are for water in (6:6) SWCNs. Solid symbols are for simulation results along the adsorption path of the isotherm and open symbols are for desorption. Lines are guides for the eye, symbols are larger than statistical uncertainty. Figure from Striolo *et al.* (2005) with permission.

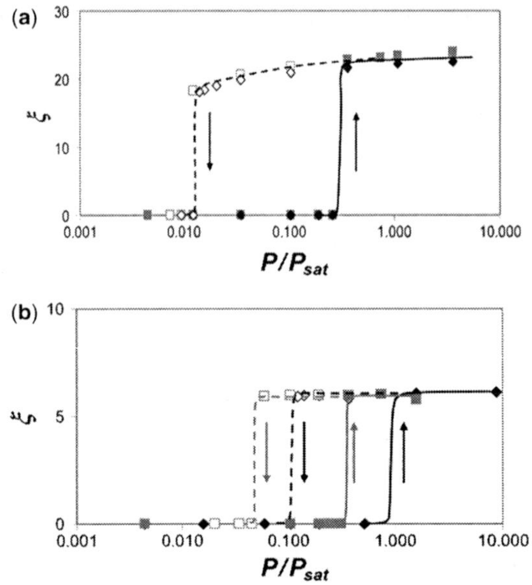

Figure 9.4 Coverage of the porous surface, ξ (number of adsorbed molecules per square nanometer of porous surface), as a function of the relative pressure for water adsorbed in carbon-slit pores at 298 K. Part a is for $H = 1.6$ nm. Part b is for $H = 0.6$ nm. Full symbols are for adsorption; open symbols are for desorption. Gray squares are obtained by using the Steele potential to account for pore-water interactions, while black diamonds are for results obtained by considering each carbon water interaction potential. Lines are guides to the eye. Symbols are larger than the simulation uncertainty. Arrows indicate the sequence of results obtained at increasing or decreasing pressure (adsorption and desorption, respectively). Figure from Striolo *et al.* (2003) with permission.

Despite the extensive study concerning the water flow through CNTs, a great number of computational studies have been performed focused in Carbon-Slit Nanopores (Striolo *et al.* 2003; Liu & Monson, 2005; Striolo *et al.* 2004). In case of wide pores (>1 nm), water does not develop condensed behaviour properties until the pressure acted on the fluid reaches values higher than the bulk in vapour phase (P_0). The above can be clarified by the absence of H-bonding with the carbon surfaces result the dense water to destabilize. As the pore diameters approach the molecular diameters, the strong interactions originated from both the graphite surfaces balance the dense water (Figures 9.4 and 9.5). However, the appearance of an additional destabilization for $P < P_0$, occurs if the pore width decrease to distances smaller than 0.6 nm (Liu & Monson, 2005). Striolo *et al.* (2003) examined the procedure of pore-filling showing that it arises by a capillary-condensation operation. The computed adsorption isotherms indicate once again a hysteresis between the cycles of adsorption and desorption. In a later study, Striolo *et al.* (2004) supported the latter, adding that the size of the aforementioned hysteresis along with the relative pressure, P/P_0, at which the pore-filling is being triggered decrease as the pore width decreases simultaneously. In particular, during the process of pore-filling in case of a hydrophobic nanopore, water undergoes a capillary condensation emanated by the development of clusters caused by the presence of H-bonds between the studied molecules. On the contrary, during the pore-emptying, a number of low density regions are exposed, demonstrating a clear tendency to approach the hydrophobic C*-surfaces. The above suggest that the interfacial surfaces are larger during desorption cycle compared with the adsorption one, giving a possible explanation for the observed hysteresis behaviour widely observed in water adsorption isotherms. Concerning the statics of the confined water, the computation of radial distribution functions showed that water molecules lay in a vapour phase before the pore-filling, while after their structure is typified with success as a liquid. Consequently, by using Carbon based pores to confine water in their interior, an analogous to vapour-liquid transition is obtained by pore-filling process.

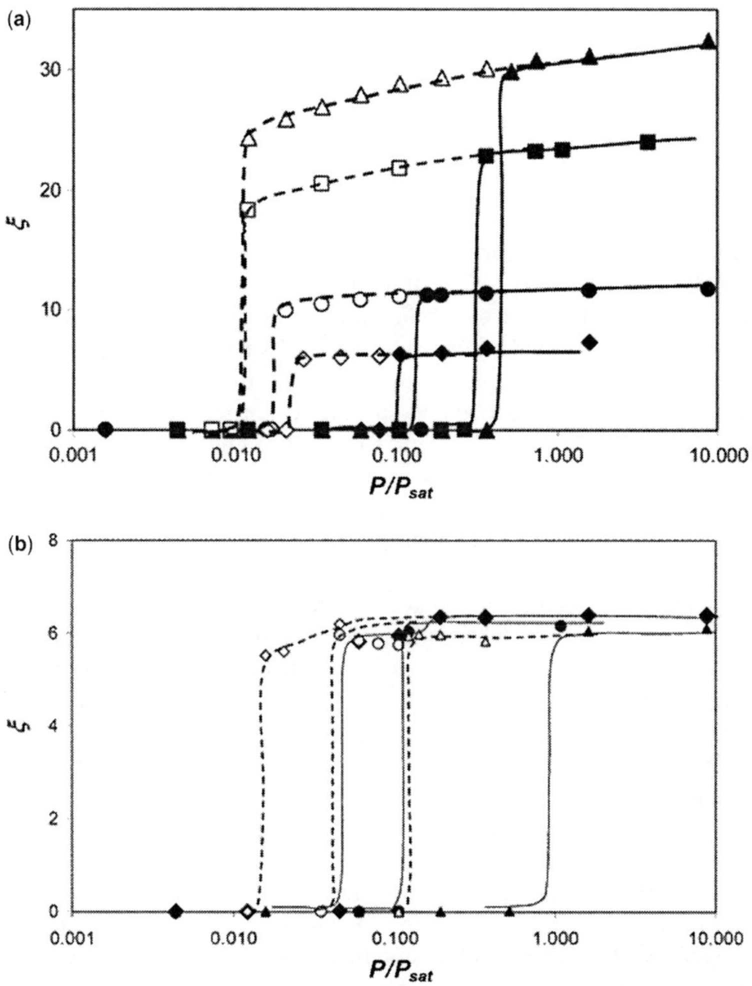

Figure 9.5 Coverage of the porous surface, ê, as a function of the relative pressure for water adsorbed in carbon-slit pores at 298 K. Part a shows the results obtained in pores of width 0.8 (diamonds), 1.0 (circles), 1.6 (squares), and 2.0 (triangles) nm. Part b shows the results obtained in pores of width 0.7 (diamonds), 0.625 (circles), and 0.6 (triangles) nm. Full symbols represent results for adsorption, while open symbols are for the desorption loops. Lines are guides to the eye. Symbols are larger than the simulation uncertainty. Pore-water interactions are computed using the Steele potential when H = 2.0 nm and by considering each carbon-water pair interaction in all other cases. Figure from Striolo *et al.* (2003) with permission.

Pressure-driven flows

A variety of different modes of flow in nanochannels have been proposed in the literature, with pressure-, electric field- and thermally-driven flows to attract the majority of the investigations (Li *et al.* 2010). A general flow mode in nanochannels is the pressure induced flow; with the two physical mechanisms involved in to the infiltration of fluid into the nanochamber and the transport of fluid inside it. Considering the infiltration process the resistance exercised by the flow and the structure change of the liquid are crucial factors. Additionally, the molecular structure combined with the bulk average velocity influence the transport process. The infiltration and defiltration mechanisms in nanoflows differ from that of in bulk. In case a flow in a nanochannel is studied, it is critical to work on how the molecular structure and the velocity of the flow are related with. Thomas and McGautney (2009) used MD to explore the pressure-driven water flow through 0.83–1.66 nm diameter SWNTs. The authors presumed that in subcontinuum scales the decrease in flow areas may lead to a non-monotonically increase for the flow enhancement. A reduction of the latter is observed when the size of liquid molecules are comparable to the cross-sectional area, which can be accredited to the structure changes due to the confinement of them. The diagnosed a critical diameter for the cross-section of the nanochannel which ranges in 1.25–1.39 nm. The authors pinpointed, when the characteristic flow time scale is equivalent with the relaxation time of the molecular structure, a dependence of the liquid structure and flow. Qiao *et al.* (2009) simulated the pressure-driven water flow in a silicon dioxide nanochannels by MD computations. They investigated the resistance that is acted on the water molecules as and when they penetrate to the nanochannel. In order for the water molecules to enter the nanochamber an energy barrier must be overwhelmed. The latter constitute of two parts; the first represents the change of the free energy of the infiltrated water molecules because each one of them looses two H-bonds as it enters the nanochannel and the second comprises the van der Waals and electrostatic interactions between the confined liquid molecules and the solid atoms. Apart the resistance previously described, an additional mode of resistance has been observed known as column resistance. The authors suggested that the continued infiltration has to overcome apart the aforementioned energy barrier a certain resistance which varies with the infiltration volume. Column resistance is mainly exhibited in cases where the infiltrated molecules are confined in nanopores with width comparable to the molecule size which drives them to interact directly with the solid wall.

Most of the studies so far have been devoted in the investigation of nanofluidics to nanochannels characterized by a constant circular-shaped cross section in the direction of the their axis. The literature however, is insufficient concerning nanochannels with variant cross-section and inclined solid walls, such as conical nanopores. Conical nanopores have been used for various applications in different domains of science such as bioengineering with DNA sensing, nanosequestration with particle capture and cooling techniques with pumped cooling loops (Harrel *et al.* 2006; Kovarik *et al.* 2008; Jung *et al.* 2008). Liu *et al.* (2009) attempted to further investigate the capabilities of the above nanostructures, so to understand the behaviour of nanoflows with pivotal position to be obtained by the infiltration mechanisms. To analyze the flow-mechanisms through a conical nanopore two geometrical factors must be defined; the inclination angle of the solid wall and the size effects such as the contact angle of various contributing variables. The Young's equation adjusted by the nanoscale size effects has been employed to analyze the unique mechanisms observed into cone-shaped nanopores. Under ambient conditions, despite the hydrophobic nature of the solid surface, if the apex angle is higher that a specific value spontaneous infiltration occurs. Moreover, they depict that in order to have an infiltration the applied pressure that has to be exercised must be higher that a critical infiltration pressure at which a flow occurs in case of a nanotube. The former entails the existence of a transition regime between the spontaneous infiltration (hydrophilic) and pressure originated (hydrophobic) infiltration processes. Studies for the amplitude of the access into the nanopore, showed that higher pressure is required for water molecules to travel further into the nanocone. Finally, if the apex angle takes larger values it leads to lower pressures for infiltration to occur.

Electric and thermally–driven flows

Although the physics in case of pressure-driven flows in nanochannels is rather simple, in order to be achieved large pressure gradients are required in experimental techniques. For that reason, other options as driving forces, have been scrutinized such as an electrical field (Kuo *et al.* 2001; Qiao & Aluru, 2003) and a temperature gradient (Linke *et al.* 2006). Considering the electric field-driven flows, the most important transport mechanism is the electroosmotic flow. When a charged solid surface approaches an ionic solution an electric double layer is observed in the proximity of solid-liquid interface. An electroosmotic flow is developed by applying an external electrical field directed tangentially to the solid surface, driving the ions in the layer. In MD terms, the Coulombic potential has to be taken into consideration by adding an electric field force in the system's equation of motion. The most common technique in modelling an electroosmotic flow is by implementing a combination of the Navier-Stokes equations (NSE) and the Poisson-Boltzmann equation (PBE). The former is used to simulate the fluid transport while the latter models the ion distribution (Ermakov *et al.* 1998). However, Kuo *et al.* (Kuo *et al.* 2001) illustrated several issues that arise by the aforementioned technique. By performing continuum calculations physical process dominating in nanoscale flows are ignored such as the modelling of water as a dielectric continuum in the PB equation instead of a point charged configuration (Li *et al.* 2010). To that direction MD present a more thorough understanding in nanoscale physics by discrediting the assumptions of the continuum theory. For instance, Qiao and Aluru (2003, 2005) presented a group of studies focusing in the electroosmotic flows by modifying the PBE through MD techniques. They appointed hydrogen bonding for the different ionic distribution computed in positively and negatively charged surfaces. Their work also expanded by proving that the direction of an electroosmotic flow is the opposite in contrast to what has been suggested by continuum models provided that the walls and the ions are signed with the same charge.

Thermophoresis is a phenomenon that bases its origin, in the movement of added molecules into a motionless fluid which is subjected to a temperature gradient causing the aforementioned molecules to travel directed to the colder regions. The theory of thermophoresis in liquids has been well established however a significant difficulty arises since other mechanisms too may be responsible apart the kinetic interaction (Li *et al.* 2010). For this reason Han (2005) performed a MD simulation to investigate thermophoresis. He supported that the finite thermophoretic velocity detected in liquids, cannot be computed through the kinetic theory of rarefied gases. He also showed that the intermolecular interactions have a vital role in case liquid states are studied.

Recently, the thorough investigation that is taking part in flows through confined geometries, showed that a possible handling in the geometry of the nanostructures, that is, a Carbon Nanoscroll (CNS), may lead to a more controlled fluid transport. To fully take advantage the possibility to manipulate the geometries of the nanostructures, it is inevitable to have an understanding for the physical and chemical principles that govern the fluid mechanics in nanoscale. The latter will lead to the development of nanodevices taking advantage the properties of the confined flows. During the fabrication of several carbon based nanostructures, it is rare to produce them in isolated forms but only in bundles. The agglomeration comes as a consequence of the van der Waals attractive interactions that keep them in distances equal to 3.4 Å. To override the latter obstacle several techniques have been proposed. Tube solubilisation is considered among those, providing access to regions that it wouldn't be feasible to exploit otherwise. Moreover a recent rising technique is the formation of tube scaffolds by introducing various types of linkers increasing and holding the tubes in a certain distance. The term of nanofluids is used for fluids composed by particles of the order of nanometers. Among others interesting properties, their improved heat transfer capabilities is remarkable. The thermal conductivity along with the convective heat transfer capability of particles in small volumes fractions are amplified, deprived of the obstacles observed in common insoluble in water cases. It is widely known that, in industrial world, fluids' heating and cooling in an efficient way is among the top priorities. As a result their thermal conductivity is a critical factor for the fabrication of devices characterized by higher energy efficiency focused in heat transfer. Nanofluids are being investigated both numerically and experimentally so to exercise advantage of the exceptionally favourable thermal transfer properties compared to traditional thermal transfer via fluids.

Additionally, the understanding of nanofluidics inside confined structures, further enlarges the envelope of the applications located this time in gas storage and adsorption devices. CNTs and CNSs are considered among the most promising nanomaterials; several studies have been made concerning the interactions of H_2 and H_2O with them. The reason why the above elements are more studied than others originates in the fact that Hydrogen is the key for the development of hydrogen based fuels with the water to be a side product with its regeneration appears to agree with the general term of environmental friendly fuels. Although a great effort has been devoted in the field of nanostorage, the use of Hydrogen energy is restricted to a limited set of applications because of the inability to produce devices with low weight storage combined with fast kinetics for loading and unloading the stored/adsorbed liquid.

Understanding the dynamics is vital for many applications in a nanoscale frame. Considering the dynamics of water in nanoconfined structures the diffusion has taken most of the attention in the molecular dynamics studies. It measures the mobility of water molecules and it is an indicative factor about how the hydrogen bond influence the translational motions, explaining the mechanisms developed in the nanochannels. Several properties may be investigated for water molecules in confined states such as the dipole correlation the velocity distribution, the residence time and the reorientation dynamics. A thorough understanding of the latter properties, can explain in detail the diffusion transport of the water molecules that experience a restricted motion. Although diffusion dominates the transport mechanisms in nanoscales and characterize water's nanofluidics, most of the potential models have not been parametrized to reproduce the values of self diffusion coefficients. Several studies have been devoted in the investigation of water in various kinds of nanopores such as carbon nanotubes, silica and slit nanopores. Theoretical works have investigated the diffusion coefficients in cylindrical shaped pores with different surface behaviours such as hydrophobic or hydrophilic channels. Allen *et al.* (1999) showed that in case of hydrophobic pores the self-diffusion coefficient of water molecules decreases for radius between 3.6 and 4.1 Å with its value to be beyond the bulk value explained by the hydrogen-bonding network. As the radius increases the self-diffusion values tend to the bulk value. Water molecules possess a preferable dipole orientation which can explain the fact that in case of hydrophilic channels such as Carbon Nanotubes, the water diffusion is slower that what is computed in hydrophobic. Hummer *et al.* (2001) however, concluded in their work that even in case of hydrophobic nanochannels such as CNTs, the water occupancy can be noticeable despite the noticeable degradation in number of Hydrogen bonds compared with bulk phase.

Focusing in water flow through CNTs Mashl *et al.* (2003) illustrated that there is a critical pore size where the mobility of the water molecules is minimized; showing a condensation and an ice behaviour of the water even in ambient temperatures. As described indirectly in the previous lines, the diffusion mechanism is strongly dependent by the radius of the cylindrical pore; for narrow pores the diffusion of the confined water molecules can be characterized as Single-file Diffusion (SDF) and in case of wider pores molecules can move to three dimensions giving rise to Fickian diffusion. Thomas and McGaughey

(2008b) showed that the density of unconfined water in the exterior of CNTs is independent of their radius with a nonuniform distribution as approaching the channel's surface. In the interior of the nanotubes the curvature of the carbon wall affects both the distribution and the water density.

9.5 MOLECULAR MODELLING OF WATER

Water is the most investigated liquid by numerical studies since the beginning of molecular simulations five decades ago. Its importance in nature and in general in human activities, lead the scientific world to actively study water and its properties in various forms that it can be observed. The ubiquity that water possesses mainly in bulk, designates it as one of the most studied element in literature. However, despite the great amount of studies concerning water, its properties are far from conceived. The advances in nanotechnology, have emerged the fabrication of nanochannels composed of various materials which may lead to the understanding the properties of water in restricted volumes and to the validation of molecular simulations. The present chapter, the brief presentation of molecular models for water will follow the overall behaviour of water in constrained nanochannels concerning both its statics and dynamics. At this point, it should be wise to describe the basic information required for the water molecule in MD theory. It constitutes of three atoms, two hydrogen atoms and one oxygen atom. An isolated water molecule, in gas phase, has two bonds with their length, r_{OH} to be 0.95718 Å and a formed bond angle θ_{HOH} equals to 104.474°. In liquid phase the aforementioned values are slightly adapted by the additional interactions that appear (water–water or/and water-ion interactions) suggesting $r_{OH} = 0.97$ Å and $\theta_{HOH} = 106.0°$ (Ichikawa *et al.* 1991). The charge separation that it is observed between the higher electronegative oxygen atom compared with hydrogen atom, arises an electric dipole, which can characterize the water molecule in cases of how it behaves in electrical fields (Karniadakis *et al.* 2005). In the frame of the efforts to discover the properties of water, a large number of theoretical models have been established. The models involve the placing of the electrostatic sites along with the Lennard–Jones (LJ) sites which in some cases may coexist with one or more of the charged sites. In general, each of the models has been formulated to match an individual property or in some cases a set of properties failing on the other hand to confirm the rest of physical structures and parameters. The majority of the models are empirical with their basis to be the widely recognized opinion that the hydrogen bond mainly resulted in the co-presence of the attractive interaction potential energy, estimated by the classical electrostatic interactions, and the repulsive electronic energy. To model the charge distribution of the water molecule point charges on the nuclei are used (SPC model) and seldom the use of several artificial sites located either in the plane (TIP4P) defined by the water molecule or out of it (ST2). The LJ potential is used to compute the electronic repulsion between two water molecules including the dispersion energy as well, located in the oxygen site. A brief description is followed of the most used water molecular models.

The Simple Point Charge model (SPC) is treated as a 3-site model, consisting of a tetrahedral water model with an OH distance equal to 1 Å, a H–O–H angle equal to 109.47°. Two point charges are introduced; on the oxygen and the hydrogen equal to $-0.82e$ and $+0.41e$ respectively. The total interaction energy between two water molecules modelled as SPC, consists of a LJ and a Coulombic potentials. Concerning the geometry of a SPC water molecule, its centre of mass is situated in the position of the oxygen atom compared with the SSD model where it is an artificial site. The model shows a satisfactory degree of behaviour for many purposes, however it needs improvement based in the density, the diffusivity, the radial distribution function and the critical properties. The extended Simple Point Charge model (SPC/E) is typified by three point masses with the interatomic distance to be 1 Å, and the angle equal to 109.47°, with the charges positioned in the oxygen and the hydrogen sites equal to $-0.8476e$ and $+0.4238e$

respectively. The SPC/E shows a considerable improvement in the diffusivity and its structure compared with the SPC. A water molecule modelled as Soft Sticky Dipole (SSD) is treated as a LJ sphere with an embedded point dipole accompanied by a tetrahedral potential, both of them placed in the molecular centre of mass (M) located in a distance of 0.654 Å standing in the bisector of the H—O—H angle which is 104.52° with a bond length 0.9572 Å. The H—M—H angle is 109.47° and the centre of mass to be the single interaction site of the model. The SSD model for the calculation of the total interaction potential energy between two molecules includes three terms; the LJ potential, the point dipole–dipole point potential and finally the third term to be the tetrahedral sticky potential.

The TIPnP models are widely used with the basic variants to be the TIP3P, TIP4P and the TIP5P. For all the aforementioned models the bond length and the H—O—H bond angle have been chosen to be equal with the experimental values in gas phase; 0.9572 Å and 104.52°. For a TIP5P model, the negatively charged sites are positioned in a symmetric way along the lone-pair directions with an intervening angle of 109.47°. A charge equal with +0.241e is taken into consideration on each hydrogen atom along with the lone pair interaction sites, on the contrary the oxygen atom is not considered as charged. The TIP4P as expected, is model with a four interaction sites positioned in a planar architecture. All the details for the TIPnP models are summarized in Karniadakis *et al.* (2005). The ST2 model is a rigid molecule described as a four-site charged molecule. The interaction between two ST2 molecules is modelled with a LJ central potential developed by the oxygen atoms, with the Coulombic potential interactions acting on the point charges. Many empirical potentials as described above, are pairwise additive although they take into consideration many body effects effectively, the dipole mechanism has a vital role in the many body contributions. To achieve it, a practical way is to develop models, rigid or flexible, approached by several point charges with the choice of their magnitudes to be based in a way to reproduce the dipole moment in gas phase assigning a polarizability point to the oxygen site (Guilot, 2002). The polarisable point charge (PPC model) is considered as an effective model maintaining the characteristics of three site models whereas it keeps its reaction to an electric field as computed from ab initio studies (Guilot, 2002). Finally, there is a Six-Site model proposed by Nada and van der Eerden (2003), based again in a rigid molecule with six interaction sites this time reproducing successfully the properties of ice and water approaching the melting point.

REFERENCES

Allen M. P. and Tildesley D. J. (1987). Computer Simulation of Liquids. Oxford University Press, Oxford.

Anderson J. D., Jr. (1995). Computational Fluid Dynamics: The Basics with Applications. McGrawhill Inc, London.

Asproulis N. and Drikakis D. (2010a). Surface roughness effects in micro and nanofluidic devices. *J. Comput. Theor. Nanosci.*, **7**(9), 1825–1830.

Asproulis N. and Drikakis D. (2010b). Boundary slip dependency on surface stiffness. *Phys. Rev. E.*, **81**(6), 061503.

Bahukudumbi P. and Beskok A. (2003). A phenomenological lubrication model for the entire Knudsen regime. *J. Micromech. Microeng.*, **13**, 873.

Baudry J., Charlaix E., Tonck A. and Mazuyer D. (2001). Experimental evidence for a large slip effect at a nonwetting fluid-solid interface. *Langmuir*, **17**, 5232–5236.

Benke M., Shapiro E. and Drikakis D. (2008). An efficient multi-scale modelling approach for ssDNA motion in fluid flow. *J. Bionic Eng.*, **5**, 299–307.

Beskok A. (2001). Validation of a new velocity-slip model for separated gas microflows. *Numer. Heat Transf. Part B: Fundam.*, **40**(6), 451–471.

Bitsanis I., Somers S. A., Davis H. T. and Tirrell M. (1990). Microscopic dynamics of flow in molecularly narrow pores. *J. Chem. Phys.*, **93**(5), 3427–3431.

Blazek J. (2005). Computational Fluid Dynamics: Principles and Applications. Elsevier Science Ltd, Oxford.

Bocquet L. and Barrat J. L. (1993). Hydrodynamic boundary conditions and correlation functions of confined fluids. *Phys. Rev. Lett.*, **70**, 2726–2729.

Bocquet L. and Barrat J. L. (1994). Hydrodynamic boundary conditions, correlation functions, and Kubo relations for confined fluids. *Phys. Rev. E*, **49**, 3079–3092.

Boehnke U. C., Remmler T., Motschmann H., Wurlitzer S., Hauwede J. and Fischer M. Th. (1999). Partial air wetting on solvophobic surfaces in polar liquids. *J. Colloid Int. Sci.*, **211**, 243–251.

Bonaccurso E., Butt H. S. and Craig V. S. J. (2003). Surface roughness and hydrodynamic boundary slip of a Newtonian fluid in a completely wetting system. *Phys. Rev. Lett.*, **90**, 144501.

Bonaccurso E., Kappl M. and Butt H. S. (2002). Hydrodynamic force measurements: boundary slip of water on hydrophilic surfaces and electrokinetics effects. *Phys. Rev. Lett.*, **88**, 076103.

Brenner H. and Ganesan V. (2000). Molecular wall effects: are conditions at a boundary 'boundary conditions'? *Phys. Rev. E*, **61**, 6879–6897.

Cheikh C. and Koper G. (2003). Stick-slip transition at the nanometer scale. *Phys. Rev. Lett.*, **91**, 156102.

Cho J. H., Law B. M. and Rieutord F. (2004). Dipole-dependent slip on Newtonian liquids at smooth solid hydrophobic surfaces. *Phys. Rev. Lett.*, **92**, 166102.

Churaev N. V., Sobolev V. D. and Somov A. N. (1984). Slippage of liquids over lyophobic solid surfaces. *J. Colloid Int. Sci.*, **97**, 574–581.

Darden T., York D. and Pedersen L. (1993). Particle Mesh Ewald-an N.Log(N) method for Ewald sums in large systems. *J. Chem. Phys.*, **98**, 10089–10092.

Das S., Das T. and Chakraborty S. (2006a). Analytical solutions for the rate of DNA hybridization in a microchannel. *Sensors and Actuators*, **114**, 957–963.

Das S., Das T. and Chakraborty S. (2006b). Modeling of coupled momentum heat and solute transport during DNA hybridization in a microchannel. *Microfluid Nanofluid*, **2**, 37–49.

De Gennes P. G. (1985). Wetting – statics and dynamics. *Rev. Mod. Phys.*, **57**, 827–863.

De Gennes P.-G., Brochard-Wyart F. and Quéré D. (2004). Capillarity and Wetting Phenomena: Drops, Bu4bbles, Pearls, Waves. Springer, Berlin, Heidelberg.

Drikakis D. and Rider W. (2005). High-Resolution Methods for Incompressible and Low-Speed Flows. Springer Verlag, Berlin.

Ermakov S. V., Jacobson S. C. and Ramsey J. M. (1998). Computer simulations of electrokinetic transport in microfabricated channel structures. *Anal. Chem.*, **70**, 4494–4504.

Ewald P. (1921). Die Berechnung optischer und elektrostatischer Gitterpotentiale. *Ann. Phys.*, **369**, 253–287.

Ferziger J. H. and Peric M. (1999). Computational Methods for Fluid Dynamics. Springer, Berlin.

Gad-el-Hak M. (2001). The MEMS Handbook. CRC Press, Boca Raton, Florida.

Gad-el-Hak M. (2002). Use of continuum and molecular approaches in microfluidics. 3rd Theoretical Fluid Mechanics Meeting 24–26 June, St. Louis, Missouri, AIAA 2002–2868.

Goldstein S. (1938). Note on the condition at the surface of contact of a fluid with a solid body. In: Modern Development in Fluid Dynamics, S. Goldstein (ed.), Vol. 2, Clarendon, Oxford, 676–680.

Guillot B. (2002). A reappraisal of what we have learnt during three decades of computer simulations on water. *J. Mol. Liquids*, **101**(1–3), 219–260.

Han M. (2005). Thermophoresis in liquids: a molecular dynamics simulation study. *J. Colloid Int. Sci.*, **284**, 339–348.

Harrell C. C., Choi Y., Horne L. P., Baker L. A., Siwy Z. S. and Martin C. R. (2006). Resistive-pulse DNA detection with a conical nanopore sensor. *Langmuir*, **22**, 10837–10843.

Ho T. A., Papavassiliou D. V., Lee L. L. and Striolo A. (2011). Liquid water can slip on a hydrophilic surface. *PNAS*, **108** (39), 16170–16175.

Hockney R. W. and Eastwood J. W. (1989). Computer Simulation Using Particles. Adam Hilger, NY.

Hockney R. W., Goel S. P. and Eastwood J. W. (1973). A 10,000 particle molecular dynamics model with long range forces. *Chem. Phys. Let.*, **21**(3), 589–591.

Hoffmann K. A. (1989). Computational Fluid Dynamics for Engineers. Engineering Education System, Austin, TX.

Hummer G., Rasaiah J. C. and Noworyta J. P. (2001). Water conduction through the hydrophobic channel of a carbon nanotube. *Nature*, **414**, 188–190.

Ichikawa K., Kameda Y., Yamaguchi T., Wakita H. and Misawa M. (1991). Neutron-diffraction investigation of the intramolecular structure of a water molecule in the liquid-phase at high-temperatures. *Mol. Phys.*, **73**, 79–86.

Jung J.-Y., Oh H.-S., Lee D. K., Choi K. B., Dong S. K. and Kwak H.-Y. (2008). A capillary-pumped loop (CPL) with microconeshaped capillary structure for cooling electronic devices. *J. Micromech. Microeng.*, **18**, 017002.

Kalra A., Garde S. and Hummer G. (2003). Osmotic water transport through carbon nanotube membranes. *Proc. Natl. Acad. Sci. USA*, **100**(8), 10175–10180.

Karniadakis G., Beskok A. and Aluru N. R. (2005). Microflows and Nanoflows: Fundamentals and Simulation. Springer Verlag, New York.

Kovarik M. L. and Jacobson S. C. (2008). Integrated nanopore/microchannel devices for AC electrokinetic trapping of particles. *Anal. Chem.*, **80**, 657–664.

Kuo T. C., Sloan L. A., Sweedler J. V. and Bohn P. W. (2001). Manipulating molecular transport through nanoporous membranes by control of electrokinetic flow: effect of surface charge density and debye length. *Langmuir.*, **17**, 6298–6303.

Laney C. B. (1998). *Computational Gas Dynamics.* Cambridge University Press, Cambridge.

LeVeque R. J. (2002). Finite volume methods for hyperbolic problems. Cambridge University Press, Cambridge.

Li Y., Xu J. and Li D. (2010). Molecular dynamics simulation of nanoscale liquid flows. *Microfluid Nanofluid*, **9**, 1011–1031.

Linke H., Aleman B. J., Melling L. D., Taormina M. J., Francis M. J., Dow-Hygelund C. C., Narayanan V., Taylor R. P. and Stout A. (2006). Self-propelled Leidenfrost droplets. *Phys. Rev. Lett.*, **96**, 154502.

Liu J.-C. and Monson P. A. (2007). Studies of a lattice model of water confined in a slit pore. *J. Phys. Chem. C*, **111**(43), 15976–15981.

Liu L., Yin C.-Y., Culligan P. J. and Chen X. (2009). Mechanisms of water infiltration into conical hydrophobic nanopores. *Phys. Chem. Chem. Phys.*, **11**, 6520–6524.

Mashl R. J., Joseph S., Aluru N. R. and Jakobsson E. (2003). Anomalously immobilized water: a new water phase induced by confinement in nanotubes. *Nano Lett.*, **3**(5), 589–592.

Nada H. and van der Eerden J. P. (2003). An intermolecular potential model for the simulation of ice and water near the melting point: a six-site model of water. *J. Chem. Phys.*, **118**, 7401–7413.

Petersen H. G. (1995). Accuracy and efficiency of the particle mesh Ewald method. *J. Chem. Phys.*, **103**, 3668–3679.

Pit R., Hervert H. and Léger L. (2000). Direct experimental evidence of slip in hexadecane: solid interfaces. *Phys. Rev. Lett.*, **85**, 980–983.

Pratt L. R. and Pohorille A. (2002). Hydrophobic effects and modeling of biophysical aqueous solution interfaces. *Chem. Rev.*, **102**, 2671–2692.

Priezjev N. V. (2007). Effect of surface roughness on rate-dependent slip on simple fluids. *J. Chem. Phys.*, **127**, 144708.

Priezjev N. V., Darhuber A. A. and Troian S. M. (2005). Slip behavior in liquid films on surfaces of patterned wettability: comparison between continuum and molecular dynamics simulations. *Phys. Rev. E.*, **4**(71), 041608.

Qiao R. and Aluru N. R. (2003). Ion concentrations and velocity profiles in nanochannel electroosmotic flows. *J. Chem. Phys.*, **118**, 4692–4701.

Qiao R. and Aluru N. R. (2005). Atomistic simulation of KCI transport in charged silicon nanochannels: interfacial effects. *Colloids Surf. A.*, **267**, 103–109.

Ruckenstein E. and Rajora P. (1983). On the no-slip boundary condition of hydrodynamics. *J. Colloid Int. Sci.*, **96**, 488–491.

Striolo A., Chialvo A. A., Cummings P. T. and Gubbins K. E. (2003). Water adsorption in carbon-slit nanopores. *Langmuir*, **19**(20), 8583–8591.

Striolo A., Chialvo A. A., Gubbins K. E. and Cummings P. T. (2005). Water in carbon nanotubes: adsorption isotherms and thermodynamic properties from molecular simulation. *J. Chem. Phys.*, **122**(23), 1–14.

Striolo A., Chialvo A. A., Cummings P. T. and Gubbins K. E. (2006). Simulated water adsorption in chemically heterogeneous carbon nanotubes. *J. Chem. Phys.*, **124**(7), 074710.

Striolo A., Gubbins K. E., Chialvo A. A. and Cummings P. T. (2004). Simulated water adsorption isotherms in carbon nanopores. *Mol. Phys.*, **102**(3), PART II, 243–251.

Thomas J. A. and McGaughey A. J. H. (2008a). Reassessing fast water transport through carbon nanotubes. *Nano Lett.*, **8**, 2788–2793.

Thomas J. A. and McGaughey A. J. H. (2008b). Density, distribution, and orientation of water molecules inside and outside carbon nanotubes. *J. Chem. Phys.*, **128**, 084715.

Thompson P. A. and Robbins M. O. (1989). Simulations of contact line motion – slip and the dynamic contact-angle. *Phys. Rev. Lett.*, **63**, 766–769.

Thompson P. A. and Troian S. M. (1997). A general boundary condition for liquid flow at solid surfaces. *Nature*, **389**, 360–362.

Toro E. F. (2009). Riemann Solvers and Numerical Methods for Fluid Dynamics: A Practical Introduction. Springer Verlag, London.

Tretheway D., Stone S. and Meinhart C. (2004). Effects of absolute pressure and dissolved gases on apparent fluid slip in hydrophobic microchannels. *Bull. Am. Phys. Soc.*, **49**, 215.

Tropea C., Yarin A. L. and Foss J. F. (2007). Spinger Handbook of Experimental Fluid Mechanics. Springer, Berlin.

Tuzun R. E., Noid D. W., Sumpter B. G. and Merkle R. C. (1996). Dynamics of fluid flow inside carbon nanotubes. *Nanotechnology*, **7**(3), 241–246.

Verlet L. (1967). Computer 'experiments' on classical fluids. I. Thermodynamical properties of Lennard–Jones molecules. *Phys. Rev.*, **159**, 98–103.

Versteeg H. K. and Malalasekera W. (2007). An Introduction to Computational Fluid Dynamics: The Finite Volume Method. Prentice Hall, London.

Voronov R. S. and Papavassiliou D. V. (2008). Review of fluid slip over superhydrophobic surfaces and its dependence on the contact angle. *Ind. Eng. Chem. Res.*, **47**, 2455–2477.

Werder T., Walther J. H., Jaffe R. L., Halicioglu T., Noca F. and Koumoutsakos P. (2001). Molecular dynamics simulation of contact angles of water droplets in carbon nanotubes. *Nano Lett.*, **1**(12), 697–702.

Zhu Y. and Granick S. (2001). Rate-dependent slip of Newtonian liquid at smooth surfaces. *Phys. Rev. Lett.*, **87**, 096105.

Zhu Y. and Granick S. (2002). Limits of the hydrodynamic no-slip boundary condition. *Phys. Rev. Lett.*, **88**, 106102.

Chapter 10

Computational recipes of transport phenomena in micro and nanofluidics

N. Asproulis

10.1 INTRODUCTION

Micro and Nanofluidic Devices (MNFDs) have experienced a rapid development over the past decade. These devices offer numerous benefits including improved accuracy, reduced operating volume and decreasing analysis time in comparison to conventional macro scale devices. This makes them especially advantageous for a wide range of applications spanning from the field of chemical engineering to bioengineering (Freemantle, 1999; Nguyen & Wu, 2005; Whitesides, 2006; Ehrfeld *et al.* 2000). The latest technological developments in micro- and nanoscale manufacturing allow construction of high precision microfluidic devices in a variety of materials. Polydimethylsiloxane (PDMS), due to manufacturing benefits, became the most common material for microfluidics fabrication (McDonald *et al.* 2000; Thorsen *et al.* 2002). Glass and silicone are also widely used in applications where chemical and thermal stability is essential or, where rigid walls might be necessary (Whitesides, 2006). In order to provide the desired surface properties of the fluidic channels, the base material is often covered with various thin film coatings. Due to their unique features and adjustable surface properties, micro- and nanofluidic devices are widely used in many biomedical systems (Ehrfeld *et al.* 2000; Weigl & Yager, 1999). Particularly in bio-analysis systems, where the main purpose of these devices are to detect the presence of specific target materials in the investigated fluid sample. The target materials are mostly synthetic or organic macromolecules such as drugs, lipids, enzymes and DNA. Detection efficiency is determined by the sensitivity of the sensor and the local concentration of sample molecules around the sensor area. If local sample concentration in the sensor region is insufficient, efficiency can drop dramatically which might easily lead to system malfunction. In most bio-detection applications, the main purpose of the employed MNFDs is to ensure that the sample material is delivered to the sensor area and to enable sufficient residence time of the sample in the sensor zone for the detection to take place. Understanding and modelling the fundamental transport processes in MNFDs is essential in order to achieve these objectives. The present study aims to give a comprehensive overview of the existing mathematical models used to describe the transport of sample macromolecules on a variety of scales.

10.2 MODELLING APPROACHES

From the onset it is possible to identify two main modelling approaches which can be used to consider transport processes in MNFDs based on the length scales separation. Namely the continuum level and the molecular level modelling. The first approach is based on the Navier-Stokes system of equations with additional models for the sample transport. For the second approach, molecular models based on the motion of molecules governed by the laws of classical mechanics are employed. The choice of the most suitable approach is determined by the characteristic length scale of the flow and the level of detail required from the physical model (see e.g. Karniadakis *et al.* 2005; Drikakis & Kalweit, 2006; Gad-el-Hak, 2005).

At the continuum level, transport and mixing of the sample in a fluid flow is governed by two main mechanisms – convection and diffusion (see e.g. Clusser (1997)). Convection is associated with the transport due to velocity of the carrier fluid flow and diffusion is governed by the stochastic molecular motion. While convection provides the dominant transport mechanism in high-speed flows, the typical fluid flows in MNFDs are relatively low-speed. For flow regimes which are relevant to microfluidics, the two mechanisms are of equal importance. In nanoscale devices, diffusive mechanism is the primary transport mechanism in the cross-stream direction At the nanoscale, the motion of the sample as well as that of the carrier liquid is governed by molecular interactions only. There are two separate mechanisms, the interaction between particles of comparable sizes leading to Fick's law of diffusion and the interaction of small carrier liquid particles with bigger particles of the sample resulting in Brownian motion related diffusion. The first mechanism describes the behaviour of a system with considerably large number of particles of the same size, therefore this mechanism is particularly relevant to nanoscale motion of the solvent molecules. The latter one describes the displacements of larger particles due to non-zero net momentum resulting from a number of random collisions of the smaller solvent molecules. The scope of our study is related to the motion of the molecules which are sufficiently large in comparison with these of the carrier liquid, therefore we restrict our investigation to the Brownian-related diffusion occurring on this scale.

10.2.1 Modelling multiple scales

Fundamental processes occurring in many MEMS applications cover a wide range of time and length scales. Although continuum level models describe macroscopic behaviour of systems, the flow in a variety of micromechanical (MEMS) systems, like microreactors, μTAS applications, drug delivery systems, fuel cells, separators, and so on, cannot be fully predicted by using only continuum flow models. Continuum approaches average fine time- and length scale processes and therefore fundamental phenomenon occurring on these scales can only be resolved in statistical sense. In cases when the continuum models are unable to capture detailed physics of the system, molecular models have to be employed (Figure 10.1). Atomistic models provide detailed description of processes on the characteristic scales of 10 nm in space and 10^{-9}s, in time. Pure molecular models are based on the solution of Newton equations of motion written for individual particles. The acting forces are resulted from pre-defined inter-molecular potentials. Solution of the Newton equations of motion is obtained by numerical integration of the governing equations for molecular positions. Despite the apparent simplicity of the model, calculations are extremely demanding computationally. The main shortcoming of molecular models is their high computational cost which restricts their application to simulations of nanoscale system to very short time periods. Thus, modelling phenomena at micro and nanofluidic devices presents significant difficulties due to the inaccuracy of the continuum models and inefficiency of the molecular ones. In order to confront this dilemma, multiscale methods have been developed to couple microscopic and macroscopic descriptions of the investigated system and to facilitate the exchange of information. Hybrid methods bridge the gap between the macroscopic and microscopic length scales and provide a unifying description of liquid flows from nanoscale to larger scales. In multiscale simulations, molecular and continuum-level models are applied simultaneously. The two-way information exchange between the molecular and continuum domains is achieved by a hybrid solution interface (HSI) connecting the corresponding interfaces. Multiscale methods provide atomic level description of the physics in selected regions of the computational domain, while the remaining part of the domain is described by the continuum model. Therefore local fine scale representation of fundamental processes avoids limitations of pure molecular models. In the continuum level framework, it is possible to study the transport of individual macromolecules via meta-modelling approaches which rely on a mechanical model representation of molecular structures. Meta-models draw information from molecular techniques in order to define model parameters but are essentially continuum-scale (see, e.g. Doi & Edwards, 1986; Trebotich *et al.* 2005). This approach is especially advantageous, if the size of the investigated molecule is large compared to the solvent molecules. For some regimes the transport cannot be wholly addressed within the scope of either continuum or molecular approach alone, which calls for the multi-scale modelling approach which allows selective or blended application of both approaches simultaneously (Drikakis & Kalweit, 2006). In general case, multiscale models include continuous two-way exchange of information between continuum level and molecular level models.

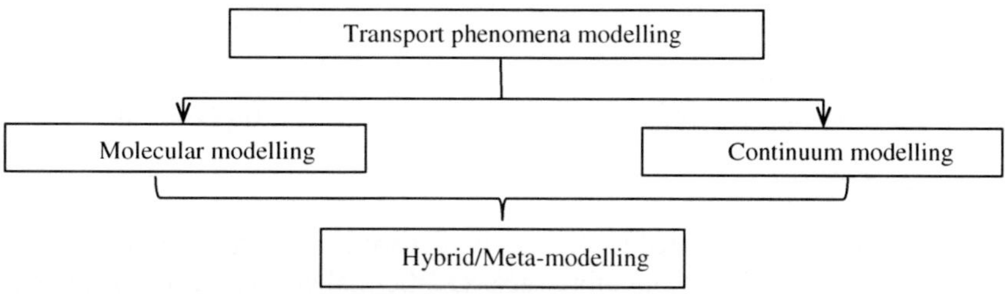

Figure 10.1 Modelling approaches across the scales.

10.2.2 Brownian motion

Brownian motion is caused by carrier liquid molecules colliding against the particle, and transferring momentum to it (Einstein, 1905). Due to the high but limited number of collisions, the net momentum transferred to the particle per unit time will not be zero (R. Kubo, 1978), which results in the acceleration of the particle. The final effect is that bigger particles immersed in liquid are subject to erratic motion, similar to the thermal agitation but on a different (slower) time scale, that is responsible for diffusive process. Brownian motion forms the basis of the meta-models of macromolecule motion. The mobility of the particles is primarily related to their mass and size. For example, micron sized objects motion can be tracked on a time scale of microseconds. In the very last years, however, the evolution of microscale visualising techniques has allowed some authors to report subtle discrepancies with the existing theories. The purely "thermal" description of Brownian motion has been questioned, and although the existing models still offer a good qualitative and quantitative description, its exact physical characterisation is currently a re-opened problem (S. N. Bagayev & Panov, 2007).

From the point of view of modelling Brownian motion, the widely adopted solution is a random-walk path, implemented as a Gaussian (Weiner) stochastic process (Karatzas & Shreve, 2000). The whole process is synthesised in a force of uniformly distributed random direction, and Gaussian distributed random intensity over time. Although apparently very simplified, this approach is made legitimate under the assumption of statistically independent random collision with a high number of molecules, by the central limit theorem (see, e.g. Frey & Kroy, 2005). The central limit theorem ensures that a random variable x which is the sum of "many" (ideally, infinite) identical and independent random variables x_i will always behave as Gaussian variable of opportune mean and variance, whatever the actual statistical law of the x_i. In our case, the x_i variables represent contribution of single collisions, and their sum is the total momentum transferred to the particle or molecule over a small time interval δ t, that is the mean force acting on the particle (or molecule) over δ t. In order to apply the central limit theorem, we usually write governing equations of motion for every particle in the form introduced by Langevin, including explicitly the random force and the viscous drag due to the carrier liquid as:

$$m_i \mathbf{a}_i = \mathbf{F}_i(t) - \zeta \mathbf{v}_i \tag{10.1}$$

where $\mathbf{a}_i$ is the acceleration, $\mathbf{v}_i$ the velocity and $\zeta = 6\pi\eta_s$ a is the Stokes friction coefficient for a particle of hydrodynamic radius a in a carrier liquid with a η_s dynamic viscosity (Kundu & Cohen, 2002). $\mathbf{F}_i$ (t) is the Gaussian-distributed random Brownian force due to collisions. The amplitude of the oscillations of the random force is related to the friction coefficient by the fluctuation-dissipation theorem:

$$\langle \vec{F}_i(t) \rangle = 0 \tag{10.2}$$

$$\int \langle \vec{F}_i(t) \cdot \vec{F}_i(t') \rangle dt = 6 k_B T \zeta \tag{10.3}$$

$\langle \mathbf{F}_i(t) \rangle$ where t' is the reference time. For the purpose of computations, it is often assumed that the random force is completely non-correlated at different time steps, so that Equation (10.3) becomes

$$\langle \mathbf{F}_i(t) \cdot \mathbf{F}_i(t') \rangle = 6 \zeta k_B T \delta(t - t') \tag{10.4}$$

In case of Langevin equations implemented for Langevin Dynamics numerical simulations, this relationship between dissipative friction force and random force acts as a thermal bath and ensures the preservation of the kinetic temperature of the system of particles. The Brownian motion theory is applied to meta-models which represent polymer molecules as chains of constrained particles moving under the influence of Brownian motion and friction drag (Doi & Edwards, 1986; Trebotich *et al.* 2005).

Link between molecule motion and transport coefficients

The calculation of the diffusion coefficient D for homogeneous and equilibrium systems can be performed following either of the two widely used approaches – Einstein equation approach or Green-Kubo approach. The results obtained with these are usually in good agreement (22). The Einstein approach computes D directly from the atom displacements using the Einstein relation:

$$D = \frac{1}{6} \lim_{t \to 0} \frac{\langle |\mathbf{r}(t_0 + t) - \mathbf{r}(t)|^2 \rangle}{t} \tag{10.5}$$

where $\mathbf{r}$ denotes atom positions and $\langle \cdot \rangle$ represents the average over the trajectory. The Green-Kubo approach relies on the computation of the diffusion coefficient D from equilibrium velocity fluctuations according to

$$D = \int_0^{\infty} \langle \mathbf{v}(0) \cdot \mathbf{v}(t) \rangle dt \tag{10.6}$$

where **v** is the velocity, represents the ensemble average and is the velocity autocorrelation function. Equation (10.6) is only valid for homogeneous systems at equilibrium; however practically equations (10.5) and (10.6) lead to the same results (Karniadakis *et al.* 2005). The computation of the quantities involved in the above mentioned equations has become feasible in recent years for some chemical species (subject to constraints imposed by the computational cost), by means of molecular dynamics techniques. It has to be underlined that if the diffusion is computed under flow conditions, only the relative velocities (with respect to the macroscopic velocity) must be considered. In principle, the techniques briefly outlined can be also used for the simulation of (not too long) polymer chains, in order to estimate the diffusion coefficient of their center of mass. In practice, only short chains are computationally feasible. The knowledge of scaling laws for polymer properties (Doi & Edwards, 1986) can ease this difficulty, and allow careful extrapolations. However, it is also necessary to include corrections due to finite-size effects of the simulation box: in particular, in the case of simulations with periodic boundary conditions, the simulated hydrodynamic radius is consistently underestimated; the reason has been attributed to the screening of the hydrodynamic interactions between mirror images. The needed corrections are inversely proportional to the linear box size (Dunweg & Kremer, 1993). Another element is the unrealistically low viscosity of most water models adopted for example in biomolecular simulation, an effect that can be taken into consideration applying an opportune scaling coefficient (Yeh & Hummer, 2004).

10.2.4 Continuum scale diffusion

Diffusion is caused by thermally induced random motion of particles. However on macroscale diffusion is typically characterised as the migration of the solute in the solvent from regions of high to low concentrations of the solute. Pure diffusion occurs when the velocity field of the solvent is zero, whereas in case of non-zero velocity field the motion of the solute is partially convective, since dissolved particles are carried along by the solvent. In the following, we review the historical approach to diffusion modelling at macroscopic scale, discuss approaches used to obtain binary diffusion coefficients and describe multi-component diffusion.

Different diffusion equations

Generally, diffusion can be considered as a process of relative thermal motion of same or different species. Therefore proper macroscale description of the diffusion process should be based on the notion of relative velocity. There are several diffusion equations currently used in the literature. These different formulations are summarised in Table 1. The complete description of the mass transfer requires separating the convection and diffusion contributions.

(total mass transported) = (mass transported by diffusion) + (mass transported by convection)

Particularly, the total mass flux is defined as the mass transported per area per time relative to fixed coordinates. Based on the mass flux, an average solute velocity is defined as $\mathbf{n}_1 = c_1\mathbf{v}_1$, where c_1 is the local concentration. Then the velocity $\mathbf{v}_1$ can be divided into two parts as follows:

$$\mathbf{n}_1 = c_1(\mathbf{v}_1 - \mathbf{v}^a) + c_1\mathbf{v}^a = \mathbf{j}_1^a + c_1\mathbf{v}^a \quad (10.7)$$

where $\mathbf{v}^a$ is the convective reference velocity. In Eq. 10.7 the first part $\mathbf{j}_1^a$ represents the diffusion flux whereas the second term $c_1\ \mathbf{v}^a$ represents the convection. The selection of the convective reference velocity is a debatable issue. There is no uniquely correct a priori selection, for instance it can be the mass average velocity or the velocity of the solvent. Particularly, based on the concentration of the diluted substance in the fluid, one can define the following types of solutions:

- *Negligible number density*: In this case classical diffusion theory – determination of species concentrations based on continuum-level species transport equations – cannot be applied to describe the physics of diffusion. Instead, Lagrangian description of individual particles pathways relative to a fixed Eulerian grid is a more feasible approach.
- *Dilute solutions*: The concentration of the substance in the fluid is very low. The presence of the solute material does not affect the physical (macroscopic) properties of the carrier fluid (density, viscosity, thermal properties). Classical diffusion theory can be applied and concentration field behaves effectively as a passive scalar field.
- *Semi-dilute solutions*: Sample concentration in the solution is considerable. Material properties of the solution may differ significantly from these of the carrier fluid. Due to the dissolved substance, non-Newtonian effects may appear in Newtonian fluids. Description of diffusion in semi-dilute solutions can be based on the classical diffusion theory.
- *Concentrated solutions*: Concentration of the dissolved material is near the solubility limit. Material properties of the solution are governed by the concentration and properties of the dissolved substance. Strong non-Newtonian effects can be observed. Diffusion phenomena description in concentrated solutions can be based on the classical diffusion theory.

Multicomponent diffusion

The diagonal terms D_{ii} are usually similar to the corresponding binary counterparts. The off-diagonal cross-terms $D_{ij,i} \neq j$ are usually ten percent or less, in magnitude, of the diagonal terms.

In addition to binary diffusion, diffusion processes often include transport of many solutes. In most cases multi-component diffusion can be described by generalising the Fick's law equation to an n-component system (Clusser, 1997), which leads to the following flux

$$\mathbf{j}_i = -\sum_{j=1}^{n} D_{ij} \cdot \nabla c_j \tag{10.8}$$

where $\mathbf{j}$ is the flux of substance i, ∇c_j is the spatial concentration gradient of the substance j, and D_{ij} are the components of the diffusion coefficient tensor. The tensor is in general asymmetric, that is $D_{ij} \neq D_{ji}$. The diagonal terms D_{ii} are usually similar to the corresponding binary counterparts. The off-diagonal cross-terms $D_{ij,i} \neq j$, are usually ten percent or less, in magnitude, of the diagonal terms.

A n-component system is described therefore by diffusion tensor, with one substance arbitrarily chosen as a solvent or carrier. The approximation of D_{ij} for simple gas molecules can be achieved directly. In case of complicated mixtures, diffusion coefficients and even the relation between binary coefficients in general is not known. Multicomponent effects, however are usually small in diluted solution. In most situations a sound application of Fick's law for binary diffusion yields correct results (Clusser, 1997).

Diffusion coefficients

At the macroscale level diffusion coefficients in liquids are mainly estimated from the Stokes-Einstein equation given by

$$D = \frac{k_B T}{f} = \frac{k_B T}{6\pi\mu R_0} \tag{10.9}$$

where f is the friction coefficient of the solute, k_B is Boltzmann's constant, μ is the solvent viscosity and R_0 is the solute radius. The above equation is derived by assuming a rigid solute sphere diffusing in a continuum of solvent. Hence, Equation (10.9) should provide better approximation when the size of the sample particles is large in comparison with that of carrier liquid particles. However, when the size of the sample particles is five times the size of the carrier liquid particles the Stokes–Einstein equation fails to predict the diffusion coefficient correctly. The values predicted by the Stokes–Einstein equation are only accurate to ~20%, but this approach still remains a common practice despite the availability of a number of alternatives.

In all cases summarised in this table the diffusion coefficient is inversely proportional to the viscosity coefficient, however since vary with temperature, this proportionality might be misleading and it is preferable to consider the variation of diffusion with temperature using an appropriate expression for the variation of viscosity with temperature. For example, for water one may consider (see, e.g. Likhachev, 2003).

$$\mu = \mu_0 \exp\left[ap + \frac{E - bp}{R(T - \theta - cp)}\right] \qquad (10.10)$$

where p denotes pressure, T denotes temperature and a, b, c, E, R and θ are tabulated constants.

10.3 META-MODELLING FOR MACROMOLECULES

Meta-models are based on the direct mechanical modelling of the macromolecule motion in the fluid flow (Figure 10.2). In this case the diffusion of the modelled large molecules is resolved directly by tracking each individual structure. In order to build a meta-model, it is necessary to establish a continuum-scale mechanical representation of the molecule of interest. The model must represent accurately both static and dynamical behaviour of the investigated real macromolecules and must be able to describe the interaction between the flowing fluid and the travelling large molecule. In order for meta-models to be applicable from the computational efficiency point of view, it is necessary to have very low number density of tracked macromolecules in the fluid flow. Based on experimental observations, it is possible to create a mechanical representation of many flexible polymer types using a number of connected beads along the chain. A simple but widely used Freely Jointed Chain (FJC) model (Doi & Edwards, 1986; Janshoff *et al.* 2000) constructs the molecule from beads connected by flexible joints oriented independently of each other. In case of beads connected with rigid bonds, the interaction potential cancels out from the model equation and bond lengths are maintained by additional geometrical constraints. The motion of these beads is governed by the hydrodynamic forces exerted by the flow on a bead and Brownian motion of individual beads (Doi & Edwards, 1986; Trebotich *et al.* 2005). The hydrodynamic forces can be described by the Stokes drag force (see, e.g. Happel & Brenner, 1983). For particle, the equation of motion is written as (Trebotich *et al.* 2005):

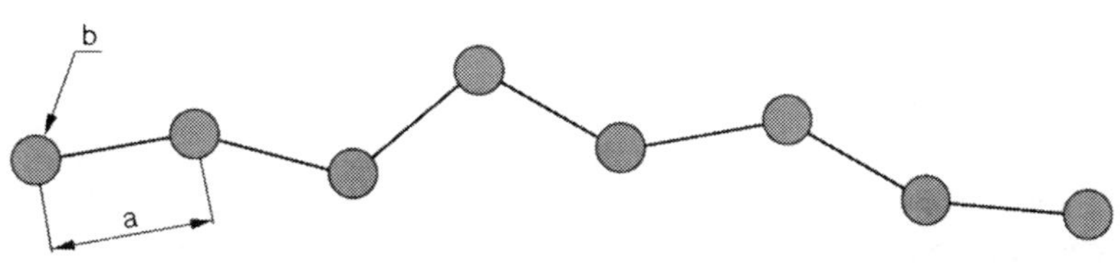

Figure 10.2 Mechanical representation of a biomolecule.

$$m_n \frac{d^2\mathbf{r}_n}{dt^2} = \mathbf{f}_n \tag{10.11}$$

where is the mass and is the radius of bead. The total force acting on particle is as follows:

$$\mathbf{f}_n = \underbrace{m_n \gamma_n (\mathbf{v}(\mathbf{r}_n) - \mathbf{v}_n)}_{\text{Stokes drag}} + \underbrace{\boldsymbol{\phi}_n(t)}_{\text{Random force}} \tag{10.12}$$

where and is the friction coefficient arising from the Stokes drag of a sphere with radius is the fluid velocity at particle position, is the particle velocity and is the random force. This random force represents the effect of solvent molecules stochastically bumping to the beads and exchanging impulse. The random force adheres to Gaussian distribution characterised as (Doi & Edwards, 1986; Trebotich *et al.* 2005):

$$< \boldsymbol{\phi}_n(t) > = 0 \tag{10.13}$$

$$< \boldsymbol{\phi}_n(t)\boldsymbol{\phi}_n(t') > = 2m_n \gamma_n k_B T \delta(t - t') \tag{10.14}$$

Model equations of this bead-rod mechanical structure are tightly coupled with the governing equations of the carrier fluid. For scales larger than few molecules (>10 nm) a liquid can be generally treated as a continuum medium (Gad-el-Hak, 2005). Governing equations of fluid flow are therefore more appropriate to microfluidic rather than nanofluidic devices. The equations that govern this regime are the Navier-Stokes equations. The tight coupling of the macromolecule motion with the flow is achieved via the addition of a forcing term into the governing equations of fluid flow, for example

$$\frac{\partial u_i}{\partial t} + u_j \frac{\partial u_i}{\partial x_j} = -\frac{1}{\rho}\frac{\partial p}{\partial x_i} + \mu \frac{\partial u_i}{\partial x_j \partial x_j} + \frac{1}{\rho} F_i \tag{10.15}$$

The main challenge then lies in the implementation of data exchange between the moving particles and the fixed Eulerian cells. A series of technical issues arises because of the fact that particle coordinates vary in each time step and do not necessarily coincide with cell centres.

From the total force acting on each particle, the additional volume force in the momentum equation is calculated as:

$$\mathbf{F}(\mathbf{r}) = (F_1, F_2, F_3) = -\sum_n \mathbf{f}_n \delta_\varepsilon (\mathbf{r} - \mathbf{r}_n) \tag{10.16}$$

with smoothed Dirac function of length scale. This function incorporates a division with unit volume, therefore transforms the concerning point-like force into a volumetric force field. To complete the mathematical model, additional equation is necessary to model rigid bonds between the beads: $\left\| \mathbf{r}_n - \mathbf{r}_{n+1} \right\| = a$, and index n runs from 0 to N−1.

10.4 HYBRID CONTINUUM-MOLECULAR MODELS

To overcome the limitations of pure molecular models and resolve the inaccuracy of continuum scale description, hybrid continuum-molecular methods have been developed. These hybrid methods bring in the balance between the accuracy of the multiscale phenomena description and the computationally efficiency, bridging the gap between the macroscopic and microscopic length scales and providing a unifying description from nanoscale to larger scales. The hybrid methods can be broadly classified into three groups:

- *Domain decomposition techniques* (DDT) (Flekkoy *et al.* 2000; Drikakis & Kalweit, 2006; Oconnell & Thompson, 1995)
- *Embedding based techniques* (EBT) (Ren & Weinan, 2005; Karniadakis *et al.* 2005)
- *Equation free approach* (EFA) (Samaey *et al.* 2006; Gear, 1971; Hyman, 2005)

Domain decomposition is appropriate for problems where continuum equations are still valid in large regions of the system, but fail to fully describe the phenomenon in a particular area. In this case two regions are defined, where the one is solved by the continuum solver and the other one, that needs molecular modelling, is solved by molecular dynamics (Drikakis & Kalweit, 2006; Oconnell & Thompson, 1995). The advantage of this approach is that computationally slow molecular dynamics technique is employed in a small region, which is essential, whereas the rest of the domain is treated with the several orders faster CFD solvers. The idea of the domain decomposition was introduced in 1995 by O'Connell (Oconnell & Thompson, 1995). Since then several coupling approaches have been developed based on the idea of the domain decomposition. These include the relaxation method (Wang & G, 2007; Oconnell & Thompson, 1995), coupling through state- Schwarz Method (Hadjiconstantinou, 1999; Hadjiconstantinou & Patera, 1997; Hadjiconstantinou, 2005; Werder *et al.* 2005) and coupling through fluxes (Wagner *et al.* 2002; Flekkoy *et al.* 2005). In the embedding multiscale methods, the whole domain is covered with the macroscopic solver and the microscale model, which enters as a refinement, is used to obtain macroscopic properties. An additional property is that the time steps for the macroscale and the microscale are naturally decoupled. These schemes were introduced to handle the time scale constraints introduced by geometrical coupling. The Heterogeneous Multiscale Method when applied to the moving contact line and the Marangoni flow problems inherits the characteristics of the embedding based framework (Ren & Weinan, 2005) Patch Dynamics and equation free approach are techniques initially developed by Yiannis Kevrekidis and James Hyman. The goal of Patch Dynamics is to bridge the time and length scales and predict the macroscale dynamics by performing only microscopic simulation over small batches. More particularly, Patch Dynamics use locally averaged properties for a short period of time and for a small region, in order to advance and predict long space-time scale dynamics. The general framework of the patch dynamics circumvents the need for a closed analytical description of the macroscale systems and delivers macroscopic information.

REFERENCES

Bagayev S. N., Orlov V. A. and Panov S. V. (2007). Observation of the Brownian motion of individual microobjects in liquids for small time and spatial scales. *Doklady Physics*, **52**, 473–477.

Clusser E. L. (1997). Diffusion: Mass Transfer in Fluid Systems. Cambridge University Press, Cambridge.

Doi M. and Edwards S. F. (1986). The Theory of Polymer Dynamics. Oxford Science Publications.

Drikakis D. and Kalweit M. (2006). First Handbook in Theoretical and Computational Nanotechnology. American Scientific Publishers.

Dunweg B. and Kremer K. (1993). Molecular dynamics simulation of a polymer chain in solution. *The Journal of Chemical Physics*, **99**, 6983–6997.

Ehrfeld W., Hessel V. and Lowe H. (2000). Microreactors: New Technology for Modern Chemistry. Wiley–VHC, Weinheim.

Einstein A. (1905). On the motion of small particles suspended in liquids at rest required by the molecular-kinetic theory of heat. *Annalen der Physik*, **17**, 549–560.

Flekkoy E. G., Delgado-Buscalioni R. and Coveney P. V. (2005). Flux boundary conditions in particle simulations. *Physical Review E – Statistical, Nonlinear, and Soft Matter Physics*, **72**, 1–9.

Flekkoy E. G., Wagner G. and Feder J. G. (2000). Hybrid model for combined particle and continuum dynamics. *Europhysics Letters*, **52**, 271.

Freemantle M. (1999). Microscale technology. *Chemical and Engineering News*, 27–36.

Frey E. and Kroy K. (2005). Brownian motion: a paradigm of soft matter and biological physics. *Annalen der Physik (Leipzig)*, **14**, 20–50.

Gad-el-Hak M. (2005). Liquids: the holy grail of microfluidic modeling. *Physics of Fluids*, **17**, 100612.

Gear C. W. (1971). Numerical Initial Value Problems for Ordinary Differential Equations. Prentice–Hall.

Hadjiconstantinou N. G. (1999). Hybrid atomistic-continuum formulations and the moving contact-line problem. *Journal of Computational Physics*, **154**, 245–265.

Hadjiconstantinou N. G. (2005). Discussion of recent developments in hybrid atomistic-continuum methods for multiscale hydrodynamics. *Bulletin of the Polish Academy of Sciences: Technical Sciences*, **53**, 335–342.

Hadjiconstantinou N. G. and Patera A. T. (1997). Heterogeneous atomistic-continuum representations for dense fluid systems. *International Journal of Modern Physics C*, **8**, 967–976.

Happel J. and Brenner H. (1983). Low Reynolds Number Hydrodynamics. Martinus Nijhoff Publishers, Hague.

Hyman J. M. (2005). Patch dynamics for multiscale problems. *Computing in Science and Engineering*, **7**, 47–53.

Janshoff A., Neitzert M., Oberdorfer Y. and Fuchs H. (2000). Force spectroscopy of molecular systems – single molecule spectroscopy of polymers and biomolecules. *Angewandte Chemie International Edition*, **39**, 3212–3237.

Karatzas I. and Shreve S. (2000). Brownian Motion and Stochastic Calculus. Springer, New York.

Karniadakis G., Beskok A. and Aluru N. (2005). Microflows and Nanoflows: Fundamentals and Simulation. Springer.

Kubo R. and Toda M. (1978). Statistical Physics II – Nonequilibrium Statistical Mechanics. Springer–Verlag.

Kundu P. and Cohen I. (2002). Fluid Mechanics. Academic Press.

Likhachev E. (2003). Dependence of water viscosity on temperature and pressure. *Technical Physics*, **48**(4), 514–515.

McDonald J. C. *et al.* (2000). Fabrication of microfluidic systems in poly(dimethylsiloxane). *Electrophoresis*, **21**, 27–40.

Nguyen N. and Wu Z. (2005). Micromixers – a review. *Journal of Micromechanics and Microengineering*, **15**, R1–R16.

Oconnell S. T. and Thompson P. A. (1995). Molecular dynamics-continuum hybrid computations: a tool for studying complex fluid flows. *Physical Review E – Statistical, Nonlinear, and Soft Matter Physics*, **52**, R5792–R5795.

Ren W. and Weinan E. (2005). Heterogeneous multiscale method for the modeling of complex fluids and micro-fluidics. *Journal of Computational Physics*, **204**, 1–26.

Samaey G., Kevrekidis I. G. and Roose D. (2006). Patch dynamics with buffers for homogenization problems. *Journal of Computational Physics*, **213**, 264–287.

Thorsen T., Maerkl S. J. and Quake S. R. (2002). Microfluidic large-scale integration. *Science*, **298**, 580–584.

Trebotich D., Miller G. H., Colellac P., Graves D. T., Martin D. F. and Schwartz P. O. (2005). *A tightly coupled particle-fluid model for DNA-laden flows in complex microscale geometries*, 3rd M.I.T. Conference on Computational Fluid and Solid Mechanics. p. 1018.

Wagner G., Flekkoy E., Feder J. and Jessang T. (2002). Coupling molecular dynamics and continuum dynamics. *Computer Physics Communications*, **147**, 670–673.

Wang Y. and He G. (2007). A dynamic coupling model for hybrid atomistic-continuum computations. *Chemical Engineering Science*, **62**, 3574–3579.

Weigl B. H. and Yager P. (1999). Microfluidic diffusion – based separation and detection. *Science*, **283**, 346–347.

Werder T., Walther J. and Koumoutsakos P. (2005). Hybrid atomistic-continuum method for the simulation of dense fluid flows. *Jounral of Computational Physics*, **205**, 373–390.

Whitesides G. (2006). The origins and the future of microfluidics. *Nature*, **442**, 368–373.

Yeh I. C. and Hummer G. (2004). Diffusion and electrophoretic mobility of single-stranded RNA from molecular dynamics simulations. *Biophysical Journal*, **86**, 681–689.

Chapter 11

Multi-detection of waterborne pathogens in raw and treated water samples by using ultrafiltration concentration and DNA array technology

Sophie Courtois, Anne Cajon, Aurore Romey, Fanny Poyet and Claude Mabilat

11.1 INTRODUCTION

Control of microbiological water quality (drinking and domestic water) is a key issue not only because of the health impact of potential contamination, but also because water resources must be protected to ensure long-term sustainability. However, there is a rather limited knowledge on the microbiological principles governing the prevalence and pathogenesis of emerging microbial pathogens in drinking water (Leclerc *et al.* 2002; Albinana-Gimenez *et al.* 2006). One of the main reasons for the lack of knowledge is that accurate detection, identification and quantification of microbial pathogens in water is difficult and only possible with a combination of conventional and molecular biology methods (Purohit *et al.* 2002; Bej, 2003).

There is a need for more appropriate methodologies to track specific pathogens both for routine monitoring of water and to investigate disease outbreaks. The challenges involved in developing a universal pathogen detection method are such that there is currently no single method to simultaneously collect, process and analyze a water sample for all pathogenic bacteria, viruses and parasites (Straub *et al.* 2003).

Recent developments in molecular detection technology, such as quantitative real-time PCR and DNA arrays, offer a great opportunity to obtain valid data on the occurrence of emerging pathogens. The development of molecular fingerprints based on environmental nucleic acids bear the potential of new screening tools for emerging bacterial, viral and protozoan pathogens. DNA chip technology holds great potential for microbiological diagnostic applications due to its powerful capacity to simultaneously analyze a large number of nucleic sequences (Ramsay, 1998). For example, the photolithography chip developed by Affymetrix is a particularly powerful tool, allowing simultaneous testing of up to 500,000 unique oligonucleotide probes. By confirming the presence of DNA sequences, this technology has been used in various applications, firstly within the context of clinical applications such as gene expression profiling (Chee *et al.* 1996), DNA sequencing (Pease *et al.* 1994), disease screening (Khan *et al.* 1999), but also in microbial community analysis (Rudi *et al.* 2002; Brodie *et al.* 2006), strains or species identification (Troesch *et al.* 1999; van Leeuwen *et al.* 2003) and bacteria detection (Peplies *et al.* 2003).

The overall goal of the "Healthy Water" European project (http://www.helmholtz-hzi.de/en/healthy_water/) was to develop an integrated research approach by combining molecular and conventional detection techniques and activity assessment of emerging pathogens in a broad range of drinking water systems from different European regions. In the framework of this project, a new detection technology was developed, based on the use of DNA microarray which allows the simultaneous screening of bacteria, parasites, and viruses in raw and treated water samples. The ultimate objective of this work was to develop an innovative approach for detecting multiple waterborne pathogens utilizing large volume ultrafiltration as a universal concentration technique, direct extraction of nucleic acids, Reverse-Transcription PCR and hybridization to a multi-pathogen, water quality microarray. The most important aspects in applying molecular pathogen-detection techniques to water samples are the "front-end" processes of sample concentration and nucleic acid extraction. The most sensitive and specific detection system available will be virtually useless if presented with low concentrations of poor quality DNA. This proposal addresses these critical issues by focusing on methods for concentrating pathogens from large volumes of water and extracting high quality nucleic acids so that the full potential of advanced molecular detection technologies can be realized.

11.2 IMPROVED AND SIMPLIFIED METHOD FOR CONCENTRATING VIRAL, BACTERIAL, AND PROTOZOAN PATHOGENS

11.2.1 Technical challenges for a universal concentration protocol

Several steps in the development and application of rapid molecular techniques are necessary and essential for successful detection of bacteria, viruses or parasites in environmental samples. One of the most critical amongst them is the concentration of the sample in order to analyze a large volume of the initial sample that is representative of the matrix to be analyzed. Currently, the microbiological monitoring of the different types or genus of microorganisms generally requires specific filter systems for each type of microbial target, with different ranges of volume (from 100 mL to 1000 L). With increasing interest to monitor for all types of pathogens that may be present in a water sample and the additional costs of performing separate analyses for each organism of interest, there is a fresh need in developing a single and universal method to simultaneously concentrate and detect all waterborne pathogens.

Effective monitoring of drinking and resource water requires the assessment of volumes on a scale of 10–100 L for detecting the presence of target organisms with low infectious doses. The development of an efficient concentration stage is thus very important to enable efficient and reliable detection of low numbers of target pathogens and their genomic sequences, which are significantly diluted in high numbers of non-target sequences. Ultrafiltration (UF) membranes have pore sizes small enough to remove molecules having molecular masses in the order of 10,000 to 100,000 Da, allowing the simultaneous concentration of various waterborne microbes, including viruses, bacteria, and parasites. Sequential filtration is deemed necessary to improve the ability to detect very low numbers of target pathogens since the hold-up volume after primary UF concentration is about 0.1 to 1.5 L.

Primary concentration based on hollow-fiber ultrafiltration (HFUF)

Hollow fiber ultrafilters are cartridge packed with many individual strands of hollow membranes. They offer an alternative concentration method that can simultaneously concentrate all types of pathogenic organisms (Payment *et al.* 1989; Juliano *et al.* 1998; Morales-Morales *et al.* 2003). This technique is increasingly viewed as a time-efficient and cost-effective alternative for simultaneously recovering various microbes from large volumes of drinking water. Due to high flow rate capability, HFUF-based procedures are rapid; 100 L water samples can be filtered in less than one hour.

Ultrafilters are commonly composed of non-cellulosic synthetics designed for low-protein binding (polysulfone, polyvinylidene fluoride, polyacrylonitrile). They are commercially available (generally sold for medical uses e.g. dialysis, hemoconcentration) and are reusable (Morales-Morales *et al.* 2003; Oshima *et al.* 2007). However the use of commercial ultrafilters require the use of many additional adapters, connectors and clamps (Smith *et al.* 2009).

Table 11.1 Comparison of UF techniques for universal primary concentration of large water volumes.

Type of water and initial volume	HFUF type & name (surface area, manufacturer)	Retentate volume (eluant)	Average recovery of target organisms	Reference
TP, WW, SW 10 L	Tangential Microza (0.2 m^2; Pall Corp)	250 mL (glycine 0.05 M)	In surface water: *C. parvum* (10 oocysts/L): 19–54%* *E. coli* (9.10^5 cfu/L) : 87–96% T1 phage (10^5 pfu/L): 31–74%	(Morales-Morales *et al.* 2003)
TP, 100 L	Tangential Microza (1 m^2, Pall Corp)	1 L (glycine 0.05 M)	*C. parvum* (100 oocysts/L): 26%* *E. coli* (10^6 cfu/L): 99% *Bacillus* spores (10^6 cfu/L): 87% PP7 phage (10^6 pfu/L): 62%	(Oshima *et al.* 2007)
SW, 100L	Tangential Microza (1 m^2, Pall Corp)	1.5 L (glycine 0.05 M/ NaOH (pH 7.0)/ Tween 80 0.1%)	PP7 phage (10^{8-} 10^9 pfu/L): 64%	(Rajal *et al.* 2007)
TP, SW, 100 L	Tangential Exceltra Plus210 (2.1 m^2, Baxter)	70 mL	*E. coli* (10–100 cfu/L): 70% (SW) ; *C. perfringens* spores (10–100 cfu/L): 30% (SW) MS2 phage (10–100 pfu/L): 84% (SW) Poliovirus (10–100 pfu/L): 16% (SW)	(Gibson *et al.* 2011)
RecW, 100 L	DEUF, F80A Hemoflow (1.8 m^2, Fresenius)	350 mL (Urea 4 M/lysine 50 mM pH 9.0)	Enterococci (4–2500 cfu/L): 251% (range from 7% to 708%)**	(Leskinen *et al.* 2008)
TP, 100 L	DEUF, REXEED 25S (2.1 m^2, Asahi Kasei)	533 mL (Tween 80 0.05%, Na polyphosphate and 0.001% Y30 antifoam)	Different turbidity ranges: *C. parvum* (10^4 oocysts/L): 63–87% *E. faecalis* (10 cfu/L): 71–93% *C. perfringens* spores (15–38 cfu/L): 57–94% MS2 phage (10^3 pfu/L) : 57–82%	(Smith *et al.* 2009)

TP: tap water; WW: well water; SW: surface water, RecW: recreational water; DEUF: dead-end ultrafiltration.
*: determined after IMS-IFA (Morales-Morales *et al.* 2003) or centrifugation (Oshima *et al.* 2007).
**: quantification of enterococci naturally present in RecW. Total CFU in 100 L water samples (extrapolated from analysis of 100 mL membrane filtration) may have been underestimated or overestimated.

Hollow-fiber UF techniques for concentrating bacteria, viruses and parasites from water samples can be based either on a tangential flow (i.e. recirculating flow) or dead-end approach that have both been shown to be effective for microbe recovery, although recovery efficiencies for the filtration stage can vary and be sample dependent (Table 11.1). Generally, tangential (or cross-flow) mode reduces filter membrane fouling (accumulation of a gel layer) by providing a constant flow across and through the membrane, while dead-end filtration systems are considered to be more susceptible to membrane fouling because of the accumulation of suspended organic material and particulates. However, concentrating by tangential-flow UF requires comprehensive operator training which is generally not conducive to rapid-response implementation for field sampling. For emergency response, outbreak investigations, or other field investigations performed by personnel with limited training in water sampling, a dead-end UF (DEUF) technique would be useful for capturing and recovering multiple microbe classes (Kearns *et al.* 2008; Leskinen *et al.* 2008; Smith *et al.* 2009; Leskinen *et al.* 2010).

Unlike other filtration systems, recovery of viruses using HFUF is largely unaffected by complex chemical constituents found in natural water, as shown in studies where viral recovery has been determined by conventional plaque assay methods and qualitative PCR (Oshima *et al.* 1995; Winona *et al.* 2001; Morales-Morales *et al.* 2003). For increasing virus recoveries, some authors (Morales-Morales *et al.* 2003; Oshima *et al.* 2007; Gibson *et al.* 2011) reported that the use of a blocking agent prior to filtration (consisting of overnight incubation in a solution of 5% calf serum) reduced the attachment of viruses to the filters. Elution of the concentrate from the filter can either be made by backflushing or by adding eluant solution (glycine 0.05 M or Tween 80 0.1%) for increasing the viruses recovery.

Evaluation studies of HFUF recovery efficiencies were mainly based on spiking experiments (with often, very high loads of seeded agents) and the efficiencies can be affected by the experimental conditions (risk of attachment of virus to the plastic feed tank used for seeding the initial volume of water sample for example) (Rajal *et al.* 2007). Recovery efficiencies determined from naturally occurring organisms in water samples (fecal indicators, adenovirus) showed to be much more variable (Leskinen *et al.* 2008; Knappett *et al.* 2011), probably due to the contribution of water quality or to the variability in microbe distribution in 100 L samples, since initial target concentrations are usually determined in much smaller volumes (i.e. 100 mL).

To summarize, the current challenges for increasing the use of HFUF as a first step in the concentration process are:

- allowing the recovery of viruses, parasites and bacteria using the same initial concentration method. The recovery rates should be determined with a realistic inoculum level.
- the compatibility of eluants with the molecular methods (i.e. PCR-based methods) that will be used for the detection of organisms
- ease of on-site use with no need for adaptors.

Secondary concentration devices

The ultimate goal of integrating concentration steps and molecular detection of waterborne pathogens present in low level in the initial volume is being able to analyze the highest original sample volume during the molecular analysis. As retentate volumes after primary UF step are about 0.1–1L, secondary concentrations are needed. To date, few studies have described a complete process from concentration of 100 L water sample to simultaneous detection of bacteria, parasites and viruses.

In their study, Oshima and Smith (2007) used centrifugation at 1240 g for 20 min, separating the pellet fraction containing bacteria and protozoa from the supernatant containing viruses. The supernatant was concentrated by a secondary UF step, using spin columns. The average recoveries of the whole concentration process from 100 L of drinking water were lower than those obtained after the primary UF concentration: 50.2% for *E. coli*, 57% for *Bacillus* spores, 26% for *C. parvum*, 55% for phages PP7 and 85.8% for poliovirus. After this process the final concentrate volume was 10 mL for the pellet fraction and 20mL for the secondary UF step.

The first UF step concentrate can also be directly concentrated through a secondary UF step. For this purpose, different commercial UF systems can be used, such as Centricon Plus-70 (Millipore) centrifugal filtration device with molecular cut-off at 100 kDa described by Gibson and Schwab (2011).

11.2.2 Protocol

The method described here is based on a two-step ultrafiltration protocol using a HFUF cartridge specifically designed for water concentration (prototype) and commercial UF centrifugal devices, allowing the concentration to be obtained from 30 L of water samples to less than 300 μL for drinking water samples and 300 μL–1 mL for surface water samples. The HFUF prototype cartridge was designed by Aquasource (Degremont, Toulouse, France). It measures 25 cm long and 6 cm in diameter and is comprised of 200 acetate cellulose hollow fibers, corresponding to a surface area of 0.7 m^2; the volume inside the cartridge is about 150 mL (Figure 11.1a).

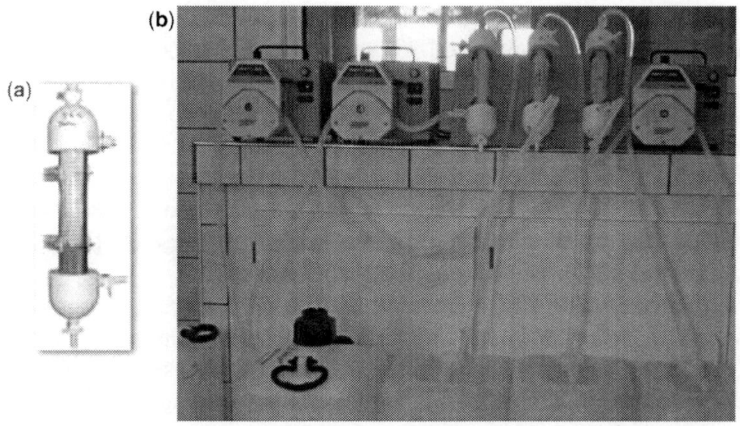

Figure 11.1 Dead-End Hollow Fiber Ultrafiltration Prototype (a) in use for concentrating 30 L of tap water samples (b).

30 L of tap water samples were collected in a tank and de-chlorinated by adding 20 mg/L of sodium thiosulfate. During filtration, the water sample is forced inside the fibers from the outside using a peristaltic pump at a flow rate of 0.8 L/min., with an inner pressure of 1.5 bars. No extra connectors or adaptors were needed for performing the first-step concentration on the sampling site (Figure 11.1b). In the case of surface water concentration, pre-filtration (100 μm, Prefilter Arkal type 100 μm, Pall Gelman) is needed in order to remove the largest organic and mineral particles which could interfere with the filtration. For treated water concentration, pre-treatment of the UF1 cartridge with 0.1% BSA (Bovine Serum Albumin) is carried out by shaking the module for 1 hour (using a wrist action shaker with arms).

After filtration, particles and micro-organisms retained on the fibers are detached by mechanical stirring (10 min at 600 oscillations/min with wrist shaker) and the liquid remaining in the cartridge is concentrated by applying compressed air, in order to recover a first concentrate (35 mL). A second backflushing-based elution is performed as described above after injecting sterile de-ionized water (raw water concentration) or NaOH 1 mM (tap water concentration). This procedure is repeated twice for raw water samples permitting the initial volume of 30 L to 70 mL (treated water) or 140 mL (raw water) to be decreased.

When pathogen spiking experiments were performed for evaluating recovery efficiencies, target microorganism load was first added to a 30 mL, de-chlorinated tap water sample and was then steadily and securely injected into the sample during filtration using the second peristaltic pump (Figure 11.2).

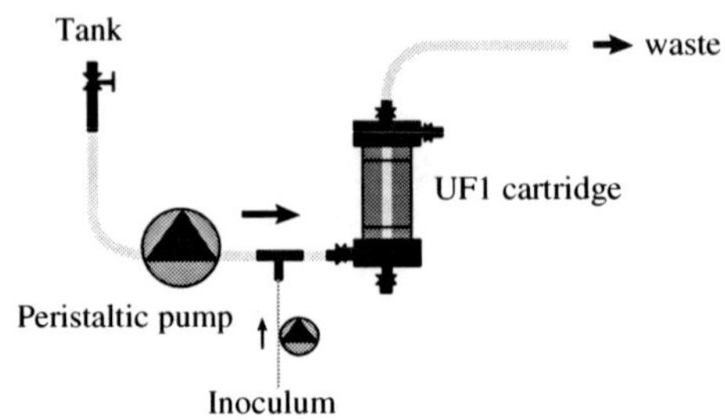

Figure 11.2 Diagram of the Dead-End HFUF prototype setup for spiking and filtering water samples.

From the 70 ml concentrate of treated water, a secondary ultrafiltration (UF2) concentration was obtained using a commercial centrifugal device (Vivacell 70, Sartorius), with a molecular weight cut-off at 10 kDa. First, the Vivacell filter was blocked with 0.1% BSA, shaken for one hour and then quickly rinsed with sterile de-ionized water. The 70 mL eluate was ultra-filtered by centrifugation for 13 min at 1100 g using a centrifuge with a swinging bucket rotor. After filtration, a volume of 2 mM NaOH equal to the volume remaining inside the cartridge was added and each side of the filter was vortexed for 30 sec. If necessary the volume was adjusted to 300 μL by adding a saline solution and if the final volume was above 300 μL, a centrifugation step for 5 min. at 1100 g was added. At the end, 10 μL H_2SO_4 (100 mM) is added to the 300 μL concentrate for neutralizing the NaOH elution.

From the 140 mL recovered after HFUF filtration of 30 L of raw water, a centrifugation at 4800 g for 60 min was performed in order to collect bacteria and protozoa in pellet fraction and viruses in supernatant. The supernatant was then concentrated into 300 μL by using 30 kDa Vivacell filter. RNA extraction is directly performed on the pellet fraction after re-suspending in 1 mL.

Following these concentration steps, concentrates from raw water and treated water were ready to be analyzed by molecular tools (through DNA or RNA extraction before PCR amplification).

11.3 INTEGRATED PROTOCOL FOR NUCLEIC ACID EXTRACTION, AMPLIFICATION AND SEQUENCE IDENTIFICATION THROUGH HIGH DENSITY MICROARRAY

DNA chip technology holds great potential for microbiological diagnostic applications due to its powerful capacity to simultaneously analyze a large number of nucleic sequences (Lin *et al.* 2006; Loy *et al.* 2006). In this study, the photolithography chip developed by Affymetrix was used as a particularly powerful tool allowing the simultaneous identification of a large set of waterborne microbes (*E. coli*, *Salmonella* spp., *Pseudomonas aeruginosa*, *E. coli* O157:H7, *Legionella* spp., *L. pneumophila*, *Cryptosporidium* spp., *C. parvum*, *Giardia* spp., *G. lamblia*, Enteroviruses, Hepatitis A viruses and Noroviruses). For bacteria and protozoa targets, ribosomal and messenger RNA sequences (rRNA and mRNA) were selected in order to detect viable pathogens through the presence of RNA markers.

Universal protocols were successfully optimized and validated for lysis, RNA extraction and purification from all targeted micro-organisms recovered in UF concentrates. Briefly, 300 µl concentrates were equally divided into three tubes and 10 µL of lyzozyme (100 mg/ml), 2 µL of 100xTrisEDTA buffer and 1µL of β-Mercaptoethanol (14.3 M) were added into each tube and incubated for 15 min. at 37°C. Then, 25 µL of Proteinase K (>600 mAU/ml), and 385 µL of Qiagen lysis solution (RLT solution) containing 1% β-mercaptoethanol were added and incubated for 15 min. at 65°C. Purification was performed using Qiashredder (Qiagen) and Rneasy Mini Kit (Qiagen) columns, adding an additional wash step with the Qiagen RW1 washing buffer. Total RNA was eluted in 25 µL of RNAse-free water.

Multiplex one-step RT-PCR amplifications were then carried out to amplify the extracted RNA for 11 microbial targets in only 4 reaction tubes (two for bacteria, one for viruses and one for parasites). RT-PCR primer sequences (described in Table 11.2) were designed specifically for this study (P. Renaud, E. Guillot, C. Mabilat, C. Vachon, B. Lacroix, G. Venret, M-A Charvieu, P. Laffaire, April 2004, France, patent number WO0202811), except for *Cryptosporidium* (Xiao *et al.* 1999) and Noroviruses (Vinje *et al.* 1996) primers. 25 µl of sample RNA was added to a 25 µL RT-PCR mix. RT-PCR reactions were performed using Access one-step RT-PCR Kit (Promega) according to the manufacturer's instructions with some modifications (Table 11.2). Thermal cycling conditions were optimized and were the same for all microorganisms tested : 45 min. at 48°C for reverse transcription; 5 min. at 94°C for initial denaturation; 40 cycles of 30 s at 94°C, 1 min. at 55°C, 1 min. at 68°C; and a final extension step of 7 min. at 68°C.

All RT-PCR products were mixed and labeled by incorporation of a biotinylated marker (*meta*-biotinphenylmethyldiazomethyl (bioMerieux) at 95°C for 25 min. and cleaved into smaller fragments with 36 mM of HCl (Korimbocus *et al.* 2005). Fragmented labeled DNA was then purified (using QiaQuick PCR purification kit) and denaturated by heating, in order to obtain biotin-labeled single DNA strands.

For the design of high-density DNA probe array specifically dedicated to waterborne pathogens, oligonucleotide sequences which allowed the identification of a total of 47 parameters (38 bacterial taxons, 5 viral taxons, and 4 parasitic taxons) were determined as follows: for each parameter, sequence databases were retrieved from Genebank, in order to determine specific sequences of 20 to 40 bases by comparing sequence alignments. Each sequence was then compared to Genebank, in order to confirm the specificity, and was selected as a reference sequence. The repertoire of selected probes was then synthesized on the array using 4-L or 2-L tiling array strategy such as previously described (Troesch *et al.* 1999; Korimbocus *et al.* 2005). For each relevant base of a given sequence, the chip contained four (or two) probes of equal lengths (20-mer). One probe represented the perfect match, while the others corresponded to the possible mismatches at the interrogating base position, centrally located within the probes. For the 2-L tiling strategy, the perfect match and the most unlikely mismatch were considered.

DNA probe array hybridization was done at 45°C for 45 min on Affymetrix GeneChip Fluidics Station 400, as described by Korimbocus *et al.* (2005). After the array was washed, the fluorescent signal emitted by the target bound to the probes was detected by a GeneArray scanner at a wavelength of 570 nm and with a pixel resolution of 3 µm. The highest signals came from the probes that best matched the target viral sequence. Probe array fluorescence intensities, nucleotide base call, sequence determinations, and reports were generated by functions available on the GeneChip 3.2 software. For each target, the percentage base-call (BC%), as determined by the percentage of homology between the experimentally derived sequence and the reference sequence, tiled on the array, as well the ratio between the median intensity obtained for the tiled sequence and the median background intensity were considered as criteria for identification.

Table 11.2 Primer sequences and multiplex RT-PCR conditions.

Multiplex (targets)	Primer (final conc. in 50 μl)	Sequence (5′-3′)	RT-PCR mix (final concentration in 50 μl)
1			
E. coli &	ENTR-F (0.2 μM)	GGAAGAAGCTTGCTTTGCTGAC	PCR buffer (1X), dNTP (0.4 mM), $MgSO_4$ (2.5 mM), AMV-rt (5 U), TFL-pol (5 U), Rnasin (5 U, Promega)
Salmonella sp.	ENTR-R (0.2 μM)	CCAGTATCAGATGCAGTTCC	
P. aeruginosa	PYO-F (0.05 μM)	GGATAACGTCCGGAAACGGG	
	FAB5RT7 (0.05 μM)	TAATACGACTCACTATAGGGAGGAGGA TTACGACTTATCGCGTTAGCTGCGCCA	
2			
Legionella sp.	LGPF-1 (0.05 μM)	CTTTAAGATTAGCCTGCGTCCG	PCR buffer (1X), dNTP (0.6 mM), $MgSO_4$ (2.5 mM), AMV-rt (5 U), TFL-pol (5 U), Rnasin (5 U, Promega)
L. pneumophila	LGPR-1 (0.05 μM)	GCACCTGTATCAGTGTTCCCGA	
E. coli	57U19 (0.5 μM)	GGCATTCAGTCTGGATCGC	
O157 :H7	278L21 (0.5 μM)	TGACCCACACTTTGCCGTAAT	
3			
Cryptosporidium	XIA2F (0.1 μM)	GGAAGGGTTGTATTTATTAGATAAAG	PCR buffer (1X), dNTP (0.6 mM), $MgSO_4$ (2.5 mM), AMV-rt (5 U), TFL-pol (5 U), Rnasin (5 U, Promega), Betain 1 M
	XIA2RT7 (0.1 μM)	TAATACGACTCACTATAGGGAGGAGGA TTAAAGGAGTAAGGAACAACCTCCA	
Giardia	GIARF1 (0.2 μM)	CGGTAATTCCAGCTCGGC	
	GIARR2 (0.2 μM)	GGTCTCGCTCGTTGTCGG	
4			
Enterovirus	EntB (0.08 μM)	GGTACCTTTGTRCGCCTG	PCR buffer (1X), dNTP (0.6 mM), $MgSO_4$ (2 mM), AMV-rt (5 U), TFL-pol (5 U), Rnasin (5 U, Promega), T4gene32 protein (Roche, 2.5 μg)
	EntC (0.08 μM)	CCAAGGTAGTCGGTTCCG	
HAV	H1 (0.08 μM)	GGAAATGTCTCAGGTACTTTCTTTG	
	H2 (0.08 μM)	GTTTTGCTCCTCTTTATCATGCTATG	
Norovirus	JV12 (0.15 μM)	ATACCACTATGATGCAGATTA	
	JV13 (0.15 μM)	TCATCATCACCATAGAAAGAG	

The Figure 11.3 summarizes the whole multi-detection process. Less than 10 hours are necessary to perform it, leading to results within one day.

To prevent false-positive or false-negative results, different process controls were systematically processed for each analysis. Two negative controls consisting in 300 μl sterile distilled water (analyzed from the RNA extraction step) and a negative RT-PCR control in which template RNA was replaced by sterile distilled water were performed. Positive controls were also performed: one process control in which 300 µl of sterile distilled water was spiked with *E. coli*, *Enterovirus* and *C. parvum* mix (respectively of 50 cfu/50 pfu/5 oocysts) and positive RT-PCR amplification controls by amplifying target RNA from bacteria, virus or parasite (1 pg in 50 µl RT-PCR mix).

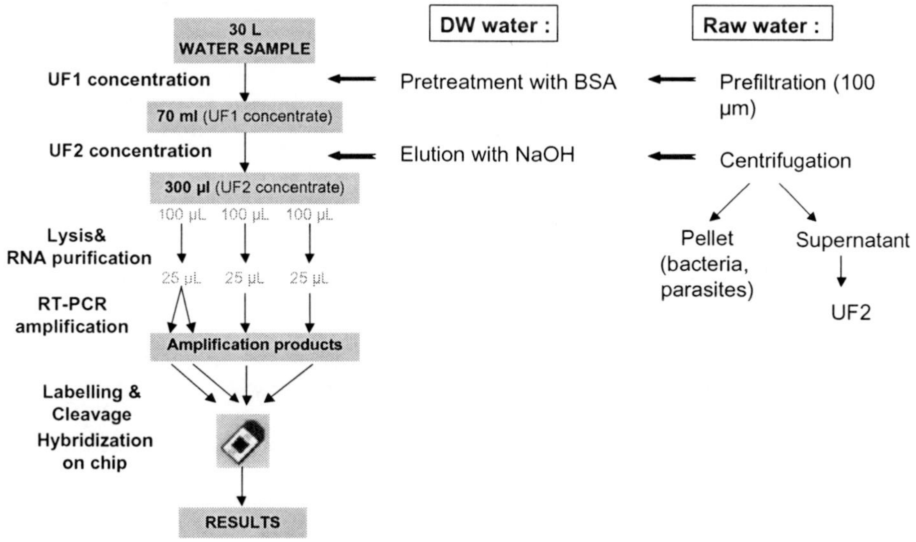

Figure 11.3 Integrated protocol for the multi-detection of waterborne pathogens by DNA chip in drinking water (DW) and raw water samples.

Detecting inhibition is also particularly important in environmental samples with low levels of pathogen contamination, as this minimal microbial load may be hidden during analysis. Improvements in RNA purification and PCR amplification were performed as follows:

- By adding a supplementary step to the Qiagen Rneasy purification protocol such as a phenol-chloroform purification
- By adding a PCR facilitator, the T4 gene 32 protein (Monpoeho *et al.* 2000; Jiang *et al.* 2005) for the relief of inhibitors of RT-PCR amplification
- By diluting the RNA extract (1/2 or 1/5) producing an ultimate solution before amplification.

11.4 RESULTS

11.4.1 Recovery from 30 L-initial volume to final concentrate

Recovery efficiencies (Table 11.3) for one or two-step UF concentration were quantified by culture-based standard methods for *E. coli* (Colilert-18 with QuantiTray 2000, Idexx) and MS2 bacteriophage (by a double-agar layer technique according to ISO 10705–2,.2001), and by IMS-IFA (Immuno-Magnetic Separation – Immuno-Fluorescence Assay, NF T90-455 standard method) for *Cryptosporidium parvum*. Microbial targets were inoculated into the initial volume of water samples in the lowest range possible (in order to be directly countable). For surface water, indigenous MS2 phages and *E. coli* cells were used. Except for MS2 phage, similar recovery efficiencies were obtained from surface and drinking water. They were also into the same order of range as those described by Smith and Hill (2009).

Table 11.3 Average recovery of organisms from 30 L of surface water (*Turbidity = 13 NTU) or drinking water using an one or two-step UF protocol.

Type of water	Microbial agents	Average recovery	Retentate from UF1 step % (SD)	Centrifugation pellet % (SD)	Centrifugation supernatant % (SD)	Retentate from UF2 step % (SD)
Surface water *	Phage MS2 (500–5000 pfu/L)	**Total** Stepwise	**34.6%** 34.6% (7.1)	NA	**35.6%** 103.7% (8.8)	**28.5%** 78.9% (21.0)
	E. coli (500–5000 cfu/L)	**Total** Stepwise	**83.3%** 83.3% (10.0)	**74.1%** 88.9% (10.2)	NA	NA
	C. parvum 10 oocysts/L	**Total** Stepwise	**42.9%** 42.9% (12.0)	**39.9%** 92.7% (5.1)	NA	NA
Drinking Water	Phage MS2 (500 pfu/L)	**Total** Stepwise	**65.2%** 65.2% (25.1)	NA	NA	NP
	E. coli (500 cfu/L)	**Total** Stepwise	**74.5%** 74.5% (13)	NA	NA	NP
	C. parvum 10 oocysts/L	**Total** Stepwise	**48.7%** 48.7% (4.5)	NA	NA	NP

N = 3 independent assays. UF1: first ultrafiltration step, UF2: second ultrafiltration step; SD: standard deviation; NA: not applicable; NP: not performed.

11.4.2 Impact of BSA blocking and elution agents on waterborne pathogen recovery using two-step ultrafiltration protocol from 30 L-drinking water

Recovery of pathogens for low seed levels was also studied for the following targets: *Legionella pneumophila* (ATCC 33152, *C. parvum* (Iowa isolate) and Poliovirus Type 3 Sabin vaccine strain. For these targets, quantification was made, using real-time PCR after DNA (or RNA) extraction on the 300 μl concentrate and following the respective methods: NF T90-471 standard method (iQCheck *Legionella pneumophila* quantification kit, Biorad); Fontaine and Guillot (2002) and Monpoeho *et al.* (2000).

First, it was shown that UF2 protocol including BSA pre-blocking and elution with 1 mM hydroxide sodium solution was essential for a better recovery of enteric viruses (Table 11.4). Similarly, these same pre-treatment and elution conditions significantly improved the total recovery for concentrating 30 L of drinking water into 300 μL (corresponding to a 5-Log concentration). The final protocol, including, for both UF steps, pre-treatment with BSA and NaOH elution, allowed small levels of waterborne pathogen targets to be detected, since about 23 copies of enterovirus (Poliovirus), 10 genome units of *L. pneumophila* and 0.3 *C. parvum* occysts could be detected per one liter of drinking water.

Table 11.4 Average total recovery of waterborne pathogens as quantified by real-time PCR-based methods using a one or two-step UF protocol from drinking water samples.

		Average total recovery % (Standard Deviation)			
		Initial volume: 70 ml Drinking water		Initial volume: 30 L Drinking water	
Waterborne Pathogen	**Initial number of target**	**UF2 protocol Water elution**	**BSA blocking UF2 protocol NaOH elution**	**UF1 protocol Water elution UF2 protocol***	**BSA blocking UF1 protocol NaOH elution UF2 protocol***
Poliovirus[a]	700 copies	10.3% (3.3)	95.4% (5.1)	Not detected	10.5% (0.7)
	1800 copies	9.5% (4.8)	85.4% (16.5)	Not detected	8.7% (0.1)
L. pneumophila[b]	300 genome units	Not performed	Not performed	Not Detected	9.8% (3.5)
	1800 genome units	Not performed	Not performed	3.4% (3.5)	7.0% (2.6)
C. parvum[c]	3 oocysts	Not performed	Not performed	29.7% (8.5)	21.3% (15.3)
	40 oocysts	Not performed	Not performed	53.4% (3.5)	65.0% (2.8)

N = 3 independant assays. UF1: first ultrafiltration step, UF2: second ultrafiltration step, BSA: Bovine Serum Albumin
[a](Monpoeho *et al.* 2002)
[b]AFNOR NF T 90-471
[c]Fontaine *et al.* 2002

The concentration efficiencies of microfiltration-based methods for bacteria and parasites (respectively 0.45 μm-membrane filtration (ISO 11731) and Envirochek filtration procedure (USEPA Method 1623) were determined in order to compare them to the recovery rates from the two-step UF protocol. The concentration efficiencies (determined after spiking of tap water samples) were similar to those obtained by the UF approach: 69($\pm$15) and 23($\pm$17)% for respectively *C. parvum* and *L. pneumophila* respectively. The two-step UF protocol was also applied for concentrating Human Adenovirus type 2 and JC polyomavirus in 10L-surface water samples. For these two viral targets, total recovery efficiencies were respectively in the range of 3.2–6.0% and 13.4–32.9% and were similar or even greater that efficiencies from glass wool-based filtration methods (Albinana-Gimenez *et al.* 2009).

Moreover, the volumes of water samples that can be filtered using the UF-protocol are similar to those that can be concentrated by the traditional methods for viruses and protozoa (i.e. glass wool and microfiltration cartridges). By contrast, the volume of water that can be analyzed for the detection of bacteria was strongly improved (more than 10 L compared to 100 mL – 1 L), increasing the capacity to detect very low levels of pathogens.

The UF concentration can be carried out on-site (without eluting) by shipping the cartridge at 4°C to the laboratory for analysis. The elution of water concentrates must be processed in less than 48 hours after sampling. Indeed, after having stored the cartridge at 4°C during 0, 24 and 48 hours, following concentration, the detection and quantification of microbial pathogens were studied by spiking 30 L of tap water with the following model microorganisms: *Legionella pneumophila* ATCC 33152, Poliovirus type 3 Sabin vaccine strain and *Cryptosporidium parvum* Iowa isolate at an initial concentrations of 3.10^4 genome units/L; 5.10^4 copies/L and 250 oocysts/L respectively. Quantification after elution was performed with real-time quantitative (RT)-PCR and no statistically significant change in the quantified concentration, even after 48 h at 4°C, was observed.

11.4.3 Back volume calculation and comparison with other available detection methods

Back volume calculations were also made in order to demonstrate the estimated original sample volume that was actually analyzed during molecular analyses and to compare these calculations to other detection methods. These back calculations are often not reported in published studies involving UF concentration from large volumes of water combined with real-time PCR (Hill *et al.* 2007; Rajal *et al.* 2007; Leskinen *et al.* 2010). The ability to relate a detection signal (in case of DNA chip-based multi-detection) or a CT value (in case of real-time PCR quantification) to a certain volume of water is important when reporting the levels (or detection results) of waterborne pathogens potentially present in the original water samples.

Table 11.5 shows the back volume calculations from the method described in this study, as well as two other studies describing multi-detection of enteric microbes by hollow-fiber ultrafiltration and real-time PCR. It shows that even if UF-based concentration allows to concentrate very large volumes of water (up to 100 L), the final sample volume present in the detection tube rarely exceeds 1 L. The back calculated volumes do not include the dilution of nucleic acid (NA) extract before PCR amplification in order to remove PCR inhibitors.

Table 11.5 Back volume calculations to determine average final sample volume in final concentrates (after UF and secondary concentration), in Nucleic Acid (NA) extracts and in PCR tubes, for bacteria (b), parasite (p) and virus (v) detection in surface water (SW) and drinking water (DW) samples.

Sample type (initial volume)	Type(s) of target	Sample volume in 100 μl of final concentrate	Volume of final concentrate (μl)	Volume used for NA extraction (μl)	NA elution volume (μl)	NA volume used for PCR (μl)	Sample volume finally analyzed	Reference
DW (30 L)	b, p, v	10 L	300	100	25	25	10 L	This study
SW (30 L)	b, p	1–3 L	1000–3000	100	25	25	1–3 L	This study
SW (30 L)	v	10 L	300	100	25	25	10 L	This study
DW (100 L)	v	5.3 L	870	200	100	5	0.5 L	(Gibson *et al.* 2011)
SW (100 L)	v	1.4 L	4 820	200	100	5	0.14 L	(Gibson *et al.* 2011)
DW (100 L)	b, p	1.7 L	1–4	500	160	10–20	0.5–1 L	(Hill *et al.* 2007)
DW (100 L)	v	1–3.3 L	1000–3000	500	160	10	0.3–1 L	(Hill *et al.* 2007)

11.4.4 Multi-detection of waterborne pathogens by DNA Chip hybridization

Whole process performances of the multi-detection of 11 microbial targets were assessed for each parameter by spiking respectively 300 μL of sterile water [in order to determine the sensitivity of the molecular detection (including RNA extraction, RT-PCR amplification and detection on DNA chip steps)] and 30-L of drinking water. The results of such spiking experiments with a single parameter are shown in Table 11.6. The sensitivity data indicate the smallest tested quantity of cells which gave positive results for all the replicates (n = 3).

Table 11.6 Sensitivity and specificity of detection on high-density DNA array for waterborne pathogens (A: bacteria targets, B: protozoan and viral targets).

A

Bacteria:	***E. coli***	***E. coli* O157:H7**	***Salmonella* spp.**	***Pseudomonas aeruginosa***	***Legionella* spp**	***L. pneumophila***
RNA target	16S rRNA	uidA mRNA	16S rRNA	16S rRNA	16S rRNA	16S rRNA
Sensitivity in 300 μl of sterile water	2 cfu	6 cfu	3 cfu	3 cfu	3 cfu	3 cfu
Sensitivity in 30 L of drinking water	500 cfu	750 cfu	350 cfu	390 cfu	580 cfu	580 cfu
Inclusivity: Number of strains tested	2	1	2	3	2	1
Exclusivity: Number of strains tested	28	28	28	28	28	28

B

Protozoa & viruses:	***C. parvum***	***Giardia lamblia***	**Hepatitis A**	**Enterovirus**	**Norovirus**
RNA target	18S rRNA	18S rRNA	VP1 conserved region	5′non-coding region	Polymerase coding region
Sensitivity in 300 μl of sterile water	5 oocysts	5 kysts	5 pfu	5 pfu	Not quantified
Sensitivity in 30 L of drinking water	5 oocysts	50 kysts	500 pfu	500 pfu	Not quantified

In order to validate the specificity of bacteria identification by DNA probe array, the inclusivity and exclusivity of the detection were assessed by testing up to 31 RNA extracted from bacterial strains that can be recovered in the water environment, whether or not belonging to the corresponding target. The results showed that all the bacteria target strains (*E. coli*, *E. coli* O157:H7, *Salmonella enteritidis*, *Salmonella typhimurium*, *Ps. aeruginosa* and *L. pneumophila*) were unambiguously identified by the probe array with a Base-Call% near 100% for all targets. However, both *Enterobacter cloacae* and *Citrobacter freundii* -which are fecal coliform bacteria- were identified as *Salmonella* and *Burkholderia gladioli* was identified as *Ps. aeruginosa* (data not shown). For parasites, noroviruses and Hepatitis A virus, the study of the specificity of the identification could not be achieved due to the limited number of available strains. For enterovirus identification, Poliovirus, echovirus 7, coxsackieviruses B1 and A9 were also tested and gave expected results. A further specificity study on a broader range of bacterial strains as well as on parasites and viruses would be necessary in order to validate the specificity of DNA chip detection.

The high sensitivity of this detection process was demonstrated since less than 25 cfu/liter of pathogenic bacteria as well as 0.2 to 2 (oo)kysts/liter of protozoa and 20 pfu/liter of enteroviruses or Hepatitis A can be successfully detected in drinking water.

This work illustrated the fact that high density micro-arrays (using in particular 16S rDNA probes) are suitable for distinguishing between different taxons and microbial groups for waterborne pathogens (Call *et al.* 2003; Vora *et al.* 2004). Moreover, the viability of bacteria and protozoa can be assessed since the methodology is based on specific detection of RNA fragments; it is well established that rRNA is less stable than rDNA and may be more closely related to cellular viability than DNA. Information about micro-organism viability is of paramount importance in estimating the human health risk as well as the effectiveness of water treatment. Finally, by targeting rRNA for bacteria and parasites, the protocol was consistent for the three retroviral groups in which RNA was the target.

11.5 CONCLUSIONS

Integrating the different technologies for collecting, processing and analyzing water samples for all pathogenic microorganisms of interest will represent a significant advance for research in water microbiology. In particular, by this consolidated methodology, emergent waterborne pathogens, that are today poorly documented, could be better analyzed in order to document their occurrence in the environment.

The methodology developed in this study is simple and the easily transportable design of the ultrafiltration-based concentration offers the advantage of performing the concentration step in the field. This reduces the risks and the costs of transportation and handling of large volumes of sample. Moreover, it was also demonstrated that real-time PCR quantification could also be integrated by using the same steps as for sample preparation (UF concentration and nucleic acids extraction/purification).

This research demonstrated the potential of using a single sample preparation procedure based on ultrafiltration for the simultaneous quantification waterborne pathogens instead of performing separate and costly analyses (Morales-Morales *et al.* 2003; Hill *et al.* 2007). The main difficulty of simultaneous detection of microorganisms of different regna (bacteria, parasites, viruses) was in developing a universal protocol that would be relatively easy to implement, involving successive steps effective for all three types of microorganisms which are highly different in terms of size, physical and chemical characteristics. The first concentration step was critical since pathogens, if present in treated water samples, can be found in very low concentrations against a high background of heterotrophic bacteria natural flora. Indeed, as the results demonstrated that RT-PCR amplifications with subsequent DNA chip detection were highly sensitive for all target microorganisms, optimization of upstream steps (concentration, elution and purification) is necessary in order to increase the sensitivity of the whole process.

In addition, a representative volume of water had to be sampled and analyzed in order to make the results statistically relevant. Ultra-filtration membranes, which provide an absolute barrier for the physical separation of all particles larger than 0.01 μm at ambient pH, can then concentrate viral, bacterial, and protozoan organisms in a single process from large volumes of water. We have developed and optimized two successive ultra-filtration steps: UF1 and UF2,which concentrated treated water samples of 30 L (and up to 100 L, data not shown) with a 5-log concentration factor in a small volume (under 500 μL) compatible with subsequent molecular detection. For all the types of targets, the concentration efficiencies were comparable to those in published molecular methods, but with the main advantage of its multi-detection properties.

For the first time, a single solution, from sampling to results, allows the multi-detection of bacteria, viruses and parasites in a treated water sample in less than 8 hours, compared to several weeks for conventional culture methods alone, in particular for viruses. This process offers a rapid, non-quantitative screening method that will help to draw the attention of the health authorities on the existence of a potential issue. In the event of a positive result, quantification by real-time PCR can be performed in order to provide data for microbial risk assessment. Instead of using indicators, this precise multi-detection method for water pathogens, together with assessment of their viability, will make it possible to better assess the microbiological safety of drinking water.

Acknowledgements

We are grateful to Stephane Bulteau, Marie-Astrid Armand, Carole Vachon and Emmanuelle Guillot for their previous contribution to the development of the detection protocol of the pathogen targets on the Affymetrix DNA chip.

Part of this work was supported by funds from the European Commission for the Healthy Water Project (FOOD-CT-2006-036306).

REFERENCES

Albinana-Gimenez N., Clemente-Casares P., Bofill-Mas S., Hundesa A., Ribas F. and Girones R. (2006). Distribution of human polyomaviruses, adenoviruses, and hepatitis E virus in the environment and in a drinking-water treatment plant. *Environ. Sci. Technol.*, **40**(23), 7416–7422.

Albinana-Gimenez N., Clemente-Casares P., Calgua B., Huguet J. M., Courtois S. and Girones R. (2009). Comparison of methods for concentrating human adenoviruses, polyomavirus JC and noroviruses in source waters and drinking water using quantitative PCR. *J. Virol. Methods*, **158**(1–2), 104–109.

Bej A. K. (2003). Molecular based methods for the detection of microbial pathogens in the environment. *J. Microbiol. Methods*, **53**(2), 139–140.

Brodie E. L., Desantis T. Z., Joyner D. C., Baek S. M., Larsen J. T., Andersen G. L., Hazen T. C., Richardson P. M., Herman D. J., Tokunaga T. K., Wan J. M. and Firestone M. K. (2006). Application of a high-density oligonucleotide microarray approach to study bacterial population dynamics during uranium reduction and reoxidation. *Appl. Environ. Microbiol.*, **72**(9), 6288–6298.

Call D. R., Borucki M. K. and Loge F. J. (2003). Detection of bacterial pathogens in environmental samples using DNA microarrays. *J. Microbiol. Methods*, **53**(2), 235–243.

Chee M., Yang R., Hubbell E., Berno A., Huang X. C., Stern D., Winkler J., Lockhart D. J., Morris M. S. and Fodor S. P. (1996). Accessing genetic information with high-density DNA arrays. *Science*, **274**(5287), 610–614.

Fontaine M. and Guillot E. (2002). Development of a TaqMan quantitative PCR assay specific for Cryptosporidium parvum. *FEMS Microbiol. Lett.*, **214**(1), 13–17.

Gibson K. E. and Schwab K. J. (2011). Tangential-flow ultrafiltration with integrated inhibition detection for recovery of surrogates and human pathogens from large-volume source water and finished drinking water. *Appl. Environ. Microbiol.*, **77**(1), 385–391.

Hill V. R., Kahler A. M., Jothikumar N., Johnson T. B., Hahn D. and Cromeans T. L. (2007). Multistate evaluation of an ultrafiltration-based procedure for simultaneous recovery of enteric microbes in 100-liter tap water samples. *Appl. Environ. Microbiol.*, **73**(13), 4218–4225.

Jiang J., Alderisio K. A., Singh A. and Xiao L. (2005). Development of procedures for direct extraction of cryptosporidium DNA from water concentrates and for relief of PCR inhibitors. *Appl. Environ. Microbiol.*, **71**(3), 1135–1141.

Juliano J. and Sobsey M. (1998). Simultaneous concentration of Cryptospridium, bacteria and viruses by hollow-fiber ultrafiltration. In: *Proc. of the Water Quality Technology Conference*, 1998.

Kearns E. A., Magana S. and Lim D. V. (2008). Automated concentration and recovery of micro-organisms from drinking water using dead-end ultrafiltration. *J. Appl. Microbiol.*, **105**(2), 432–442.

Khan J., Bittner M. L., Chen Y., Meltzer P. S. and Trent J. M. (1999). DNA microarray technology: the anticipated impact on the study of human disease. *Biochim Biophys Acta*, **1423**(2), M17–M28.

Knappett P. S., Layton A., McKay L. D., Williams D., Mailloux B. J., Huq M. R., Alam M. J., Ahmed K. M., Akita Y., Serre M. L., Sayler G. S. and van Geen A. (2011). Efficacy of hollow-fiber ultrafiltration for microbial sampling in groundwater. *Ground Water*, **49**(1), 53–65.

Korimbocus J., Scaramozzino N., Lacroix B., Crance J. M., Garin D. and Vernet G. (2005). DNA probe array for the simultaneous identification of herpesviruses, enteroviruses, and flaviviruses. *J. Clin. Microbiol.*, **43**(8), 3779–3787.

Leclerc H., Schwartzbrod L. and Dei-Cas E. (2002). Microbial agents associated with waterborne diseases. *Crit. Rev. Microbiol.*, **28**(4), 371–409.

Leskinen S. D., Brownell M., Lim D. V. and Harwood V. J. (2010). Hollow-fiber ultrafiltration and PCR detection of human-associated genetic markers from various types of surface water in Florida. *Appl. Environ. Microbiol.*, **76**(12), 4116–4117.

Leskinen S. D. and Lim D. V. (2008). Rapid ultrafiltration concentration and biosensor detection of enterococci from large volumes of Florida recreational water. *Appl. Environ. Microbiol.*, **74**(15), 4792–4798.

Lin B., Wang Z., Vora G. J., Thornton J. A., Schnur J. M., Thach D. C., Blaney K. M., Ligler A. G., Malanoski A. P., Santiago J., Walter E. A., Agan B. K., Metzgar D., Seto D., Daum L. T., Kruzelock R., Rowley R. K., Hanson E. H., Tibbetts C. and Stenger D. A. (2006). Broad-spectrum respiratory tract pathogen identification using resequencing DNA microarrays. *Genome Res.*, **16**(4), 527–535.

Loy A. and Bodrossy L. (2006). Highly parallel microbial diagnostics using oligonucleotide microarrays. *Clin Chim Acta*, **363**(1–2), 106–119.

Monpoeho S., Coste-Burel M., Costa-Mattioli M., Besse B., Chomel J. J., Billaudel S. and Ferre V. (2002). Application of a real-time polymerase chain reaction with internal positive control for detection and quantification of enterovirus in cerebrospinal fluid. *Eur. J. Clin. Microbiol. Infect. Dis.*, **21**(7), 532–536.

Monpoeho S., Dehee A., Mignotte B., Schwartzbrod L., Marechal V., Nicolas J. C., Billaudel S. and Ferre V. (2000). Quantification of enterovirus RNA in sludge samples using single tube real-time RT-PCR. *Biotechniques*, **29**(1), 88–93.

Morales-Morales H. A., Vidal G., Olszewski J., Rock C. M., Dasgupta D., Oshima K. H. and Smith G. B. (2003). Optimization of a reusable hollow-fiber ultrafilter for simultaneous concentration of enteric bacteria, protozoa, and viruses from water. *Appl. Environ. Microbiol.*, **69**(7), 4098–4102.

Oshima K. H., Evans-Strickfaden T. T., Highsmith A. K. and Ades E. W. (1995). The removal of phages T1 and PP7, and poliovirus from fluids with hollow-fiber ultrafilters with molecular weight cut-offs of 50,000, 13,000, and 6000. *Can. J. Microbiol.*, **41**(4–5), 316–322.

Oshima K. H. and Smith G. B. (2007). Ultra-filtration based extraction for biological agents in early warning systems. Awwarf Report Series 91142, published by Awwa Research Foundation, Denver, US.

Payment P., Bérubé A., Perreault D., Armon R. and Trudel M. (1989). Concentration of *Giardia lamblia* cysts, *Legionella pneumophila*, *Clostridium perfringens*, human enteric viruses, and coliphages from large volumes of drinking water, using a single filtration. *Can. J. Microbiol.*, **35**, 932–935.

Pease A. C., Solas D., Sullivan E. J., Cronin M. T., Holmes C. P. and Fodor S. P. (1994). Light-generated oligonucleotide arrays for rapid DNA sequence analysis. *Proc. Natl. Acad. Sci. USA*, **91**(11), 5022–5026.

Peplies J., Glockner F. O. and Amann R. (2003). Optimization strategies for DNA microarray-based detection of bacteria with 16S rRNA-targeting oligonucleotide probes. *Appl. Environ. Microbiol.*, **69**(3), 1397–1407.

Purohit H. J. and Kapley A. (2002). PCR as an emerging option in the microbial quality control of drinking water. *Trends Biotechnol.*, **20**(8), 325–326.

Rajal V. B., McSwain B. S., Thompson D. E., Leutenegger C. M., Kildare B. J. and Wuertz S. (2007). Validation of hollow fiber ultrafiltration and real-time PCR using bacteriophage PP7 as surrogate for the quantification of viruses from water samples. *Water Res.*, **41**(7), 1411–1422.

Ramsay G. (1998). DNA chips: state-of-the art. *Nat. Biotechnol.*, **16**(1), 40–44.

Rudi K., Flateland S. L., Hanssen J. F., Bengtsson G. and Nissen H. (2002). Development and evaluation of a 16S ribosomal DNA array-based approach for describing complex microbial communities in ready-to-eat vegetable salads packed in a modified atmosphere. *Appl. Environ. Microbiol.*, **68**(3), 1146–1156.

Smith C. M. and Hill V. R. (2009). Dead-end hollow-fiber ultrafiltration for recovery of diverse microbes from water. *Appl. Environ. Microbiol.*, **75**(16), 5284–5289.

Straub T. M. and Chandler D. P. (2003). Towards a unified system for detecting waterborne pathogens. *J. Microbiol. Methods*, **53**(2), 185–197.

Troesch A., Nguyen H., Miyada C. G., Desvarenne S., Gingeras T. R., Kaplan P. M., Cros P. and Mabilat C. (1999). Mycobacterium species identification and rifampin resistance testing with high-density DNA probe arrays. *J. Clin. Microbiol.*, **37**(1), 49–55.

van Leeuwen W. B., Jay C., Snijders S., Durin N., Lacroix B., Verbrugh H. A., Enright M. C., Troesch A. and van Belkum A. (2003). Multilocus sequence typing of *Staphylococcus aureus* with DNA array technology. *J. Clin. Microbiol.*, **41**(7), 3323–3326.

Vinje J. and Koopmans M. P. (1996). Molecular detection and epidemiology of small round-structured viruses in outbreaks of gastroenteritis in the Netherlands. *J. Infect. Dis.*, **174**(3), 610–615.

Vora G. J., Meador C. E., Stenger D. A. and Andreadis J. D. (2004). Nucleic acid amplification strategies for DNA microarray-based pathogen detection. *Appl. Environ. Microbiol.*, **70**(5), 3047–3054.

Winona L. J., Ommani A. W., Olszewski J., Nuzzo J. B. and Oshima K. H. (2001). Efficient and predictable recovery of viruses from water by small scale ultrafiltration systems. *Can. J. Microbiol.*, **47**(11), 1033–1041.

Xiao L., Morgan U. M., Limor J., Escalante A., Arrowood M., Shulaw W., Thompson R. C., Fayer R. and Lal A. A. (1999). Genetic diversity within Cryptosporidium parvum and related Cryptosporidium species. *Appl. Environ. Microbiol.*, **65**(8), 3386–3391.

Chapter 12

Detection and enumeration of waterborne mycobacteria

Joseph O. Falkinham III

12.1 ECOLOGY OF WATERBORNE MYCOBACTERIA

The majority of *Mycobacterium* species are opportunistic pathogens that are normal inhabitants of natural waters, engineered water systems, and household plumbing. The current number of *Mycobacterium* species isolated from the environment is over 150 and continues to increase (Tortoli, 2003). They cause a variety of infections in humans that are usually associated with at least one of a number of risk factors. In as much as the environmental mycobacteria, called nontuberculous mycobacteria (NTM), are present in habitats that overlap with those of humans; individuals are continually exposed to the nontuberculous mycobacteria. Transmission can occur through drinking water and inhalation of aerosols from showers or flowing water.

12.1.1 Mycobacterial diseases

Environmental nontuberculous mycobacteria are opportunistic human pathogens (Marras & Daley, 2002). The major representatives include the slowly growing *Mycobacterium avium, Mycobacterium intracellulare* (Marras & Daley, 2002), *Mycobacterium xenopi* (Costrini *et al.* 1981), *Mycobacterium malmoense* (Zaugg *et al.* 1993), *Mycobacterium simiae* (Conger *et al.* 2004) and the rapidly growing *Mycobacterium abscessus, Mycobacterium chelonae,* and *Mycobacterium fortuitum* many of which are hospital-acquired (Wallace *et al.* 1998; Huitt & De Groote, 2006). Some species cause dermal and subcutaneous infections in humans and animals, including *Mycobacterium marinum* (Iredell *et al.* 1992; Aubry *et al.* 2002), *Mycobacterium ulcerans* (Johnson *et al.* 1996; Johnson *et al.* 2005) and *Mycobacterium haemophilum* (Saubolle *et al.* 1996).

The incidence of infections (principally pulmonary disease) caused by the environmental mycobacteria appears to be increasing from 1–2 per 100,000 in 1997 to 8–10 per 100,000 in 2003 (Marras *et al.* 2007). In the United States, that translates to 30,000 new cases per year. Disease presentations included: pulmonary disease and skin infections (Marras & Daley, 2002; Huitt & De Groote, 2006). Cervical lymphadenitis is found in children (Wolinsky, 1995). Disseminated disease (bacteremia) is found in individuals with AIDS or those that are immunosuppressed due to cancer, chemotherapy, or transplantation. Guidelines for the diagnosis and treatment of individuals with environmental mycobacterial infections have been published (Griffith *et al.* 2007).

Risk factors for mycobacterial pulmonary disease amongst immunocompetent individuals include lung damage, such as pneumoconiosis, black lung, smoking, or alcoholism (Marras & Daley, 2002). Patients with cystic fibrosis can be infected with the rapidly growing mycobacterium, *Mycobacterium abscessus* (Jonsson *et al.* 2007). Not just individuals with cystic fibrosis, but also carriers of mutations in the chloride membrane transport protein, CFTR (Cystic Fibrosis Transmembrane Regulator) are also at increased risk of pulmonary mycobacterial disease (Kim *et al.* 2005). Individuals with gastroesophaegeal reflux disease (GERD) are also at higher risk for mycobacterial pulmonary disease (Koh *et al.* 2007; Thomson *et al.* 2007). In the United States, the increase in the incidence of nontuberculous mycobacterial disease is primarily due to pulmonary disease in slender, elderly women who lack any of the classic risk factors (Prince *et al.* 1989; Reich & Johnson, 1991; Kennedy & Weber, 1994).

12.1.2 Mycobacterial habitats

Humans, animals, and plants are surrounded by the waterborne mycobacteria. They are normal inhabitants of natural waters, engineered water distribution systems, drinking water, and household plumbing. Very high numbers of NTM are found in acidic, humic and fulvic acid-rich waters of southeastern United States coastal swamps (Kirschner *et al.* 1992) and boreal, pine forest waters of northern Europe (Iivanainen *et al.* 1997), Canada, and the United States. NTM are present in drinking water distribution systems (Covert *et al.* 1999; Falkinham *et al.* 2001) and household plumbing (Falkinham, 2011). As such it should not be surprising that the mycobacteria are found in hot tubs and spas (Mangione *et al.* 2001), foot baths (Winthrop *et al.* 2002), and water therapy pools (Falkinham, 2009). As NTM are hydrophobic, due to the presence of the lipid-rich outer membrane (Brennan & Nikaido, 1995; Daffe & Draper, 1998), they prefer to reside attached to surfaces or at interfaces. As a consequence, the majority of NTM are found in the biofilms of pipes in distribution systems and households (Schulze-Röbbecke & Fischeder, 1989; Falkinham *et al.* 2001; Torvinen *et al.* 2004; Falkinham, 2011). The presence of the hydrophobic outer membrane imparts resistance to chlorine and other disinfectants used in water treatment (Taylor *et al.* 2000). Due to their relative disinfectant-resistance, water disinfection results in selection of the mycobacteria, which are normally poor competitors due to their very slow growth (1 generation per day).

12.1.3 Transmission of mycobacteria

Aerosol transmission of mycobacteria has been documented in showers (Falkinham *et al.* 2008), hot tubs or spas (Mangione *et al.* 2007), and automobile workers exposed to metal recovery fluid aerosols (Moore *et al.* 2000). As mycobacteria are quite hydrophobic, they are readily aerosolized from water (Wendt *et al.* 1980; Parker *et al.* 1984; Falkinham, 2003b). Compared to the concentration of mycobacteria in water, aerosolized droplets are enriched in concentration from 1,000- to 10,000-fold (Parker *et al.* 1983). The ages of children suffering from mycobacterial cervical lymphadenitis (1.5–5 years), suggests that the most probable route of infection is via water or soil introduced into the mouth and entering the cervical lymph nodes via gum lesions caused by erupting teeth (Wolinsky, 1995). Mycobacteria in drinking water or food are a likely source of pulmonary infection in individuals with gastric reflux disease (Koh *et al.* 2007; Thomson *et al.* 2007).

12.2 PHYSIOLOGICAL ECOLOGY OF WATERBORNE MYCOBACTERIA

The term physiological ecology refers to those physiologic features of a microorganism that influence or dictate its ecology. Superficially, it would appear that mycobacteria should not be present in water. They grow too slowly (i.e. one generation per day) and are impermeable to hydrophilic metal ions and nutrients (Brennan & Nikaido, 1995; Daffe & Draper, 1998). The presence of one or two ribosomal RNA operons and the diversion of energy to synthesize the long chain lipids (mycolic acids) of the mycobacterial outer membrane are the major factors dictating slow growth of mycobacteria (Brennan & Nikaido, 1995). However, slow growth means that mycobacteria are less sensitive to conditions and agents that inhibit one aspect of macromolecular synthesis (e.g. DNA replication) and hence uncouple and thereby kill other, faster growing microorganisms. The hydrophobic membrane, while reducing the rate of transport of hydrophilic compounds needed for growth, imparts resistance to anti-microbial agents (Rastogi *et al.* 1981). It also contributes to the ability of mycobacteria to utilize a wide range of hydrocarbons, including chlorinated hydrocarbons (Falkinham, 2009b). Further, the hydrophobic outer surface drives the attachment of mycobacteria to surfaces where they can survive and not be washed from a flowing system (i.e. most water systems) and grow in the biofilm.

12.2.1 The lipid-rich mycobacterial envelope

The lipid- and wax-rich outer membrane of mycobacterial cells is a major determinant of the growth, physiology, and consequently the distribution of these waterborne pathogens (Hoffman *et al.* 2008). Synthesis of the C_{40}–C_{80} chain fatty acids (e.g. mycolic acids) diverts a great deal of the available ATP away from producing more cells. Further, the presence of that lipid-rich outer membrane reduces the rates of transport of hydrophilic compounds (Brennan & Nikaido, 1995); its presence has important benefits to mycobacterial cells. First, resistant to hydrophilic compounds means that mycobacteria are resistant to disinfectants (e.g. chlorine), toxic heavy metals (e.g. Cd and Hg), and a broad spectrum of antibiotics (Rastogi *et al.* 1981; Falkinham *et al.* 1984; Taylor *et al.* 2000). In terms of possible human exposure, mycobacteria in drinking water, distribution systems, and household plumbing survive chlorination and can grow in the absence of competition from more rapidly growing microorganisms. In fact, a review of the instances of outbreaks of mycobacterial disease associated with water exposures, the outbreaks usually follow the use of disinfectants (Falkinham, 2003b). Second, a hydrophobic surface drives mycobacteria to surface attachment where they reside and grow without being at risk of being washed out of the flowing system (Schulze-Röbbecke & Fischeder, 1989; Falkinham *et al.* 2001; Torvinen *et al.* 2004). Third, surface hydrophobicity leads to the concentration of mycobacteria at the air-water interface of any body of water where organic compounds and metals accumulate and serve as nutrient sources. Fourth, the lipid-rich outer membrane while relatively impermeable to hydrophilic compounds is conducive to the binding and transport of hydrocarbons (including chlorinated hydrocarbons) as evidenced by the ability of mycobacteria to metabolize a variety of hydrocarbon pollutants (Heitkamp *et al.* 1988; Vanderberg *et al.* 1994). It is not, thereby, surprising that mycobacteria have been recovered from sites of hydrocarbon pollution (Leys *et al.* 2005; Wang *et al.* 2006).

It is best to consider the mycobacterial outer membrane, not as a detriment to their survival, but an important factor dictating the distribution of mycobacteria in the human environment. Mycobacteria are poor competitors in nutrient-rich waters, as they grow slowly and cannot keep up with the increase in numbers of other microbes. However, under harsh conditions (e.g. acid) or after disinfection, the mycobacteria become excellent competitors and ultimate can become the major population group.

12.2.2 Consequences of the slow growth of mycobacteria

As was the case for the lipid-rich outer membrane of mycobacteria, the slow growth of mycobacteria has its own advantages and disadvantages. Cells of slowly growing microorganisms are more resistant to anti-microbial agents compared to faster growing cells of the same or different types (Falkinham, 2003a). This difference in susceptibility is likely a consequence of the greater susceptibility of rapidly growing cells to any unbalance in synthesis of macromolecules (e.g. DNA, RNA, protein, cell wall) that occurs before the cell can adapt. That is not the case for mycobacterial cells whose slow grow allows the induction of pathways which can alleviate the consequences of exposure to antimicrobial agents and harsh environmental conditions (Steed & Falkinham, 2006; Falkinham, 2007). It is important to point out that although the mycobacteria grow substantially slower than do other microorganisms (e.g. 1 generation per day); this is not a consequence of slow rates of metabolism or macromolecular syntheses. Rates of mycobacterial oxygen consumption and enzyme induction are equal to those displayed by *Escherichia coli*. Mycobacteria grow slowly, in part, due to the diversion of much ATP to the synthesis of C_{40}–C_{80} lipids.

12.2.3 Viable but unculturable mycobacteria

It had been acknowledged for quite a while that mycobacteria colony counts were usually one-tenth of the number of cells seen by microscopy. This observation has been reinforced by enumeration of mycobacteria by quantitative polymerase chain reaction (qPCR). Estimates of gene copy numbers and cell counts by qPCR usually are 10-fold higher than colony counts and more in line with microscopic cell counts. Those observations suggest that only 10% of mycobacteria are culturable. Unfortunately, to date there has been not systematic study of this topic.

The question of the existence of unculturable mycobacteria is quite relevant as the water industry wants rapid tests for enumeration of waterborne pathogens, such as *Mycobacterium avium*. In the absence of data, it is unclear whether the cells that can be enumerated by microscopy or by qPCR are truly viable or not. Several preliminary experiments in my laboratory demonstrated that 2-fold higher colony counts could be obtained on suspensions or cultures by cultivation under microaerobic conditions (i.e. 6–12% oxygen). That suggests that, in spite of a great deal of experience with mycobacteria, culture conditions for the formation of colonies from every cell have not yet been identified.

12.3 RISK ANALYSIS AND SOURCE-TRACKING ENVIRONMENTAL MYCOBACTERIA

The majority of *Mycobacterium* species are opportunistic pathogens that are normal inhabitants of soils, natural waters, engineered water systems, and household plumbing (Tortoli, 2003). They cause a variety of infections in humans that are usually associated with at least one of a number of risk factors. In as much as the environmental mycobacteria, called nontuberculous mycobacteria (NTM), are present in habitats that overlap with those of humans; individuals are continually exposed to the nontuberculous mycobacteria (Falkinham, 2009a). Transmission can occur through drinking water, exposure to dusts from soil, and inhalation of aerosols from showers or flowing water.

12.3.1 *Mycobacterium avium* and the candidate contaminant list

One of the environmental opportunistic mycobacteria, *Mycobacterium avium* complex has been placed on the 3rd edition of the US Environmental Protection Agency's Candidate Contaminant List (CCL). The *M. avium* complex (MAC) includes *M. avium* and its subspecies (*avium, paratuberculosis, silvaticum, hominissuis, chimaera, colombiense*), and *Mycobacterium intracellulare*. The major human opportunistic pathogens within the MAC are *M. avium* subspecies *hominissuis* and *M. intracellulare*. Each is responsible for approximately half of the pulmonary mycobacterial infections in the United States. Almost all HIV-infected individuals with AIDS are infected with *M. avium* subsp. *hominissuis*; almost none by *M. intracellulare* or other MAC subspecies. Inclusion on the EPA's Candidate

Contaminant List requires that the agency study the organism to determine whether standards should be developed for its monitoring in drinking water.

12.3.2 Risk analysis for mycobacteria

Although *M. avium* subsp. *hominissuis* and *M. intracellulare* are recognized as opportunistic pathogens whose source of infection is drinking water, developing a risk analysis for exposure is problematic. First, there is wide diversity within both species and it is not known whether certain types within each species are virulent or avirulent. Studies of MAC strain collections have revealed differences in virulence of different isolates in animal models (Reddy *et al.* 1994). However, it is not known whether differences in virulence in animal models are duplicated in humans. To date, no one has identified a genetic or physiologic marker of virulence that could be employed in monitoring.

12.3.3 Source-tracking and DNA fingerprinting

Difficulties in identifying potentially virulent isolates of MAC notwithstanding, there has been many efforts to identify and develop typing and fingerprinting techniques for the waterborne environmental mycobacteria and, specifically, for members of the *M. avium* complex. This has been driven by the need to prove that the environment or some compartment (e.g. water) is the source of infection. Serotyping, phage typing, and enzyme profiling were the first developed, but now have been supplanted by DNA-based typing methods (Behr & Falkinham, 2009). Pulsed field gel electrophoresis (PFGE), where large DNA fragments generated by restriction endonuclease cleavage are separated and patterns compared has been quite successful (von Reyn *et al.* 1994). Other methods involving repetitive sequence and insertion sequence-based methods have been developed, in part because of the time consuming nature of PFGE. For members of the *M. avium* complex (MAC) two techniques offer a high degree of discrimination without being overly time-consuming: IS1245/IS1311 restriction fragment length polymorphism (Falkinham *et al.* 2008) and PCR amplification of sequences between repeated sequences (*rep*-PCR, Cangelosi *et al.* 2004). A study is currently underway to measure the discriminatory power and reproducibility of PFGE, IS1245/IS1311-RFLP, *rep*-PCR, and multilocus sequencing (MLS).

12.4 SAMPLING AND SAMPLE TREATMENT STRATEGIES FOR MYCOBACTERIAL DETECTION AND ENUMERATION

Sampling strategies for waterborne mycobacteria must take into account the habitats and characteristics of these opportunistic pathogens in planning a sampling strategy and methods for sample preparation and detection or enumeration. For example, hydrophobic mycobacterial numbers in water samples is low, whereas their numbers are high in biofilm (surface samples). Mycobacterial slow growth means that mycobacterial colonies will be overgrown by colonies of other, more rapidly growing waterborne microorganisms. Recognition of that fact, has led to development and implementation of decontamination regimens, designed to reduce the numbers of other microorganisms while not reducing (or at least only slightly lowering) the number of mycobacteria.

12.4.1 Sampling strategies

For the waterborne mycobacteria, water samples are simply not useful as mycobacteria in suspension represent cells that have come off biofilms. In order to recover significant numbers of cells from water as colonies or DNA for PCR or qPCR, water samples of 500–1,000 mL are required. Although the slow growth and long term survival of mycobacterial cells means that samples can be transported over days without any change in numbers or viability, the expense of transporting litre-sized samples is too high. Fortunately, biofilm samples will yield mycobacteria in rather high numbers as pipe biofilms may harbour between 500–5,000 colony-forming units per cm^2 (Schulz-Röbbecke & Fischeder, 1989; Falkinham *et al.* 2001; Torvinen *et al.* 2004; Falkinham, 2011). If natural or engineered water sources such as lakes, ponds, or rivers need sampling, consider collection of the surface biofilm using either a glass plate (Wendt *et al.* 1980) or laying filter paper on the surface. The filter paper can be immediately transferred to agar medium suitable for mycobacterial growth or the filter paper digested to isolate DNA. Hydrophobic mycobacteria are 10 to 100-fold concentrated in surface biofilms over the concentration in the bulk suspension at the air-water interface (Wendt *et al.* 1980; Parker *et al.* 1983).

12.4.2 Sample treatment

Cultivation methods for the detection or enumeration of waterborne mycobacteria have generally relied upon addition of an antimicrobial agent to reduce the number of more rapidly growing non-mycobacteria cells (i.e. decontamination). The thick lipid-rich outer membrane makes mycobacterial cells resistant to 1–5% NaOH, 1% oxalic acid, and exposures to various antimicrobial detergents such as 1% cetylpyridinium chloride (Brooks *et al.* 1984). After decontaminant exposure and neutralization of the decontaminating agent, a large percentage of non-mycobacterial cells are killed. However, it must be understood that decontamination also reduces the number of mycobacterial cells. For example although 99.99% of heterotrophic plate count bacteria are lost upon exposure to 4% NaOH, only 95% of mycobacterial cells are killed (Brooks *et al.* 1984). In spite of the fact that some unique mycobacterial species (e.g. *Mycobacterium kansasii* is acid-sensitive) or types (e.g. opaque colony types are more sensitive to decontamination than transparent types) might be lost by decontamination, the loss of the majority of rapidly growing non-mycobacteria allows detection, enumeration, and isolation of mycobacteria. However, it is important to realize that decontamination results in loss of mycobacterial cells.

12.4.3 Sample concentration methods

In addition to selecting sample sites where waterborne mycobacteria are concentrated and grow (e.g. biofilms in pipes and showerheads), liquid samples can be concentrated by centrifugation (5,000 × g for 20 min) or filtration. Mycobacteria are quite hardy, so cells in pellets can be suspended in a small volume of tap water to prepare a concentrated suspension for either direct plating (e.g. drinking water) or decontamination (e.g. biofilm suspensions). Filters can be placed directly on agar medium suitable for mycobacterial growth or digested for DNA isolation and subsequent PCR or qPCR.

In addition to filtration and centrifugation, hydrophobic partition, immunomagnetic beads, and protozoa or amoebae can be used to concentrate mycobacterial cells. Hydrophobic partition relies upon the extreme hydrophobicity of mycobacterial cells and involves mixing a water sample with a small volume of some organic solvent; 1 part hexadecane for 50 parts water sample. Following mixing, allow the organic and aqueous phases to separate and collect the organic layer and spread directly agar medium or allow the hexadecane to evaporate, suspend the cells in tap water and perform DNA extraction (Stormer & Falkinham, 1989).

Immunomagnetic beads have been employed successfully to isolate mycobacteria from samples containing high numbers of microorganisms capable of overgrowth of mycobacterial colonies (Mazurek *et al.* 1996). Magnetic beads, coated with an anti-mycobacterial antibody are added to the sample and after mixing and sufficient time for bead: mycobacterial cell binding to occur, the magnetic beads can be captured and cells eluted. The eluted cells can be either plated directly or DNA isolated.

Based on description of immunomagnetic bead capture of mycobacteria, we successfully captured and concentrated *Mycobacterium avium* cells in a water suspension using cells of the phagocytic protozoan, *Tetrahymena pyriformis*. The motivation for such a study was to see if we could develop a method for mycobacterial concentration in a setting of limited resources. The *T. pyriformis* cells were starved to encourage phagocytosis, mixed with the *M. avium* suspension, and after 60 min, the protozoa were isolated by low speed centrifugation (1,000 × g for 10 min), washed, and the resulting suspension spread on Middlebrook 7H10 agar medium. *M. avium* cells were concentrated approximately 10-fold in the small volume of the *T. pyriformis* pellet. Further, the *T. pyriformis* cells removed almost 90% of the inoculated *M. avium* cells based on enumeration of free cells in the suspension after low speed centrifugation and colony counts.

12.5 MYCOBACTERIAL DETECTION OR ENUMERATION

12.5.1 Detection or enumeration

It is important to decide whether one needs to simply detect a waterborne pathogen or actually enumerate cells in samples. Detection is sufficient for establishing the diagnosis of *Mycobacterium tuberculosis* infection, but is insufficient for the environmental mycobacteria where the number of cells is important (Griffith *et al.* 2007). Enumeration is required for initiating a risk-analysis study. Further, if source-tracking is the objective of a sampling study, it will be necessary to isolate and enumerate colonies that, in turn, can be used from DNA fingerprint comparisons. In the absence of well-established genetic markers of virulence, PCR and qPCR can be employed for detection or enumeration, but lack the ability to couple amplification of a target DNA sequence with fingerprinting.

12.5.2 Culture, PCR, or qPCR

Ideally, mycobacterial slow growth would not be a factor if PCR or qPCR was used for detection or enumeration of mycobacteria. However, the thick outer membrane makes mycobacterial cells resistant to lysis, resulting in reduced yields of DNA and thus leading to failures to detect by PCR or underestimates of mycobacterial numbers by qPCR. Therefore, it is important to consider that not all mycobacterial cells in a sample will lyse to yield DNA leading to an underestimate of mycobacterial numbers.

12.5.3 Culture of mycobacteria

Colony counts remain the "gold standard" of mycobacterial enumeration, notwithstanding the fact that cell counts can be as 10-fold higher than colony counts. Although there has been speculation that "viable but unculturable" mycobacteria are present, no study has either documented or refuted the existence of such cells, identified conditions that trigger such a state, or provide means to "resuscitate" such cells. It is quite possible that mycobacteria in some water samples subject to stress (e.g. temperature, alkalinity) might not form colonies (i.e. false negative").

The standard medium used for colony formation, isolation, and enumeration of waterborne mycobacteria is Middlebrook 7H10 agar containing 0.5% (v/v) glycerol and 10% (v/v) oleic acid-albumin (M7H10). Its transparency allows the detection of transparent colony forms; on the egg-based Lowenstein-Jensen medium transparent colonies are almost invisible. Although the vast majority of waterborne mycobacteria can grow on standard bacteriological media (e.g. Plate Count Agar and R2A), the transparent colony form requires oleic acid for growth (Fry *et al.* 1986). As the transparent forms are isolated from patients, are virulent and antibiotic-resistant, enumeration mycobacteria should always employ M7H10. The avirulent, antibiotic-sensitive opaque form appears at a frequency of 1/1000 and is favoured by laboratory cultivation because it grows more rapidly.

12.5.4 PCR-detection and qPCR enumeration of mycobacteria

A variety of DNA sequences and primers have been described for PCR detection or qPCR enumeration of mycobacteria. Principally, these primer pairs and sequences have been developed for members of the *M. avium* complex (MAC) because of its importance causing bacteremia in immunosuppressed patients due to HIV-infection, malignancy, or transplantation therapy or because of the rapidly increasing incidence of pulmonary disease in slender, elderly women. It is important to consider ensuring that the primer pair is specific for the particular *Mycobacterium* species or subspecies target because many of the primer pairs were developed and tested in human or animal samples and not water samples. Human microflora differs from that of water and it is quite possible that a PCR-based test developed for patient samples will fail due to amplification of the DNA of a waterborne bacterium not present in human or animal microflora.

Targets for PCR-based detection and enumeration of *Mycobacterium* spp. include the 16S rRNA genes (Wilton & Cousins, 1992), the ribosomal rRNA spacers (Roth *et al.* 1998), the *hsp65* (heat-shock) gene (Telenti *et al.* 1993), the *rpoB* (Whang *et al.* 2011) and *dnaJ* (Takewaki *et al.* 1993) genes, and the gene for the 32 kD antigen (Soini *et al.* 1992). Many of the methods require the digestion of an amplified PCR fragment by one or two restriction endonucleases and analysis of patterns for identification. However, some of the methods disclose primers that can be used for either PCR-based detection or qPCR-based enumeration of specific species; specifically my lab has used the primers described by Wilton and Cousins (1992) to develop a qPCR method for enumeration of *Mycobacterium avium*.

At present, PCR-based techniques for detection or enumeration of mycobacteria have not been adapted or modified to provide data suitable for DNA fingerprint comparisons. Such an assay would provide rapid identification of sources of waterborne mycobacterial infection. Such rapid identification would be of great value to individuals with risk factors for mycobacterial infection, such as slender, elderly women or carriers of mutant cystic fibrosis gene, as they could initiate measures to reduce mycobacterial exposures in the home (in-line microbiological filters on taps and showerheads).

REFERENCES

Aubry A., Chosidow O., Caumes E., Robert J. and Cambau E. (2002). Sixty-three cases of *Mycobacterium marinum* infection – clinical features, treatment, and antibiotic susceptibility of causative isolates. *Archives of Internal Medicine*, **162**, 1746–1752.

Behr M. A. and Falkinham J. O. III. (2009). Molecular epidemiology of nontuberculous mycobacteria. *Future Microbiology* **4**, 1009–1020.

Brennan P. J. and Nikaido H. (1995). The envelope of mycobacteria. *Annual Reviews of Biochemistry*, **64**, 29–63.

Brooks R. W., Parker B. C., Gruft H. and Falkinham J. O. III. (1984). Epidemiology of infection by nontuberculous mycobacteria. V. Numbers in eastern United States soils and correlation with soil characteristics. *American Reviews of Respiratory Disease* **130**, 630–633.

Cangelosi G. A., Freeman R. J., Lewis K. N., Livingston-Rosanoff D., Shah K. S., Milan S. J. and Goldberg S. V. (2004). Evaluation of a high-throughput repetitive-sequence-based PCR system for DNA fingerprinting of *Mycobacterium tuberculosis* and *Mycobacterium avium* complex strains. *Journal of Clinical Microbiology*, **42**, 2685–2693.

Conger N. G., O'Connell R. J., Laurel V. L., Olivier K. N., Graviss E. A., Williams-Bouyer N., Zhang Y., Brown-Elliott B. A. and Wallace R. J. Jr. (2004). *Mycobacterium simiae* outbreak associated with a hospital water supply. *Infection Control and Hospital Epidemiology* **25**, 1050–1055.

Costrini A. M., Mahler D. A., Gross W. M., Hawkins J. E., Yesner R. and D'Esopo N. D. (1981). Clinical and roentgenographic features of nosocomial pulmonary disease due to *Mycobacterium xenopi*. *American Review of Respiratory Disease*, **123**, 104–109.

Covert T. C., Rodgers M. R., Reyes A. L. and Stelma G. N. Jr. (1999). Occurrence of nontuberculous mycobacteria in environmental samples. *Applied and Environmental Microbiology* **65**, 2492–2496.

Daffe M. and Draper P. (1998). The envelope layers of mycobacteria with reference to their pathogenicity. *Advances in Microbial Physiology*, **39**, 131–203.

duMoulin G. C., Stottmeier K. D., Pelletier P. A., Tsang A. Y. and Hedley-Whyte J. (1988). Concentration of *Mycobacterium avium* by hospital hot water systems. *Journal of the American Medical Association*, **260**, 1599–1601.

Falkinham J. O. III. (2003a). Factors influencing the chlorine susceptibility of *Mycobacterium avium, Mycobacterium intracellulare*, and *Mycobacterium scrofulaceum*. *Applied and Environmental Microbiology*, **69**, 5685–5689.

Falkinham J. O. III. (2003b). Mycobacterial aerosols and respiratory disease. *Emerging Infectious Diseases*, **9**, 763–767.

Falkinham J. O. III. (2007). Growth in catheter biofilms and antibiotic resistance of *Mycobacterium avium*. *Journal of Medical Microbiology*, **56**, 250–254.

Falkinham J. O. III. (2009a). Surrounded by mycobacteria: nontuberculous mycobacteria in the human environment. *Journal of Applied Microbiology*, **107**, 356–367.

Falkinham J. O. III. (2009b). The biology of environmental mycobacteria. *Environmental Microbiology Reports*, **1**, 477–487.

Falkinham J. O. III. (2011). Nontuberculous mycobacteria from household plumbing of patients with nontuberculous mycobacteria disease. *Emerging Infectious Diseases*, **17**, 419–424.

Falkinham J. O. III, George K. L., Parker B. C. and Gruft H. (1984). *In vitro* susceptibility of human and environmental isolates of *Mycobacterium avium, M. intracellulare*, and *M. scrofulaceum* to heavy metal salts and oxyanions. *Antimicrobial Agents and Chemotherapy*, **25**, 137–139.

Falkinham J. O. III, Iseman M. D., de Haas P. and van Soolingen D. (2008). *Mycobacterium avium* in a shower linked to pulmonary disease. *Journal of Water and Health*, **6**, 209–213.

Falkinham J. O. III, Norton C. D. and LeChevallier M. W. (2001). Factors influencing numbers of *Mycobacterium avium, Mycobacterium intracellulare*, and other mycobacteria in drinking water distribution systems. *Applied and Environmental Microbiology*, **67**, 1225–1231.

Fry K. L., Meissner P. S. and Falkinham J. O. III. (1986). Epidemiology of infection by nontuberculous mycobacteria. VI. Identification and use of epidemiological markers for studies of *Mycobacterium avium, M. intracellulare* and *M. scrofulaceum*. *American Review of Respiratory Disease*, **134**, 39–43.

Griffith D. E., Aksamit T., Brown-Elliott B. A., Catanzaro A., Daley C., Gordin F., Holland S. M., Horsburgh R., Huitt G., Iademarco M. F., Iseman M., Olivier K., Ruoss S., von Reyn C. F., Wallace R. J. Jr. and Winthrop K. (2007). An official ATS/IDSA statement: diagnosis, treatment, and prevention of nontuberculous mycobacteria diseases. *American Journal of Respiratory and Critical Care Medicine*, **175**, 367–416.

Heitkamp M. A., Franklin W. and Cerniglia C. E. (1988). Microbial metabolism of polycyclic aromatic hydrocarbons: isolation and characterization of a pyrene-degrading bacterium. *Applied and Environmental Microbiology*, **54**, 2549–2555.

Hoffman C. A., Leis M., Niederweis M., Plizko J. M. and Engelhardt H. (2008). Disclosure of the mycobacterial outer membrane: cryo-electron tomography and vitreous sections reveal the lipid bilayer structure. *Proceedings of the National Academy of Sciences USA*, **105**, 3963–3967.

Huitt G. and De Groote M. A. (2006). Infections due to rapidly growing mycobacteria. *Clinical Infectious Diseases*, **42**, 1756–1763.

Iivanainen E. K., Martikainen P. J., Raisanen M. L. and Katila M.-J. (1997). Mycobacteria in boreal coniferous forest soils. *FEMS Microbiology Ecology*, **23**, 325–332.

Iredell J., Whitby M. and Blacklock Z. (1992). *Mycobacterium marinum* infection – Epidemiology and presentation in Queensland 1971–1990. *Medical Journal of Australia*, **157**, 596–598.

Johnson P. D., Stinear T., Small P. L., Pluschke G., Merritt R. W. and Portaels F. (2005). Buruli ulcer (*M. ulcerans* infection): new insights, new hope for disease control. *PLoS Medicine*, **2**, e108.

Johnson P. D., Veitch M. G., Leslie D. E., Flood P. E. and Hayman J. A. (1996). The emergence of *Mycobacterium ulcerans* infection near Melbourne. *Medical Journal of Australia*, **164**, 76–78.

Jonsson B. E., Gilljam M., Lindblad A., Ridell M., Wold A. W. and Welinder-Olsson C. (2007). Molecular epidemiology of *Mycobacterium abscessus*, with focus on cystic fibrosis. *Journal of Clinical Microbiology*, **45**, 1497–1504.

Kennedy T. P. and Weber D. J. (1994). Nontuberculous mycobacteria. An underappreciated cause of geriatric lung disease. *American Journal of Respiratory and Critical Care Medicine*, **149**, 1654–1658.

Kim J. S., Tanaka N., Newell J. D., De Groote M. A., Fulton K., Huitt G. and Lynch D. A. (2005). Nontuberculous mycobacterial infection. CT scan findings, genotype, and treatment responsiveness. *Chest*, **128**, 3863–3869.

Kirschner R. A. Jr, Parker B. C. and Falkinham J. O. III. (1992). Epidemiology of infection by nontuberculous mycobacteria. X. *Mycobacterium avium, M. intracellulare*, and *M. scrofulaceum* in acid, brown-water swamps of the southeastern United States and their association with environmental variables. *American Review of Respiratory Diseases*, **145**, 271–275.

Koh W. J., Lee J. H., Kwon Y. S., Lee K. S., Suh G. Y., Chung M. P., Kim H. and Kwon O. J. (2007). Prevalence of gastroesophaegeal reflux disease in patients with nontuberculous mycobacterial disease. *Chest*, **131**, 1825–1830.

Leys N. M., Ryngaert A., Bastiaens L., Wattiau P., Top E. M., Verstraete W. and Springael D. (2005). Occurrence and community composition of fast-growing *Mycobacterium* in soils contaminated with polycyclic aromatic hydrocarbons. *FEMS Microbiology Ecology*, **51**, 375–388.

Mangione E. J., Huitt G., Lenaway D., Beebe J., Bailey A., Figoski M., Rau M. P., Albrecht K. D. and Yakrus M. A. (2001). Nontuberculous mycobacterial disease following hot tub exposure. *Emerging Infectious Diseases*, **7**, 1039–1042.

Marras T. K., Chedore P., Ying A. M. and Jamieson F. (2007). Isolation prevalence of pulmonary non-tuberculous mycobacteria in Ontario, 1997–2003. *Thorax*, **62**, 661–666.

Marras T. K. and Daley C. L. (2002). Epidemiology of human pulmonary infection with nontuberculous mycobacteria. *Clinics in Chest Medicine*, **23**, 553–567.

Mazurek G. H., Reddy V., Murphy D. and Ansari T. (1996). Detection of *Mycobacterium tuberculosis* in cerebrospinal fluid following immunomagnetic enrichment. *Journal of Clinical Microbiology*, **34**, 450–453.

Moore J. S., Christensen M., Wilson R. W., Wallace R. J. Jr, Zhang Y., Nash D. R. and Shelton B. (2000). Mycobacterial contamination of metal working fluids: involvement of a possible new taxon of rapidly growing mycobacteria. *American Industrial Hygiene Association Journal*, **61**, 205–123.

Norton C. D., LeChevallier M. W. and Falkinham J. O. III. (2004). Survival of *Mycobacterium avium* in a model distribution system. *Water Research*, **38**, 1457–1466.

Parker B. C., Ford M. A., Gruft H. and Falkinham J. O. III. (1983). Epidemiology of infection by nontuberculous mycobacteria. IV. Preferential aerosolization of *Mycobacterium intracellulare* from natural waters. *American Reviews of Respiratory Disease*, **128**, 652–656.

Prince D. S., Peterson D. D., Steiner R. M., Gottlieb J. E., Scott R., Israel H. L., Figueroa W. G. and Fish J. E. (1989). Infection with *Mycobacterium avium* complex in patients with predisposing conditions. *New England Journal of Medicine*, **321**, 863–868.

Rastogi N., Frehel C., Ryter A., Ohayon H., Lesourd M. and David H. (1981). Multiple drug resistance in *Mycobacterium avium*: is the wall architecture responsible for the exclusion of antimicrobial agents? *Antimicrobial Agents and Chemotherapy*, **20**, 666–677.

Reddy V. M., Parikh K., Luna-Herrera J., Falkinham J. O. III, Brown S. and Gangadharam P. R. J. (1994). Comparison of virulence of *Mycobacterium avium* complex (MAC) strains isolated from AIDS and non-AIDS patients. *Microbial Pathogenesis*, **16**, 121–130.

Reich J. M. and Johnson R. E. (1991). *Mycobacterium avium* complex pulmonary disease. Incidence, presentation, and response to therapy in a community setting. *American Reviews of Respiratory Disease*, **143**, 1381–1385.

Roth A., Fischer M., Hamid M. E., Michalke S., Ludwig W. and Mauch H. (1998). Differentiation of phylogenetically related slowly growing mycobacteria based on 16S-23S rRNA gene internal transcribed spacer sequences. *Journal of Clinical Microbiology*, **36**, 139–147.

Saubolle M. A., Kiehn T. E., White M. H., Rudinsky M. F. and Armstrong D. (1996). *Mycobacterium haemophilum*: microbiology and expanding clinical and geographic spectra of disease in humans. *Clinical Microbiology Reviews*, **9**, 435–447.

Schulze-Röbbecke R. and Fischeder R. (1989). Mycobacteria in biofilms. *Zentralblatt Hygiene und Umweltmedizine*, **188**, 385–390.

Soini H., Skurnik M., Liippo K., Tala E. and Viljanen M. K. (1992). Detection and identification of mycobacteria by amplification of a segment of the gene coding for the 32-kilodalton protein. *Journal of Clinical Microbiology*, **30**, 2025–2028.

Steed K. A. and Falkinham J. O. III. (2006). Effect of growth in biofilms on chlorine susceptibility of *Mycobacterium avium* and *Mycobacterium intracellulare*. *Applied and Environmental Microbiology*, **72**, 4007–4100.

Stormer R. S. and Falkinham J. O. III. (1989). Differences in antimicrobial susceptibility of pigmented and unpigmented colonial variants of *Mycobacterium avium*. *Journal of Clinical Microbiology*, **27**, 2459–2465.

Takewaki S.-I., Okuzumi K., Ishiko H., Nakahara K.-I., Ohkubo A. and Nagai R. (1993). Genus-specific polymerase chain reaction for the mycobacterial *dnaJ* gene and species-specific oligonucleotide probes. *Journal of Clinical Microbiology*, **31**, 446–450.

Taylor R. H., Falkinham J. O. III, Norton C. D. and LeChevallier M. W. (2000). Chlorine-, chloramine-, chlorine dioxide- and ozone-susceptibility of *Mycobacterium avium*. *Applied and Environmental Microbiology*, **66**, 1702–1705.

Telenti A., Marchesi F., Balz M., Bally F., Böttger E. C. and Bodmer T. (1993). Rapid identification of mycobacteria to the species level by polymerase chain reaction and restriction enzyme analysis. *Journal of Clinical Microbiology*, **31**, 175–178.

Thomson R. M., Armstrong J. G. and Looke D. F. (2007). Gastroesophageal reflux disease, acid suppression, and *Mycobacterium avium* complex pulmonary disease. *Chest*, **131**, 1166–1172.

Tortoli E. (2003). Impact of genotypic studies on mycobacterial taxonomy: the new mycobacteria of the 1990s. *Clinical Microbiology Reviews*, **16**, 319–354.

Torvinen E., Suomalainen S., Lehtola M. J., Miettinen I. T., Zacheus O., Paulin L., Katila M.-L. and Martikainen P. J. (2004). Mycobacteria in water and loose deposits of drinking water distribution systems in Finland. *Applied and Environmental Microbiology*, **70**, 1973–1981.

Uslan D. Z., Kowalski T. J., Wengemack N. L., Virk A. and Wilson J. W. (2006). Skin and soft tissue infections due to rapidly growing mycobacteria. *Archives of Dermatology*, **142**, 1287–1292.

Vanderberg L. A., Burback B. L. and Perry J. J. (1995). Biodegradation of trichloroethylene by *Mycobacterium vaccae*. *Canadian Journal of Microbiology*, **41**, 298–301.

von Reyn C. F., Maslow J. N., Barber T. W., Falkinham J. O. III and Arbeit R. D. (1994). Persistent colonisation of potable water as a source of *Mycobacterium avium* infection in AIDS. *Lancet*, **343**, 1137–1141.

Wallace R. J. Jr, Brown B. A. and Griffith D. E. (1998). Nosocomial outbreaks/pseudo-outbreaks caused by nontuberculous mycobacteria. *Annual Reviews of Microbiology*, **52**, 453–490.

Wang Y., Ogawa M., Fukuda K., Miyamoto H. and Taniguchi H. (2006). Isolation and identification of mycobacteria from soils at and illegal dumping site and landfills in Japan. *Microbiology and Immunology*, **50**, 513–524.

Wendt S. L., George K. L., Parker B. C., Gruft H. and Falkinham J. O. III. (1980). Epidemiology of infection by nontuberculous mycobacteria. III. Isolation of potentially pathogenic mycobacteria from aerosols. *American Reviews of Respiratory Disease*, **122**, 259–263.

Whang J., Lee B. S., Choi G.-E., Cho S.-N., Kil P. Y., Collins M. T. and Shin S. J. (2011). Polymerase chain reaction – restriction fragment length polymorphism of the *rpoB* gene for identification of *Mycobacterium avium* subsp. *paratuberculosis* and differentiation of *Mycobacterium avium* subspecies. *Diagnostic Microbiology and Infectious Disease*, **70**, 65–71.

Wilton S. and Cousins D. (1992). Detection and identification of multiple mycobacterial pathogens by DNA amplification in a single tube. *PCR Methods and Applications*, **1**, 269–273.

Winthrop K. L., Abrams M., Yakrus M., Schwartz I., Ely J., Gillies D. and Vugia D. J. (2002). An outbreak of mycobacterial furunculosis associated with footbaths at a nail salon. *New England Journal of Medicine*, **346**, 1366–1371.

Wolinsky E. (1995). Mycobacterial lymphadenitis in children: a prospective study of 105 nontuberculous cases with long-term follow up. *Clinical Infectious Diseases*, **20**, 954–963.

Zaugg M., Salfinger M., Opravil M. and Luthy R. (1993). Extrapulmonary and disseminated infections due to *Mycobacterium malmoense*: case report and review. *Clinical Infectious Diseases*, **16**, 540–549.

Chapter 13

New molecular technologies for the rapid detection of *Legionella* in water

E. Soria, M. A. Yáñez, R. Múrtula and V. Catalán

13.1 INTRODUCTION

Legionella pneumophila is a bacterium belonging to the family Legionellaceae. Within this family, which is continually expanding, *L. pneumophila* is the agent most frequently associated with pathologies in humans and among the 15 known serogroups, serogroup 1 is the most commonly found in clinical and environmental isolates.

The importance of studying this bacterium lies in the health implications, since it is the etiologic agent of severe pneumonia, but also for its important socio-economic implications, which can affect large communities causing outbreaks, with the consequent impact on different industrial sectors such as tourism, the petrochemical industry and others.

Many countries have developed regulations and guidelines to study the presence of *Legionella* in all the water systems in order to prevent legionellosis (Nocker *et al.* 2010). These regulations and guidelines include the necessity of testing programs and all of them establish the culture isolation based on the ISO 11731 as the gold standard. However, this monitoring technique is time-consuming, since the growth rate of the bacterium is very slow. Moreover, using this methodology there is a difficulty to detect viable nonculturable bacterium, and it is difficult to isolate *Legionella* in samples containing high levels of other microbiota (Yáñez *et al.* 2005). In clinical diagnostic, the outcome of the disease depends on the speed with which a diagnosis can be performed (Heath *et al.* 1996), so in this kind of samples the most used alternative method is urinary antigen detection by enzyme immunoassay (EIA). This method is rapid, sensitive and specific for the detection of serum antibody levels. In the case of environmental samples, different methods have been tested, such as PCR, qPCR and immunodetection, all of them trying to solve the main problem of culture isolation which is the time to obtain results. But until now, the mentioned culture isolation method is the accepted procedure for *Legionella* testing in environmental samples to comply with regulations. In this chapter, we review some rapid methodologies (some completely developed and others under development) that have been designed for *Legionella* detection in environmental samples.

In the last decade, immunological detection methods have become more sensitive, specific and reproducible, and for this reason, a lot of designs are available in the market, aiming to reduce analytical time, indicating a promising future for this kind of technologies. In general, the new detection methods compared to traditional culture techniques and the ideal detection method should allow the real-time detection of microorganisms. The need of more rapid, reliable, specific and sensitive methods for a rapid detection of pathogens with a low cost is the objective of all the research in this area, especially looking for devices for the detection of microorganisms outside the laboratory environment. In this sense, the development of biosensors offers the potential to detect pathogens in real-time, but in some cases, they still require time-consuming pre-enrichment steps to detect low concentrations of microorganisms. Nevertheless, the new advances in antibody production and in physical or physical-chemical transducers allow the detection of a broad spectrum of analytes in complex samples. This is a promising development for application in the detection of microorganism in clinical samples and environmental monitoring (Leonard *et al.* 2002). We present here the latest advances in the detection and quantification of *Legionella* using these kinds of biosensors, and their main limitations, which should be solved before their use as a routine method.

13.2 IMMUNODETECTION AND *LEGIONELLA* FAST DETECTION

Using the culture isolation standard method, the time required to obtain results for *Legionella* becomes up to 10–12 days. This delay is sometimes unacceptable, for example, in the event of an outbreak, so during the years, several approaches looking for more rapid screening methods have been developed trying to solve this important drawback. The use of immunodetection assays in the search of *Legionella* requires greater specificity and sensitivity when applied to the detection of microorganism in environmental samples than many clinical diagnostic assays provide (Kfir & Genthe, 1993). With the modern methodologies and instrumentation, these methods are again increasingly popular, mainly due to their cost-effectiveness, reliability, rapidity and simplicity of the testing. The immunoassay rapid tests available in the market have been extended to detect *Legionella* antigen in clinical samples. That is the case of *Xpect*® *Legionella* from Oxoid or *BinaxNOW*® *Legionella* from Inverness Medical Innovations that are widely used methods for diagnosing Legionnaires' disease in urine (Svarrer *et al.* 2012). These commercial kits have the disadvantages of having low or no sensitivity for other serogroups different from *L. pneumophila* serogroup 1 (sg 1) and the concentration of *Legionella* in the sample must be relatively high to be detected. For this reason, all these diagnostics kits are not useful for the detection of *Legionella* in environmental water samples.

Among the methods that can be used directly in water samples, the *FastPath*® *Legionella* Detection Test (Nalco) and the *Hydrosense*® kit (Hydrosense) can alert the presence of *L. pneumophila* sg.1 in a few minutes. Both systems use the capillary flow action technology to bind a coloured antibody with the *Legionella* antigen present in the sample, showing a red line in the "Test line" of a plastic disposable device. The limit of detection of both methods is around 100 colony forming units (cfu)/mL and the hands-on time per test is 25 min. These kits present several advantages since they are small, portable, very rapid and easy to use. The interpretation of the results is visual and very easy, it is not necessary to have trained personnel, and they can be used in all kind of waters. The main disadvantages are that they inform only about the presence/absence of the microorganism (i.e. they are not quantitative assays), need pre-concentration of the water and they are designed only to detect *L. pneumophila* sg. 1.

Another rapid test available in the market is the kit *Bioalarm Legionella*® (Rodriguez *et al.* 2010). This kit uses the *Estapor*® magnetic microspheres for the detection of *L. pneumophila* in water within 50 minutes. This technology is designed to detect intact cells, because the surface integrity of a bacterial cell can be used as a criterion to differentiate viable forms and distinguish them from damaged cells. The methodology includes the capture of the *Legionella* cells by immunomagnetic spheres labeled with an enzyme-conjugated antibody in a sandwich format, and a subsequent colorimetric analysis after a short reaction with enzyme substrates. All the reactions are performed in a small device at room temperature. The method is able to perform a semi-quantitative detection, and the colour obtained can be visually compared to a colour card to estimate the magnitude order of the *L. pneumophila* concentration in the samples. One of the main drawbacks of the method is that the samples need to be concentrated in order to get a limit of detection comparable with the culture method and that it only detects *L. pneumophila* sg. 1. In this sense, the limit of detection after the sample concentration is around 2,000 cells/L. In order to compare the results obtained by this method with those obtained by culture isolation we organized an inter-collaborative study with 15 accredited laboratories to validate the methodology, and the results indicated that the method was comparable to culture isolation in 95.8% of the samples (unpublished data).

The future of all these methodologies is promising although they need to improve the ability to perform the assay directly on site to be fast and cost-effective. Currently, these are used as screening methods for the rapid monitoring of *Legionella* in water in different environments and facilities.

13.3 *LEGIONELLA* DETECTION USING MICROFLUIDICS

The impact of microfluidic technologies has dramatically increased during the last years in areas such as biotechnology, diagnostics, medicine or pharmaceutics. Initially, the main reason for miniaturization was to enhance analytical performance, but the reduction of size also presented the advantages of reduced consumption of reagents and the ability to integrate different steps within a single device. The ability of microfluidic systems to conduct measurements from small volumes of samples with efficiency and speed, without the need for a skilled operator, has been regarded as the most powerful application of Lab-on-a-Chip (LOC) technologies (Mairhofer *et al.* 2009). Many of the developments with microfluidics that are being applied for microbiology were designed for clinical diagnostic, and are being introduced very slow in the environmental diagnostic area. Therefore the future goal of the application of microfluidic technology to monitor environmental sources such as rivers, lakes, sea, and facilities such as cooling towers and evaporative condensers is to know the concentration of the microorganism at real-time and in a short period of time.

In this section, we review the different approaches related with *Legionella* although, at the moment, none of them has been successfully applied to environmental samples, so far.

13.3.1 Microarray platforms using antibodies

Based on antibodies detection, different approaches have been used for *Legionella* detection in environmental waters. Wolter *et al.* (2008) published the first flowthrough chemiluminescence microarray developed for the rapid and simultaneous detection of *Escherichia coli* O157:H7, *Salmonella typhimurium*, and *L. pneumophila* in water samples using a semiautomated readout system. The development was based on the capture of bacteria using species-specific polyclonal antibodies that were immobilized on a polyethylene glycol (PEG) modified glass substrate. Cell identification was carried out by binding of species-specific biotinylated secondary antibodies. Chemiluminescence detection was accomplished by a streptavidin-horseradish peroxidase (HRP) catalyzed reaction of luminol and hydrogen peroxide, and recorded by a sensitive charge-coupled device (CCD) camera. The overall assay time was 13 min, enabling a fast sample analysis. In multi-analyte experiments, the limits of detection (LOD) were 3×10^6, 1×10^5, and 3×10^3 cells/mL for *S. typhimurium*, *L. pneumophila*, and *E. coli* O157:H7, respectively. Quantification of samples was possible in a wide concentration range with good recoveries. In particular, the detection of *L. pneumophila* cells with this ELISA, Enzyme-Linked Immunosorbent Assay, performed under flow conditions on the antibody microarray surface showed linear signal to concentrations that covered the range 10^5–10^7 cells/mL with very high R^2 (0.992). The assay was highly reproducible, with 7% intra-assay standard deviations for *L. pneumophila* detection. However, background noise and poor slide-to-slide reproducibility are a general problem related to disposable chips, and the achieved LOD is still far from cell concentrations found in natural water samples. Nevertheless, this sandwich ELISA platform enables very rapid detection of bacterial cells in water samples, and it is adequate for automation, and integration with concentration and purification units coupled previously to the detection system, helping to meet the sensitivity needed to detect the contamination levels demanded in the legal regulations.

A similar device was developed by the same group in the Technical University of Munchen (Karsunke *et al.* 2009). A disposable chip made of acrylonitrile – butadiene – styrene (ABS) copolymer was devised as a support for a multiplexed sandwich immunoassay with six parallel flowthrough microchannels for chemiluminescence detection of pathogenic bacteria. Antibodies were directly spotted on the microchannels of the ABS chip and were directly immobilized by adsorption. Five measuring microchannels were used for calibration (using different microorganism concentrations) and one microchannel for analysis of water samples. Polyclonal antibodies against the pathogenic bacteria *E. coli* O157:H7, *S. typhimurium*, and *L. pneumophila* were immobilized on the chip by microcontact printing in order to use them as specific receptors. Detection of the bacteria was carried out by use of specific detection antibodies labelled with biotin and horseradish peroxidase (HRP) – streptavidin conjugates. The enzyme HRP generates chemiluminescence after adding luminol and hydrogen peroxide. This signal was observed by use of a CCD camera (Colour Charge-Coupled Device). The LOD 1.8×10^4 cells/mL for *E. coli* O157:H7, 7.9×10^4 cells/mL for *L. pneumophila*, and 2.0×10^7 cells/mL for *S. typhimurium*. The overall assay time for measurement and calibration was 18 min, enabling very fast analysis. Once more, authors comment that pre-enrichment and pre-concentration of bacteria would be necessary to enable analysis of real samples of drinking water. For monitoring drinking water supply the developed detection system could be used in line with microfiltration and immunomagnetic separation.

A microfluidic device developed by Yamaguchi and coworkers for *E. coli* (Yamaguchi *et al.* 2011) has been later applied to the rapid enumeration of *L. pneumophila* in cooling tower waters. This microfluidic device (microchip) was designed for "on-chip" staining and counting, and was manufactured in polydimethylsiloxane (PDMS) and glass using rapid prototyping and replica moulding techniques. In this microchip, water samples and fluorescent dyes are injected at separate inlets, so that bacterial cells in the samples are fluorescently stained during flow through a "mixing part". In the "detecting part," flowing cells are recorded as a movie using a CCD camera mounted on a microscope, and counted by image analysis software. The system can count total bacterial cells and target bacterial species with the same microfluidic device by changing the fluorescent dye, and results can be obtained within 1 h. In this work, this system was used to determine the total bacterial number in freshwater and groundwater using DAPI (4′,6-diamidino-2-phenylindole) staining, in the range of 10^4–10^6 cells/ml, and *Escherichia coli* O157:H7 spiked into pond water samples using fluorescent antibody staining. Once again, the authors are aware of the need to integrate a bacterial concentrating system prior to the analysis. The same authors (Yamaguchi, 2011), applied this rapid system for rapid counting of *L. pneumophil*a cells in cooling tower water with anti-*L. pneumophila* fluorescent antibody in a microchannel of the device. The number of *L. pneumophila* cells in the samples was determined in parallel by this microfluidic system and conventional fluorescence microscopy, showing a good correlation and with a detection limit of 10^4 cells/ml. In order to get lower LOD, a concentration step, which can also be done using microfluidics, has to be integrated.

The IMMUNOLEGIO research project, funded by The European Commission was conducted by small and medium enterprises and research centres from Hungary, France, Spain, Czech Republic, Ireland, Italy,

Belgium, Finland and UK. This project aimed to increase the effectiveness of *Legionella* detection technology through the development of an alternative method with high sensitivity and short detection time. The main objective of the project was to develop a magneto-resistive biosensor device employing magnetic particles as biological markers which allow for real time measurements obtaining a LOD of 50 cfu/L. It ensure that the Immunolegio device passes the strict threshold level regulation fixed at 100 cfu/L and enabling a complete analysis in a total test time of 30 minutes with a minimal enrichment stage. The proposed technique was applicable in both natural and man-made aquatic systems, provided a complete analysis within 30 minutes and allowed for repetitive measurements at different timing points. Moreover, it was automated and portable and had no requirements for specialised personnel or instrumentation. The developed prototype was evaluated at laboratory scale using real industrial samples and its assessment continued after the project completion (Bamford *et al.* 2009).

13.3.2 Microarray platforms using DNA

Based on DNA detection, Zhou and coworkers (Zhou *et al.* 2011) have developed an oligonucleotide based microarray system for water-borne pathogen detection in drinking water and other water samples, particularly for quick detection of *L. pneumophila.* The PCR-based DNA microarray uses the sequences of 16S–23S rDNA internal transcribed spacer regions (ITS) and the gyrase subunit B gene (gyrB) found in the most prevalent water-borne pathogenic agents. This new diagnostic tool contains 26 specific probes and can simultaneously detect, together with *Legionella pneumophila*, other microorganisms such as *Aeromonas hydrophila*, *Klebsiella pneumoniae*, *Pseudomonas aeruginosa*, *Salmonella* spp., *Shigella* spp., *Staphylococcus aureus*, *Vibrio cholerae*, *Vibrio parahaemolyticus*, *Yersinia enterocolitica* and *Leptospira interrogans*. For each group of bacteria, one to four specific capture probes were designed. Primers and probes specificity testing was carried out against a total of 218 bacterial strains, including 53 representative strains, 103 clinical isolates and 62 strains of other bacterial species belonging to 10 genera and 48 species. In addition, a conserved region of 16S rDNA was chosen to be a positive control probe. PCR amplification and purification of products is performed previous to chip hybridization. Using this methodology, 30 samples of bottled drinking water and 12 evaporative condensers were analyzed. The results demonstrated the specificity and reproducibility of the design with sensitivity of 0.1 ng DNA or 10^4 cfu/ml. In the case of *L. pneumophila*, results from evaporative condenser water samples were in agreement using this microarray and qPCR methods used as reference. The authors considered that this developed method is faster and less labour-intensive that the culture isolation method and antiserum agglutination.

Vitens Company and TNO's (Applied Scientific Research) Quality of Life division in collaboration with the Kennemerland Regional Public Health Laboratory in the Netherlands have developed the "*Legionella* Chip"®. It is a biochip based on DNA technology that uses biomarkers to detect the presence of *L. pneumophila*. The most relevant advantage of "*Legionella* Chip"® is that it can distinguish between pathogenic and non-pathogenic strains of *Legionella* in water systems. There are dozens of species of *Legionella*, only one of which, *L. pneumophila*, is responsible for most infections, although some of the other species are occasionally pathogenic. It is also known that not all isolates within the *L. pneumophila* species are equally dangerous. In the course of the research, an accurate comparison was made between the composition of the DNA of strains of *L. pneumophil*a which have been found to be pathogenic and that of strains which have not caused and are highly unlikely to cause illness (Yzerman *et al.* 2010). The basis of the study is formed by the strain collection from the Dutch *Legionella* surveillance program, including patient-derived strains from The Netherlands and environmental strains collected in an attempt to identify the source of infection for those patients. Comparative-genome analysis of 257 strains from this collection enabled to select five markers that proved the ability to distinguish pathogenic from non pathogenic strains. These pathogenicity markers, in addition to a selection of other diagnostic *Legionella* markers, are finally the biological content of the "*Legionella* Chip"®.

The features of this new diagnostic tool consist of the combination of unique DNA markers spotted on the high-tech chip surface embedded in a standard lab device called the "Eppendorf vial". After sample preparation, DNA extraction and labeling, the assay is run within a standard lab environment within 4 hours time. Therefore, the "*Legionella* Chip"® is able to determine within four to six hours whether a water sample contains *L. pneumophila* serogroups 1 or 2–14, and *L. anisa*, and whether the detected bacteria are a health risk. The results provide actionable management information aiming for the establishment of a healthy and safe environment and striving for significant cost reduction.

We devise a possible drawback of the currently developed "*Legionella* Chip"® that is its specificity for strains isolated from a particular geographic area. It is highly probable that the appropriate genetic markers will be different, for example, in Southern Europe. That limitation should be addressed over the coming years, and may force to develop and validate different versions of this detection system.

Simultaneous detection of waterborne pathogens using a PCR microfluidic platform

In a project under development in our laboratory in collaboration with the Sensors and Biosensors Group from Autonomous University of Barcelona in the frame of the project SOSTAQUA (CENIT; CEN20071039), we are working in the development of a microfluidic platform for microbial detection by PCR. The goal of this project is the development of a microfluidic system for the simultaneous detection by PCR of faecal pollution indicators (e.g. *Escherichia coli* and Enterococci) and water-transmitted pathogens (including *Clostridium perfringens*, *Vibrio cholerae*, *Salmonella* spp. and *L. pneumophila*, among others).

A modular system is proposed, which involves the separation of microfluidic and electronic platforms constructed in different materials. A disposable microfluidic platform to handle the sample and reagents and to run the reaction was designed to be fabricated using polymer substrates. However, the electronic platform which controls the temperature of the reaction is designed to be reusable and was fabricated with Low Temperature Co-fired Ceramics (LTCC) technology. This PCR platform, eventually in a portable device in which the manipulation steps are reduced to a minimum, can be complemented with sample concentration and DNA extraction modules, and will present a highly integrated solution for microbiological diagnostic laboratories.

Amplification system design

As a first step in the development of the PCR platform, a search for specific primers and probes for each of the target microorganisms was performed. The amplification designs (primer pairs and probes) fulfil specific requirements for qPCR (primer annealing temperature, amplified fragment length, etc.), so as to use the same thermal conditions allowing for the simultaneous detection of all the target microorganisms in the same run. As an added value to the developed PCR platform, all the designed systems include an internal amplification control (IAC), which helps in the interpretation of results, allowing to easily distinguishing between the absence of target and an inhibited reaction caused by impurities in the sample.

Biocompatibility studies and PCR microfluidic platform design

The substrate for the manufacturing of the microfluidic PCR platform must be carefully chosen in order to avoid possible problems, including adherence of reagents to the microchannels walls, and interference of the material with the PCR product detection system. Most PCR microchambers or microchannels were fabricated from silicon or glass substrate. However, these materials show serious interactions between the surface and the PCR ingredients. This is a common issue in the microfluidic systems, because the increased surface-to-volume ratio, which in many cases is an advantage, will be a challenge due to an increased probability of interaction between the biochemical components and the surrounding surfaces. The PCR inhibition effect due to binding of reaction components to the microchannel walls is a particularly important phenomenon in microsystems so that sometimes a surface treatment (e.g. passive coating) is needed (Christensen *et al.* 2007). For all these reasons, in this work, different polymers which have increasingly been utilized as alternative substrates, such as polydimethyloxane (PDMS), polycarbonate (PC), polymethylmethacrylate (PMMA), COC (cycloolefin copolymer) and Low Temperature Co-fired ceramics (LTCC) were evaluated as candidates for using them as substrates for microfluidic platform.

The biocompatibility studies consisted of different assays, such as testing the binding of reagents to the substrates, the inhibition of reaction after contact with the materials, the interferences with detection systems, and so on. The analyses for all the biocompatibility assays were performed by means of qPCR quantification of a Salmonella DNA control, using the specific primers and fluorescence probe designed in this work for this target detection. Performance of the exposed amplification mixtures was always evaluated in comparison with control mixtures, and including in every PCR run positive tests with DNA and PCR reagents not exposed to the materials. In all the assays, both the amplification kinetics and the obtained Ct value (threshold cycle; number of cycles required for the fluorescent signal to cross the threshold) from each sample were analyzed. Since the Ct value allows calculating the starting amount of target DNA, the comparison of Ct values from samples and controls permitted the detection of an anomalous amplification reaction due to a negative effect of the substrate assayed.

According to the obtained results, no DNA adhesion to the microfluidic platform was observed for the COC-based device. Moreover, PCR amplification and efficiencies for COC-exposed amplification mix were comparable to those obtained with untreated reagents, indicating that this material is PCR-compatible and suitable for the fabrication of the microfluidic platform (capillaries and chambers). Then, COC was selected as the substrate to manufacture low cost and disposable microfluidic platforms.

To obtain a disposable multiparametric microfluidic platform, in this work, a device based on centrifugal propulsion is proposed. This configuration is an elegant, simple and inexpensive way to handle liquids in microfluidic structures. The application of centrifugal propulsion avoids the use of pumps, allows integrating valves and separation steps and enables a high throughput of multiple tests by highly parallel and automated liquid handling. Fluid propulsion depends on fluid properties, centrifugal force (Fc), rotation rate and geometry and location of channels and reservoirs.

For precise liquid handling and dosage, a microfluidic structure based on capillary force valves was developed. These valves are situated at the centre of rotation and their geometry has been optimized in this work so that an adequate configuration of combined capillary valves and reservoirs permits the separation of multiple aliquots from a single total inserted volume as well as the evacuation of remaining sample to a waste reservoir. The current design includes eight reaction chambers with around 20 μL capacity, although it can be easily adapted to other needs. These multilayer microfludic devices were designed using CAD (Computer Aided Design) software, and rapid prototyping was achieved by means of a computer numerically controlled (CNC) machine (Protomat C100/HF, LPKF Laser and Electronics S.L., Madrid, Spain). To attain highly hermetic multilayer devices, thermal bonding was applied. Once mechanized, individual layers were aligned and laminated by applying pressure and temperature using a hydraulic hot press. Lamination consisted on applying 2–6 bar at a temperature slightly above the polymer glass transition temperature. Using this kind of configuration two different options are possible. First, eight different targets can be investigated simultaneously in a single sample, and second, with a different fluidic platform design, a single microorganism could be assayed in eight different samples.

Thermal platform

Adequate and consistent amplification product during the PCR reaction requires a precise control of temperature. The thermocycler designed in this work simultaneously provides three temperature zones according to the values associated to a qPCR process (94°C, 50°C and 72°C), by three sensor/actuator elements fabricated with LTCC (Low Temperature Cofired Ceramics) technology. Therefore, to perform a temperature cycle, the microfluidic platform should be automatically positioned onto the corresponding zone following the proper rotation sequence (Figure 13.1). These elements include a planar resistor and a highly precise temperature sensor both integrated in the same substrate. The multilayer approach of the LTCC technology permitted to integrate the sensor/actuator set in three independent substrates and lately inserts them in a polycarbonate structure that improves the heat isolation between zones. In this way, thermal interference between zones was avoided. In this case, mechanization was performed using a laser micromachining tool. Finally, a triple PID (Proportional + Integral + Derivative) control was implemented using a microcontroller in order to keep the three zones at the desired temperature.

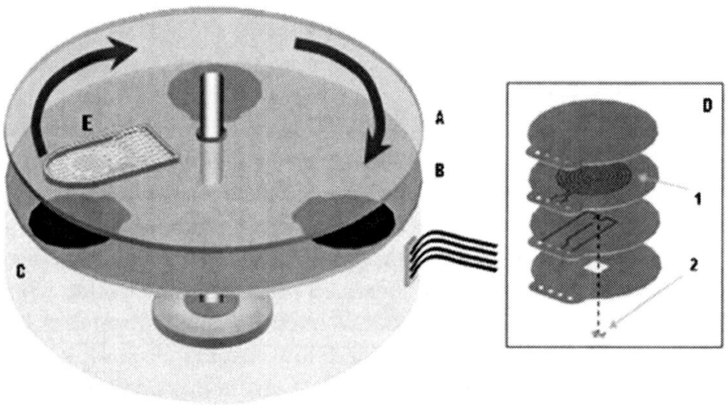

Figure 13.1 Schematic view of the PCR microfluidic platform system. A: rotating microfluidic platform. B: thermal platform with three defined temperatures. C: rotating platform. D: sensor/actuator pair for temperature control in the thermal platform; 1: resistor; 2: sensor. E: disposable microfluidic card for sample and reagents.

Freeze-dried PCR reagents

Considering the advantages that offering disposable microfluidic platforms with pre-charged reagents would provide, we evaluated the feasibility of PCR reagents lyophilization. The integration of lyophilized reagents inside the microfluidic PCR platform would provide the system with great versatility, avoiding pipetting reagents prior to PCR analysis. This would reduce the PCR preparation process complexity and would promote its attractive for later commercialization. For this purpose, different and simple microfluidic platform prototypes were fabricated for lyophilization studies.

The results of our assays showed a satisfactory lyophilization of reagents into the corresponding chambers. Once the reagents were freeze-dried, the device was sealed on top, and the rehydration of the lyophilized material into the microfluidic systems was performed by pipeting 15 uL of deionized water into the reservoir and applying centrifugal force to take it to the reaction chamber containing freeze-dried products. Subsequent PCRs assays using these reagents were satisfactory. In practice, the sample will be directly pipetted and used for hydrating the PCR reagents.

Preliminary results of PCRs performed in the microfluidic platform have been obtained by gel electrophoresis analysis after recovering the PCR mixtures from the reaction chambers, as the detection system is still under development. The detection system design is a major challenge in the project when the final goal is to build a portable device that can be used outside the laboratory, but ideally it should permit the fast detection of a low number of target copies in short time courses.

Waiting for the detection system design, we consider that the work depeloped in the frame of this project makes an important contribution to the development of microfluidic PCR systems for environmental diagnosis.

13.4 FUTURE RESEARCH DIRECTIONS

Although conventional methods for the detection and identification of *Legionella* (mainly culture isolation) are sensitive and offer both qualitative and quantitative information, they require several days to yield results. *Legionella* detection by qPCR is the alternative method of choice because its sensitivity and rapidity, but in the current format, it requires some training for analysis and results interpretation. Biosensors offer a good alternative, allowing rapid and "real-time" results which are essential for the detection of this bacterium.

New methods based on nucleic acid analysis and immunofluorescence techniques are being developed, and together with the evolving microtechnologies, will greatly contribute to the microbiological diagnostic field. Moreover, implementation of LOC systems, particularly for the analyses including concentration steps, will allow high throughput analysis. In the particular case of environmental samples, and mainly in the case of *Legionella*, where the expected concentration in the environment is low, a pre-concentration step is necessary. For monitoring drinking water supply, for example, the developed detection systems could be used after microfiltration and immunomagnetic separation. Problems associated with sample concentration, purification, and efficient recovery still remains, and the development of microfluidics and microelectromechanical Systems (MEMS) will help to overcome them, mainly providing a significant reduction in sample preparation time. Moreover, microfluidic devices will help to achieve a higher level of standardisation than current laboratory standard operating procedures, especially those based upon minute volumes of sample.

It is important to note that, in the future, instead of biosensors for detecting a single or few microorganisms, multiple-sensing instruments will be developed and implemented with the aim of detecting virtually anything of interest. This capability is expected to have a major impact on quality control of food and water, enabling fast release of batches and an increase in the product quality and safety with minimal investment. In the case of environmental samples, the simultaneous detection and quantification of *Legionella* and total heterotrophic counts could be very useful for the on-line control of cooling towers and tap water distribution systems. While many of the new molecular methods are addressed to direct detection of pathogens, regardless their infectivity capability, it could be also interesting to establish a correlation between pathogen concentration and the real health risk.

Finally, a good indicator of the growing interest in microfluidic platform technologies can be seen, apart from the number of published papers in the recent years, in the remarkable number of spin-off companies dedicated to commercialize LOC products. We expect to see very soon an increase in highly sensitive, fully automated and inexpensive biosensors for the rapid detection of microorganisms including *Legionella*, being routinely used both in the field and the laboratory. In the meantime, scientific community and particularly the applied market is waiting for regulations concerning these new methodologies, so that they become a real alternative to the conventional detection methods. Together with this, before this new analytical tools can be used in a routine way, they should meet the critical technical requirements and should be considered as standardized methods more than scientific approaches.

REFERENCES

Bamford F., De R., Dowling P., Gaboyard M., Gatti R., Krejci J., Matti V., Merret H., Muñoz F. and Clavell N. (2009). Rapid Biotechniques Based on Immunosensors for in situ Detection of *Legionella* in Industrial and Environmental Water Samples (IMMUNOLEGIO). http://www.ist-world.org (accessed 10 December 2011)

Christensen T. B., Pedersen C. M., Gröndahl K. G., Jensen T. G., Sekulovic A., Bang D. D. and Wolff A. (2007). PCR biocompatibility of lab-on-a-chip and MEMS materials. *J. Micromech. Microeng.*, **17**, 1527–1532.

Heath C. H., Grove D. I. and Looke D. F. M. (1996). Delay in appropriate therapy of *Legionella* pneumonia associated with increased mortality. *Eur. J. Clin. Microbiol. Infect. Dis.*, **15**, 286–290.

Hydrosense *Legionella* Field Kits. http://www.hydrosense.biz (accessed 10 December 2011)

International Organization Standardization. ISO 11731 (1998). Water quality-Detection and enumeration of *Legionella*.

International Organization Standardization. ISO 11731 (2004). Water quality-Detection and enumeration of *Legionella*. Part. 2: Direct membrane filtration method for waters with low bacterial counts.

Karsunke X. Y. Z., Niessner R. and Seidel M. (2009). Development of a multichannel flow-through chemiluminescence microarray chip for parallel calibration and detection of pathogenic bacteria. *Anal. Bioanal. Chem.*, **395**, 1623–1630.

Kfir R. and Genthe B. (1993). Advantages and disadvantages of the use of immunodetection techniques for the enumeration of microorganism and toxins in water. *Water Sci. Tech.*, **27**, 243–252.

Leonard P., Hearty S., Brennan J., Dunne L., Quinn J., Chakraborty T. and O'Kennedy R. (2003). Adavances in biosensors for detection of pathogens in food and water. *Enzyme Microb. Technol.*, **32**, 3–13.

Mairhofer J., Roppert K. and Ertl P. (2009). Microfluidic systems for pathogen sensing: a review. *Sensors*, **9**, 4804–4823.

NALCO an Ecolab Company FastPath™ *Legionella* Detection Test. http://www.nalco.com (accessed 10 December 2011)

Nocker A., Yáñez M. A., Soria E. and Catalán V. (2010). Waterborne Pathogens: *Legionella*. http://www.waterbornepathogens.org (accessed 18 December 2011)

Rodriguez G., Sultan F., Bendrina B. and Solis I. (2010). Fast detection of *Legionella pneumophila* in cooling towers by immunomagnetic microspheres. *Medlab Mag.*, **2**, 28–32.

Svarrer C. W., Lueck C. P., Elverdal P. L. and Uldum S. A. (2012). The immunochromatic kits Xpect® *Legionella* and BinaxNOW® *Legionella* for detection of *Legionella pneumophila* urinary antigen have low sensitivities for the diagnosis of Legionnaires' disease. *J. Med. Microbiol.*, **61**, 213–217.

Wolter W., Niessner R. and Seidel M. (2008). Detection of *Escherichia coli* O157:H7, *Salmonella typhimurium*, and *Legionella pneumophila* in water using a flow-through chemiluminescence. *Anal. Chem.*, **80**, 5854–5863.

Yamaguchi N. (2011). Rapid Enumeration of *Legionella pneumophila* in Cooling Tower by Using a Microfluidic System. ASM 2011. http://m.core-apps.com/TriStar-ASM2011/abstract/84dc19e6de1adf2c66b74fa54f6fd5c7 (accessed 12 December 2011)

Yamaguchi N., Torii M., Uebayashi Y. and Nasu M. (2011). Rapid, semiautomated quantification of bacterial cells in freshwater by using a microfluidic device for on-chip staining and counting. *Appl. Environ. Microbiol.*, **77**, 1536–1539.

Yáñez M. A., Carrasco-Serrano C., Barberá V. M. and Catalán V. (2005). Quantitative detection of *Legionella pneumophila* in water samples by immunomagnetic purification and real-time PCR amplification of the dotA gene. *Appl. Environ. Microbiol.*, **71**, 3433–3441.

Yzerman E., den Boer J., Caspers M., Almal A., Worzel B., Van der Meer W., Montijn R. and Schuren F. (2010). Comparative genome analysis of a large Dutch *Legionella pneumophila* strain collection identifies five markers highly correlated with clinical strains. *BMC Genomics*, **11**, 433.

Zhou G., Wen S., Liu Y., Li R., Zhong X., Lu F., Wang L. and Cao B. (2011). Development of a DNA microarray for detection and identification of *Legionella pneumophila* and ten other pathogens in drinking water. *Int. J. Food Microbiol.*, **145**, 293–300.

Chapter 14

Detection of virus in the water environment

Johan Nordgren and Lennart Svensson

14.1 INTRODUCTION

The water environment is a major route of transmission for many viruses causing disease in humans. Due to the development of new detection methodologies in the later years, the available information regarding waterborne viruses is steadily increasing. The analysis of virus in waters is valuable for several reasons: for evaluating the risk of virus infection from contaminated waters, to determine the efficiency of virus removal in water treatment processes, to assess drinking water quality, and for obtaining epidemiological information about circulating viruses (Bosch *et al.* 2008). In developing countries, unsafe water is recognized as a major obstacle in improving public health (Fewtrell *et al.* 2005), and numerous disease outbreaks directly or indirectly due to virus contamination in waters are reported world-wide every year (Maunula *et al.* 2005; Sinclair *et al.* 2009). Viruses are not monitored in water treatment plants routinely, nor in recreational waters or other water environments used directly or indirectly by humans. Commonly used bacterial and fecal indicators have proven unreliable in terms of virus contamination (Vivier *et al.* 2004; Ehlers *et al.* 2005; Serracca *et al.* 2010), leading to a frequent underestimation about the risk of virus infection. Thus, from a public health perspective, it is important to develop, standardize and implement sensitive methods for the detection of viruses in the water environment in order to reduce the spread of waterborne viral disease. Also, by measuring the presence of viruses in wastewater, epidemiological information about virus infections occurring in the community can be obtained. An essential component for the investigation of virus in the water environment is readily available detection techniques, which optimally should be fast, cost effective and applicable for several viruses and water types.

In this chapter, we will briefly introduce the most common human pathogenic and waterborne viruses, their different transmission routes in the water environment and their relevance to human health. We will then describe and review the most common concentration and detection methodologies used today.

14.1.1 The waterborne viruses

The viruses that have the greatest potential to spread and cause disease through the water environment are enteric viruses that infect humans via the fecal-oral route. These waterborne viruses often cause gastroenteritis in humans, but also hepatitis A and E causing liver disease are commonly transmitted through the water environment (Table 14.1). The enteric viruses (viruses that primarily infect and replicate in the gastrointestinal tract) are non-enveloped, robust and highly resistant to environmental and

Table 14.1 Waterborne viruses causing disease in humans.

Virus	Family	Genetic code	Size (nm)	Symptoms	Comments
Norovirus	*Caliciviridae*	+ssRNA	28–38	Gastroenteritis	The most important waterborne virus, causing the majority of waterborne virus outbreaks
Hepatitis A	*Picornaviridae*	+ssRNA	27–32	Hepatitis	The most common cause of hepatitis world-wide. One of the most important waterborne viruses
Hepatitis E	*Hepeviridae*	+ssRNA	27–34	Acute hepatitis	Endemic waterborne transmissions in many regions, can be fatal for pregnant women
Sapovirus	*Caliciviridae*	+ssRNA	28–35	Gastroenteritis	Generally causes milder infections as compared to norovirus, relatively rare in water related outbreaks
Enteric Adenovirus (40,41)	*Adenoviridae*	dsDNA	70–90	Gastroenteritis	Primarily infects young children, relatively rare in water related outbreaks. Suggested as an indicator for human viral contamination
Non-enteric Adenovirus	*Adenoviridae*	dsDNA	70–90	Respiratory disease, conjunctivitis, encephalitis, pneumonia	Usually causes waterborne outbreaks from recreational waters such as swimming pools
Astrovirus	*Astroviridae*	+ssRNA	28–30	Gastroenteritis	Primarily infection young children, many under 6 months. Relatively rare in water related outbreaks
Rotavirus	*Reoviridae*	dsRNA	70	Gastroenteritis	The most common cause of severe gastroenteritis in young children. Waterborne spread is not considered a high risk but often used for epidemiological investigations
Enterovirus	*Picornaviridae*	+ssRNA	27–30	Respiratory disease, fever, meningitis, paralysis, myocarditis	Usually causes waterborne outbreaks from recreational waters. Easy to propagate in invitro cultivation, thus much used for infectivity studies

disinfection factors, usually more so than bacteria (Duizer *et al.* 2004). Several studies have shown that the enteric viruses can remain infectious in environmental waters for long periods of time. For example, in a recent study, the authors were able to detect norovirus in a seeded water sample for over three years, and the virus remained infectious for at least 61 days (Seitz *et al.* 2011).

Norovirus, a member of the *Caliciviridae* family is by far the most important viral pathogen associated with waterborne transmission of gastroenteritis, and is often the focus of many studies of waterborne viruses (Maunula *et al.* 2005; Nordgren *et al.* 2009). Worldwide, it is the most common cause of acute gastroenteritis, causing disease in children and adults alike, with an estimation of 900,000 clinical visits in industrialized countries, and 200,000 deaths in children in developing countries annually (Patel *et al.* 2008). Waterborne outbreaks related to norovirus are very frequently occurring (Maunula *et al.* 2005; Sartorius *et al.* 2007; Nenonen *et al.* 2008; Maunula *et al.* 2009) (Table 14.2). Other waterborne viruses that cause gastroenteritis include sapovirus, enteritic adenovirus (serotypes 40 and 41), astrovirus and rotavirus. These viruses have been suggested or reported in waterborne outbreaks (Kukkula *et al.* 1997; Villena *et al.* 2003; Maunula *et al.* 2004; Nakagawa-Okamoto *et al.* 2009; Scarcella *et al.* 2009; Koroglu *et al.* 2011), but they occur far less frequently as compared to norovirus, and further mostly cause disease in children (Table 14.1 and 14.2).

Hepatitis A, which also spread via the fecal-oral route, is a common cause of hepatitis world-wide. The disease is self-limiting and rarely fatal, but can often incapacitate patients for several months. The hepatitis A virus has been associated with several waterborne outbreaks (Cao *et al.* 2009; Chobe *et al.* 2009), and might be considered together with norovirus and hepatitis E, the most important waterborne virus related to human health. Hepatitis E, although globally less frequent than hepatitis A, has a higher mortality rate, particularly in pregnant women. It is the most important cause of acute clinical hepatitis in adults throughout Asia, the Middle East and Africa, while it is rare in industrialized countries (FitzSimons *et al.* 2010). Hepatitis E is frequently transmitted via the water environment in developing countries where it is endemic (Guthmann *et al.* 2006; FitzSimons *et al.* 2010; Swain *et al.* 2010; Acharya *et al.* 2011), but it is also observed in the water environment of industrialized countries (Pina *et al.* 2000; Miyamura, 2011). For example, an outbreak of hepatitis E affecting ~79,000 individuals in Kanpur, India in 1991 was ascribed to contaminated drinking water (Naik *et al.* 1992).

Besides gastroenteritis and hepatitis, water borne viruses can cause many other types of diseases. Prominent among these are non-enteric adenovirus (which includes several serotypes) and enterovirus. Waterborne non-enteric adenovirus outbreaks causing conjunctivitis, respiratory disease, encephalitis and pneumonia are frequently implied in outbreaks from recreational water, such as swimming pools (Sinclair *et al.* 2009). The genus enterovirus, which includes many viruses such as poliovirus, coxsackievirus and echovirus, can cause a variety of symptoms, such as respiratory illnesses, paralysis, meningitis, myocarditis and fever. These viruses have been widely investigated in the water environment, partly due to the relative ease of propagating them in cell culture assays (Table 14.1). Waterborne outbreaks of enterovirus have been reported (Begier *et al.* 2008), but the risk of waterborne transmission is unclear. However, waterborne echovirus outbreaks seem to have increased lately (Sinclair *et al.* 2009). There are also several other potential waterborne viruses, such as coronavirus (Decaro *et al.* 2010) and aichi virus (Kitajima *et al.* 2011), although their prevalence in waterborne transmissions has not been extensively investigated (Table 14.2).

14.1.2 Transmission of virus in the water environment

A focus on bacterial and fecal indicators, as well as the lower availability of detection techniques for viruses has hampered the knowledge about transmission routes for waterborne viruses until recently. It is now

Table 14.2 Examples of waterborne virus outbreaks from different water sources.

Source of waterborne virus outbreaks	Virus	Suggested reason for outbreak	References
Drinking water	Norovirus	Sewage contamination, septic tank failure	(Hewitt *et al.* 2007)
Community tap water (surface water)	Norovirus	Sewage contamination, inadequate chlorination	(Kukkula *et al.* 1999)
Community tap water (ground water)	Norovirus	Unchlorinated water	(Kuusi *et al.* 2004)
Drinking water	Hepatitis E	Sewage contamination, inadequate chlorination	(Sarguna *et al.* 2007)
Drinking water	Hepatitis E	Interruption of water treatment plant	(Swain *et al.* 2010)
Well	Norovirus	Contamination from septic pit	(Beller *et al.* 1997)
Well	Norovirus	Overflow of sewage system	(Nygard *et al.* 2003)
Lake	Norovirus	Fecal contamination from nearby toilets	(Sartorius *et al.* 2007)
Swimming pool	Adenovirus	Inadequate chlorination	(Harley *et al.* 2001)
Swimming pool	Hepatitis A	Sewage contamination	(Mahoney *et al.* 1992)
Swimming pond	Enterovirus	Fecal contamination, inadequate disinfection	(Hauri *et al.* 2005)
Ocean	Enterovirus	Sewage contamination	(Begier *et al.* 2008)
Fountain	Norovirus	Fecal contamination, inadequate chlorination	(Hoebe *et al.* 2004)
Mollusks (oysters, mussels, clams)	Norovirus, hepatitis A, astrovirus, sapovirus, enterovirus, rotavirus	Effluent sewage contamination, heavy rains (treatment plant overflow), inadequate depurination	(Le Guyader *et al.* 2008; Polo *et al.* 2010; Halliday *et al.* 1991; Nenonen *et al.* 2008; Nakagawa-Okamoto *et al.* 2009; Nenonen *et al.* 2009)
Berries	Norovirus, hepatitis A	Irrigation with untreated water, insufficient heating	(Maunula *et al.* 2009; Niu *et al.* 1992; Hjertqvist *et al.* 2006)

recognized that waterborne viruses have many pathways to cause disease in humans (Maunula *et al.* 2005; Bosch *et al.* 2008). Infected individuals can excrete large amount of virus particles through feces (10^5–10^{12}/g feces) and vomit into sewage systems. High virus concentrations are frequently observed in both incoming and outgoing water from wastewater treatment plants (Nordgren *et al.* 2009). These viruses can subsequently reach rivers, oceans and lake systems, where they can cause disease outbreak directly through recreational activities (Figure 14.1). Viruses have also been frequently observed in drinking water sources related to disease outbreaks (Maunula *et al.* 2005). The drinking water outbreaks are often reported to occur through wastewater/septic tank leakages into groundwater, private wells or household water subsequently used as drinking water (Nygard *et al.* 2003; Maunula *et al.* 2005). Public water supply systems which have disinfected and treated surface or ground water are however generally safe with the exception of rare events, usually due to low or absent chlorination (Kukkula *et al.* 1999; Kuusi *et al.* 2004). Also, agricultural irrigation and/or fertilization are an important source of transmission for waterborne viruses (Figure 14.1). Irrigation of different food items using water that has been insufficiently treated (Katzenelson *et al.* 1976) can contaminate food products such as vegetables, and eventually lead to food-borne outbreaks (Cheong *et al.* 2009). For example, frozen imported raspberries have been a frequent source of norovirus outbreaks in many European countries (Ponka *et al.* 1999; Le Guyader *et al.* 2004; Falkenhorst *et al.* 2005), and frozen strawberries have led to outbreaks of hepatitis A (Niu *et al.* 1992). Outbreaks in relation to mollusk consumption are also well documented, particular oysters, clams and mussels (Le Guyader *et al.* 2008). The mollusks can filter large amount of water and are thus capable of accumulating large quantities of virus particles. This is of particular concern for mollusks that are cultivated in proximity of the outlet of wastewater treatment plants (Nenonen *et al.* 2008). Examples include an outbreak of hepatitis A and viral gastroenteritis after ingestion of raw clams affecting approximately 300,000 individuals in Shanghai (Halliday *et al.* 1991).

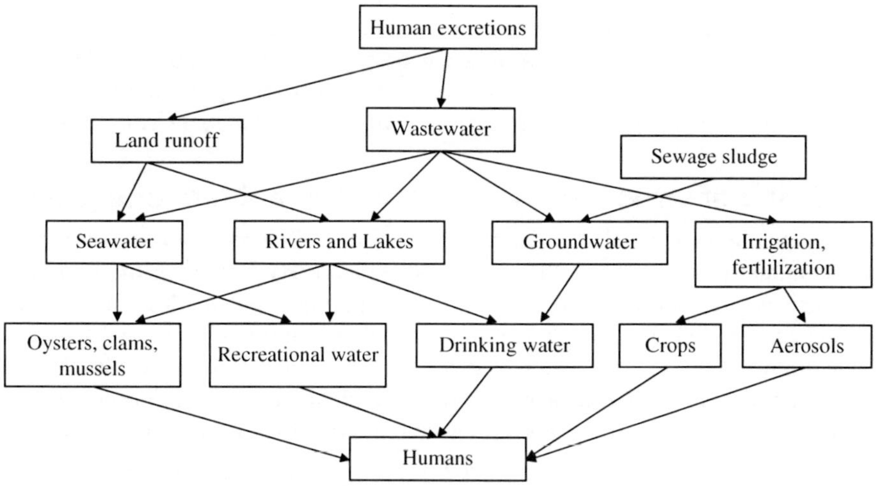

Figure 14.1 An overview of potential transmission routes for enteric viruses in the water environment. Enteric viruses can be discharged in extremely high amounts from feces or vomits from infected individuals. Through untreated/insufficiently treated wastewater the viruses can reach the oceans, lake systems and groundwater. Individuals can then subsequently be exposed directly through recreational activities or through other routes; such as mollusks from exposed cultivations, drinking water, and vegetables, berries or other crops irrigated or fertilized with contaminated water or sludge. Adapted from (Bosch, 1998).

14.1.3 Virus in wastewater treatment plants

Human enteric viruses are shed in large quantities by infected individuals, and will thus subsequently be found in high quantities in raw sewage (Lodder *et al.* 2005; Haramoto *et al.* 2006; da Silva *et al.* 2007; Nordgren *et al.* 2009). Many studies have tried to assess the reduction of viruses in different types of wastewater treatment plants. The reduction levels exhibit a large degree of variation. For norovirus, reduction has been observed to vary between 0 to 4 $\log_{10}$ units between different studies and treatment plants (Lodder *et al.* 2005; Haramoto *et al.* 2006; Nordgren *et al.* 2009). Moreover, the same studies often report a large variation in reduction in the same water treatment plant at different sampling times (da Silva *et al.* 2007; Nordgren *et al.* 2009). The measurement of virus reduction in treatment plants is inherently tricky. It is difficult to estimate the lag-time of water from incoming to outgoing water making a larger sample set necessary in order to accurately estimate the reduction. The detection techniques used today require a long analysis time, technical expertise and can usually not be performed at the treatment site, thus limiting the sample numbers. In a conventional municipal wastewater treatment system, using physical processes such as flocculation, sedimentation, activated sludge and trickling filters, approximately 50–99.9% of the incoming viruses are removed (Lodder *et al.* 2005; Haramoto *et al.* 2006; Nordgren *et al.* 2009), but the results vary significantly between studies. Owing to the high amount of virus in raw sewage this means that a high number of viruses are also present in the effluent water from wastewater treatment plants. Due to the low amount of infectious virus particles needed to cause disease in humans (for norovirus as low as 10–100 infectious virus particles (Teunis *et al.* 2008), and the fact that the enteric viruses are generally more resistant to environmental and disinfection factors as compared to bacteria, risk assessment become difficult. There is today no recognized viral indicator that is used for assessing the risk of viral contaminants in waters. Many studies have shown a low level of correlation between bacterial indicator and viruses (Hernandez-Morga *et al.* 2009) and viral contaminates have frequently been detected in waters where the bacterial and/or fecal indicators have indicated safe use (Vivier *et al.* 2004; Ehlers *et al.* 2005; Serracca *et al.* 2010). Many different viruses have been suggested as indicators (Roslev *et al.* 2011), but no general agreement has yet been reached. Human adenoviruses and polyomaviruses have been proposed as a potential candidates, due to their prevalence and stable concentrations throughout the year in different geographical locations, and that they are human specific (Girones *et al.* 2010; Serracca *et al.* 2010; Wyn-Jones *et al.* 2011).

14.1.4 Monitoring of virus in sewage water – an epidemiological tool

The sewage water, more than being a potential health risk, can also be regarded as a mirror reflecting what goes on in the community connected to the wastewater system. By following the virus prevalence and diversity in sewage waters during a period of time, important epidemiological information can be obtained. The data can be used as a complement for clinical epidemiology, especially if such data are not readily available from the community connected to the sewage plant. Previous studies have shown a correlation between viral prevalence in sewage waters, and to viral strains observed in symptomatic patients in hospitals and health care centers in the community (Iwai *et al.* 2009; Kremer *et al.* 2011). Moreover, this approach will give a more thorough understanding of viral circulation, as asymptomatic infections, or infections causing less severe symptoms, will not appear in clinical samples but will be present in sewage water (Iwai *et al.* 2009; Nordgren *et al.* 2009; Kremer *et al.* 2011). New and potential emerging strains could thus be detected before they become highly prevalent in clinical infections. Also, monitoring specific viruses in sewage waters can be used as a tool to evaluate vaccine efficacy, which is often performed as a follow up for the polio vaccine campaigns (Grabow *et al.* 1999). Two recent studies have further investigated the efficiency of rotavirus vaccination by studying the prevalence, diversity and levels of rotavirus in community and hospital wastewaters after the introduction of rotavirus vaccination into the community (Bucardo *et al.* 2011; Fumian *et al.* 2011).

14.2 CONCENTRATION OF VIRUS FROM WATER SAMPLES

Due to the low concentration of viral particles that is usually present in the water environment, especially natural waters, the techniques for detecting viruses usually first involve one or several concentration steps. This can involve sample volumes ranging from milliliters to thousands of liters (Table 14.3). An exception is wastewater samples, where virus concentrations are often, but not always, elevated enough to enable a more direct analysis. It is furthermore important that the concentration steps are also able to adequately reduce or minimize the presence of potential inhibitors that can hamper the efficacy of subsequent detection techniques. A wide variety of techniques described for the enrichment of viruses in water samples exist today (Wyn-Jones, 2007; Polaczyk *et al.* 2008). A well working concentration method should optimally comply with several criteria: it should be easy to perform, fast, cost effective, reproducible, useful for different viruses, have a high yield and a low end volume giving a high virus content in the concentrate (Bosch *et al.* 2008). No method today complies with all these criteria, and enrichment in several steps is often performed which vary widely based on the type of virus as well as the type of water that is analyzed. As a general rule, in order to sufficiently concentrate virus from surface, lake and ground waters or other relatively "clean" water environments, it is necessary to use large volumes (~1–1000 liters) since the virus quantity per volume water can be assumed low. In turbid waters such as raw sewage, a lower starting volume (~10–1000 ml) is often used in order to concentrate virus in the water sample (Table 14.3).

The techniques used today can broadly be divided into two main groups; ionic strength techniques and size exclusion techniques (Wyn-Jones, 2007). When using ionic strength techniques, the viruses are adsorbed to, and eluted from, membranes using electrostatic forces. In the size exclusion techniques, viruses are concentrated from water by size separation, generally using ultrafilters, relative to the virus particle size. Other techniques such as precipitation/flocculation and ultracentrifugation are also often used in separate or in combination with these two main strategies. Below, we will describe these techniques and list the advantages and disadvantages of each method.

14.2.1 Concentration based on ionic charge (electrostatic adsorption/elution)

The techniques based on ionic charge are generally performed for large volumes of waters, and often constitute a first concentration step which is subsequently followed by a second concentration using the smaller volume of concentrate. The basic principle is that the water sample is passed through a matrix, generally a microporous filter (often 47 mm in diameter for non-cartridge filters), to which the virus particles will adsorb under specific pH conditions (Michen *et al.* 2010). Different types of adsorption matrixes are used, but the most common are membrane filters, cartridge filters, or glass wool. The pore size of these matrixes is larger (0.45–3 μm) than the virus particles (20–100 nm); thus the need for an electrostatic force for virus adsorption. For a given pH interval which varies amongst virus particles [(having different isoelectric values (Michen *et al.* 2010)], the surface molecules will contain a charge, and will thus be attracted and adsorbed onto membranes that contain an opposite charge (Sobsey *et al.* 1979; Michen *et al.* 2010). The addition of different flocculants and/or multivalent cations to the water sample is common; which will lead to the formation of virus containing complexes which will be more readily adsorbed onto the matrix. The maximum water volumes that can be processed are related to the

Table 14.3 Commonly used concentration techniques for large volumes of water. The starting volumes are approximate and based on commonly used water volumes reported in the literature.

Principle	Methods	Starting volume (liters)	Advantages	Disadvantages
Electrostatic force	Adsorption/elution to positively charged membranes	1–150	High yield, can be used without precondition of water sample	Less efficient for turbid waters and seawater
	Adsorption/elution to negatively charged membranes	0.5–600	High yield	Require precondition of water sample. Can affect the viability of some viruses
	Adsorption/elution to glass wool	10–1500	High yield, cost-effective	Usually requires clean waters. Intra-assay variability
Size separation	Tangential/vortex flow ultrafiltration	2–100	High yield, can be used without precondition of water sample	Can be time consuming and costly. The retentate can contain inhibitors

Adapted from (Wyn-Jones, 2007; Bosch *et al.* 2008).

saturation of the membrane. Turbid waters are thus usually pre-filtered through membrane with larger pore sizes before being run through the virus adsorbing matrix. Elution is usually performed with a beef extract and/or amino acid (glycine, lysine) buffers at alkaline pH, displacing the virus from the adsorbing matrix (Wyn-Jones, 2007). The eluate is subsequently often neutralized using an acidic solution as not to affect the viability of viruses due the strong alkalinity of the elution buffer. The choice of adsorbing matrix, elution buffer, pH and other conditions is dependent on the type of virus as well as water sample.

Electrostatic adsorption/elution to electropositive membranes

The use of electropositive filters/cartridges is perhaps the most commonly used concentration technique today (Ma *et al.* 1994; Chapron *et al.* 2000; Karim *et al.* 2009; Steyer *et al.* 2011). Because most enteric viruses will be negatively charged at neutral pH, the electropositive membranes can often be used without preconditioning the water (Michen *et al.* 2010). Nylon filters (polyamide) which have a natural positive charge and 1MDS filters can be used for this effect (Gilgen *et al.* 1997; Steyer *et al.* 2011), but new types of filters steadily become available (Karim *et al.* 2009). An alkaline elution buffer is subsequently used for elution of the adsorbed virus particles. Electropositive filters can be easily saturated with turbid waters, and the presence of salt and alkalinity in seawaters can render low adsorption of viruses when concentrating these types of water samples (Lukasik *et al.* 2000; Katayama *et al.* 2002; Fong *et al.* 2005).

Electrostatic adsorption/elution to electronegative membranes

The use of electronegative filters/cartridges is also a very commonly used concentration technique for a large variety of viruses (Haramoto *et al.* 2004; Lodder *et al.* 2005; Rutjes *et al.* 2005; Haramoto *et al.* 2009). Electronegative filters, (HA membrane from Millipore is commonly used) usually give a higher virus yield from marine water and waters of high turbidity as compared to electropositive filters (Katayama *et al.* 2002; Fong *et al.* 2005). Since viruses and the filter materials are both negatively charged at neutral pH (Michen *et al.* 2010), the water sample must be conditioned to allow electrostatic adsorption of virus particles to the filter matrix. The water sample can be adjusted to an acidic pH whereas the surface charge of the virus particles will become positive and thus attracted to the electronegative filter. Alternatively, multivalent cations such as Al^{3+} or Mg^{2+} ions may be added to condition the virus particles into large positive charged complexes which will be attracted to the membrane (Katayama *et al.* 2002; Rutjes *et al.* 2005; Katayama *et al.* 2008). A further alternative is to apply a pre-charging step of the filter with cations ions such as Al^{3+}, creating an Al^{3+} coated filter which would attract the negatively charged viruses (Haramoto *et al.* 2009). In order to facilitate the elution and remove other positive ions, the filter can be pre-washed with an acidic solution, removing cations and other inhibitors while promoting direct attachment of the viruses to the membrane (Katayama *et al.* 2002), whereas the sample is subsequently eluted in an alkaline buffer, converting the virus charge and displacing it from the adsorbing matrix.

Adsorption/elution to glass wool

Treated glass wool, held together by a binding agent and coated with mineral oil, presents both hydrophobic and electropositive sites on its surface (Lambertini *et al.* 2008). When a virus suspension flows through the packed material, the fiber surface is able to attract and retain negatively charged virus particles at neutral pH (Wyn-Jones *et al.* 2001; Lambertini *et al.* 2008), thus often no precondition of the water sample with regards to pH is required. The approach is more cost effective as compared to the microporous filters and can give good recoveries, especially if used for large water volumes with low turbidity. The technique has been successfully used on surface, ground, drinking water as well as effluent wastewater (Gantzer *et al.* 1997; Vivier *et al.* 2004; Lambertini *et al.* 2008), but the technique has exhibited large intra-assay variability. To exemplify, a recent study used glass wool to recover entero, adeno and norovirus from tap and well water, using 20–1500 liters of water, with recovery rates varying between 8–98% (Lambertini *et al.* 2008).

14.2.2 Concentration based on particle size separation (ultrafiltration)

These techniques involve concentration of the virus by the virtue of virus size, using ultrafilters (pore size ~1–100 nm), rather than electrostatic force. An advantage with particle size separation techniques is that they do not require pre-conditioning of water samples regarding pH and ionic strength, and no elution step is needed, and can thus be applied on a wide variety of viruses, including those that are sensitive to pH changes, such as rotavirus (Estes *et al.* 1979). In systems where the fluid passes directly through the ultrafilter without recirculation (dead-end ultrafiltration), larger particles will quickly saturate the membranes, thus this technique is mostly used as a secondary concentration step from a lower volume of concentrate. When using larger volumes of water, ultrafiltration is usually performed with tangential or vortex flow (Table 14.3).

Tangential/vortex flow ultrafiltration

In tangential flow techniques, water is circulated in parallel to an ultrafilter maintaining a turbulent flow and recycling of the water which will prevent saturation of the membranes, thus a larger water volume can be applied (Rhodes *et al.* 2011). By applying a pressure, small particles such as ions and water will pass through the filter and large particles such as virus particles will be retained in concentrate. Vortex flow rotates a cylindrical ultrafilter inside a second cylinder and the sample is passing under pressure through the ultrafilter, whereas the virus will remain in concentrate. The vortices will keep the filter surface clean and prevent clogging (Paul *et al.* 1991). These systems can handle volumes from approximately one liter up to hundreds of liters. The larger systems will usually have to be followed by a secondary concentration step in order to further reduce the water volume (Jiang *et al.* 2001; La Rosa *et al.* 2007; Rhodes *et al.* 2011). These ultrafiltration systems can be expensive, and the time for processing the sample can be high, especially for turbid water. However, filtering of 100 liter of water in 2h has been reported (Polaczyk *et al.* 2008). These systems have been used with success for surface and tap waters samples in volumes between 20 and 100 liters of drinking water (Hernandez-Morga *et al.* 2009; Jiang *et al.* 2001; Polaczyk *et al.* 2008; Rhodes *et al.* 2011).

Dead-end ultrafiltration ("microconcentration")

This is a very common and efficient technique used as a secondary concentration step for small and relatively clean water volumes, often termed microconcentration in the literature, and has been used in a large number of studies (Kukkula *et al.* 1999; Haramoto *et al.* 2004; Liu *et al.* 2007). Generally volumes between 2 and 20 ml, can be used in a microconcentrator, such as Centricon/Amicon (Millipore) (Gilgen *et al.* 1997), although systems for larger volumes are available. The microconcentrator is basically a container with an ultrafilter having pore sizes smaller than the virus particles. The water sample is applied at the top of the container and during centrifugation, water, ions and other small particles with a molecular weight below the cut-off limit of the ultrafilter will pass through, and larger particles including virus particles will be retained in a concentrated form. There are several different microconcentrator models and a wide range of cut-off values to choose from (usually between 10–100 kDa). The advantages of microconcentration are principally that it is a fast and easy method, with no pre-conditioning of the water sample required. Thus, it can be applied on a wide range of viruses, including those sensitive to the pH modifications necessary in many adsorption/elution techniques. To prevent clogging of the ultrafilter which would reduce the concentration, it is important that the water sample is relatively clean.

14.2.3 Other concentration techniques

Besides the size exclusion and ionic strength techniques, there are a variety of other methods for concentration of virus in water samples, such as immunomagnetic capture (Tian *et al.* 2008), or lyophilisation (Villena *et al.* 2003). However, the two most commonly used techniques are ultracentrifugation and precipitation/flocculation, which has been used in several studies as a direct concentration method for virus in wastewater samples (van den Berg *et al.* 2005; Nordgren *et al.* 2009), or as a secondary concentration method (Jiang *et al.* 2001; Lodder *et al.* 2005) (Table 14.4).

Ultracentrifugation

Ultracentrifugation is a catch-all concentration method mostly used for wastewater samples where the virus content is high (Nordgren *et al.* 2009), or as a secondary concentration step (Jiang *et al.* 2001; Hundesa *et al.* 2010). Wastewater further contains a high particle content which could saturate the membranes described above in the adsorption/elution and size exclusion techniques. By applying high centrifugal force to a water sample, virus particles can be pelleted and resuspended into a small volume of buffer/water, thus concentrating the virus content. Usually, 100,000–200,000 × *g* for 1–3 hours is used in order to pellet the virus particles. The maximum sample volume is limited to the amount that can be used in an ultracentrifugation tube, thus limiting its concentration ability and use for natural waters. The initial volumes vary between 12–42 ml, and the pellet is subsequently resuspended into 100 μl to 1 ml of buffer/water. Usually, the wastewater samples are first centrifuged at low speed (3000–12000 × *g*) to remove larger particles and other inhibitors before the supernatant is used for ultracentrifugation (Nordgren *et al.* 2009). In order to prevent loss of virus aggregated to larger particles, an alternative is to perform ultracentrifugation on the entire water sample, resuspend the pellet in an alkaline buffer (usually glycine) and subsequently perform low speed centrifugation followed by ultracentrifugation of the supernatant (Puig *et al.* 1994; Pina *et al.* 2000).

Precipitation/flocculation

Precipitation and organic flocculation methods are often used as a secondary concentration method or for water samples having high particle content such as wastewater (Lodder *et al.* 2005; da Silva *et al.* 2007; Hernandez-Morga *et al.* 2009). In organic flocculation, buffered beef extract is used together with acidification of the sample (pH ~ 3.5). This will create isoelectric flocculation of proteins to which the virus can adsorb, and subsequently pelleted down with centrifugation and resuspended into smaller volumes (Katzenelson *et al.* 1976; Enriquez *et al.* 1995; Wyn-Jones, 2007; Michen *et al.* 2010). If the concentrated water sample have a high protein content (which is the case if beef extract has been used in the elution buffer of virus adsorbed to filters), it is possible to directly flocculate the virus, by adjusting the pH to acidic conditions. The polyethylene glycol (PEG) precipitation procedure consists of precipitating viral particles by addition of a PEG solution to the virus concentrate (Enriquez *et al.* 1995), or directly to wastewater (van den Berg *et al.* 2005; da Silva *et al.* 2007) followed by mixing, centrifugation or sedimentation, whereas the virus containing pellet/sedimentate is resuspended/collected (Enriquez *et al.* 1995; van den Berg *et al.* 2005). The use of beef extract and PEG in these procedures has been reported to cause inhibitory effects in subsequent molecular detection assays (PCR) (Schwab *et al.* 1995). One advantage with PEG precipitation as compared to organic

Table 14.4 Commonly used concentration methods using smaller volumes of water, often used as a secondary concentration step. The starting volumes are approximate and based on commonly used water volumes reported in the literature.

Principle	Methods	Starting Volume (ml)	Advantages	Disadvantages
Sedimentation	Ultracentrifugation	12–42	Easy and fast catch all technique. No preconditioning of water sample necessary	Limited concentration ability. Often used for wastewater samples
Size separation	Dead-end ultrafiltration (Microconcentration)	2–70	Fast and easy, no precondition necessary	Costly for larger volumes. Requires clean samples
Precipitation/flocculation	Organic flocculation, ammonium sulphate, PEG precipitation	40–1000	Inexpensive and efficient for dirty water samples such as wastewater	Beef extract and PEG can be inhibitory to RT-PCR reactions. Ammonium sulphate is cytotoxic

Adapted from (Wyn-Jones, 2007; Bosch *et al.* 2008).

flocculation is that no change of pH is necessary. An adaptation of the organic flocculation technique is the addition of ammonium sulphate, which enables virus to flocculate with beef extract at neutral pH (Shields *et al.* 1986).

14.3 DETECTION AND QUANTIFICATION METHODS

After initial virus concentration from water samples, numerous techniques exist for detecting viruses (Wyn-Jones, 2007; Hamza *et al.* 2011). Broadly, the detection methods used can be divided into two separate groups; cell culture assays and molecular assays, as well as combinations of both techniques are common (Figure 14.2). Concentrated samples can thus either be extracted for viral RNA/DNA analysis (molecular assays) or inoculated into cell lines where a specific virus can propagate (cell-culture assays) (Table 14.5) (Figure 14.2). An important issue regarding the detection of virus pathogens in the water environment is the infectivity of the viruses. Optimally, the detection methods should involve, or be coupled with, infectivity determining assays, since this would give a more accurate evaluation of potential health risks and/or efficiency of treatment. However, determining infectivity generally require the use of cell-culture, although other alternatives have been suggested such as detection of oxidative damage in the viral capsid protein (Hamza *et al.* 2011). The cell-culture assays are time-consuming and only applicable for a subset of waterborne viruses that can propagate in cell-culture, such as enterovirus and certain serotypes of adenovirus (Chonmaitree *et al.* 1988; Aslan *et al.* 2011; Okamoto, 2011). For norovirus, the most prominent waterborne virus related to human health, no readily available cell culture assays are available, although recent advances show some promise (Straub *et al.* 2007; Straub *et al.* 2011).

Molecular methods, specifically nucleic acid amplification techniques (PCR, real-time PCR), are fast and in contrast to cell-culture assays can be used to detect and quantify all known viruses (Mackay *et al.* 2002). These methods are more specific and usually more sensitive as compared to cell-culture assays and they can be used independently of whether cell culture assays exist for a particular virus. They are easy to develop and adjust to new viruses, and can thus readily be used for new emerging waterborne viruses (Table 14.5). The molecular assays, however, will only provide information about the number of virus particles present in the water sample (or more accurately, virus genes); and is not able to determine infectivity (Richards, 1999). A recent study, however, suggested that the viral copy number of enterovirus genes can be correlated to the amount of infectious viruses (Donia *et al.* 2010). Although molecular assays in principle are more sensitive than cell culture assays, a limitation is that only small sample volumes (μl range) can be used, as compared to cell-culture assays (ml range), which reduces the overall detection sensitivity.

14.3.1 Cell culture assays

Cell culture assays were the most widely used technique to determine the presence of viruses, especially enteroviruses, in environmental samples before the development of molecular methods such as PCR (Fong *et al.* 2005). It still remains the best technique to determine infectious virus particles in the water environment, but can only be applied on a subset of waterborne viruses that are able to propagate in cell-culture, such as enteroviruses and certain serotypes of adenovirus (Aslan *et al.* 2011). The techniques include plaque assay and liquid culture assays, where virus suspensions are inoculated onto susceptible cells on agar plates or in liquid media. The cells will then be evaluated visually in a light microscope for cytopathic effects (CPE) or plaques, and the amount of CPE/plaques will be counted and used to determine the amount of infectious viruses in the original water sample. Many viruses do not form plaques and thus need to be inoculated onto cells growing in liquid medium. Virus replication produces cell degeneration and the CPE can often be characteristic of the infecting virus type. There are also a number of other detection techniques using cell-culture, such as immunofluorescence and flow cytometry (Calgua *et al.* 2011; Hamza *et al.* 2011), but their use for virus in environmental waters is limited as of today.

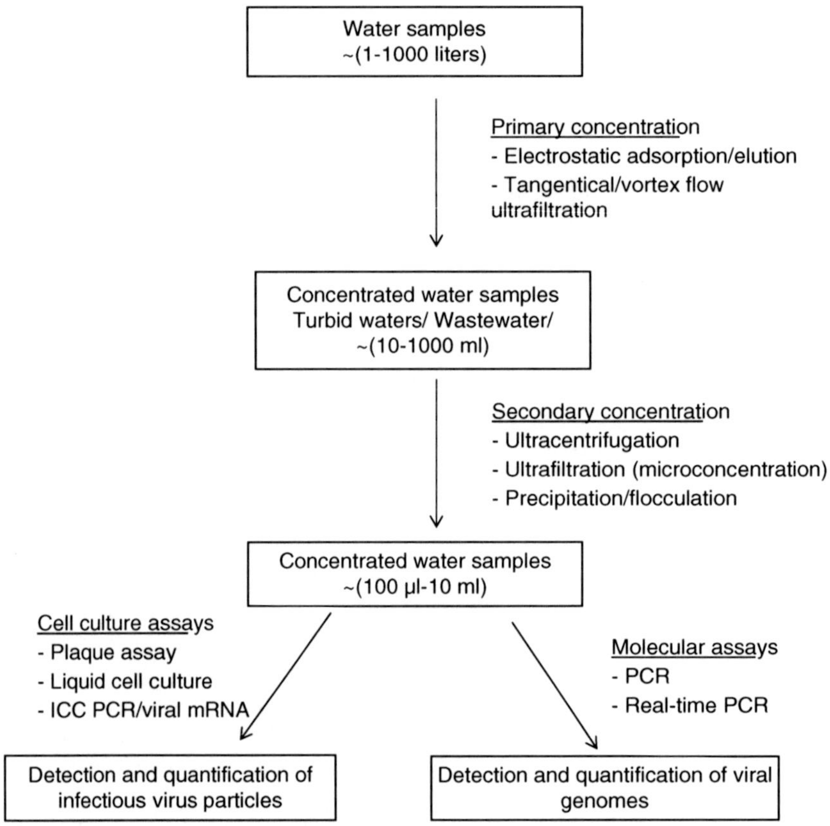

Figure 14.2 Overview of the concentration and detection methodology for virus in water samples. Water from ground, surface, river, lake, sea or drinking water sources are usually concentrated with large volumes (1–1000 liters) in a primary concentration step. Usually, this is followed by a second round of concentration to further decrease the final water volume and increase virus concentration. Virus in wastewater or other turbid waters are often concentrated directly using smaller volumes of water without a primary concentration step. From the concentrated water sample, viruses are subsequently detected and quantified by cell-culture or molecular assays.

For cell culture assays, it is important to consider that the choice of concentration methods from the initial water sample could influence the viability of viruses, thus leading to an underestimation of the amount of infectious particles. For example, pH-sensitive viruses can become less viable if the water sample was concentrated using the electrostatic adsorption/elution techniques described before. Also, the use of certain flocculants/precipitants such as ammonium sulfate, as well as compounds present in the environmental sample can be cytotoxic to the cells and thus could result in false positives when using the cell-culture assays.

Table 14.5 Advantages and disadvantages with the most common detection and quantification methods.

Detection assay	Advantage	Disadvantage
Molecular assays		
RT-PCR/PCR/real-time PCR	Fast (2–4hrs) Sensitive and specific Quantitative (esp. real-time PCR) Can be applied on all viruses	Do not measure infectivity Sensitive for inhibitors in water sample and/or compounds used in the concentration method (false negatives). Only small sample volumes can be used
Cell culture assays		
Plaque assays	Can quantitatively determine infectivity	Long processing time (days-weeks) Can only be used on viruses that replicate and form plaques in cell culture. May require multiple cell lines
Liquid cell culture	Can quantitatively determine infectivity Is applicable for a wider range of viruses as compared to plaque assays	Long processing time (days-weeks) Can only be applied on viruses that grow in cell-culture. Toxicity of water compounds and/or compounds used in the concentration method (false positives). May require multiple cell lines
Others/Combinations		
Integrated cell culture PCR/Detection of virus mRNA	Can quantitatively determine infectivity More sensitive and faster as compared to conventional cell culture Can detect viruses that do not produce CPE in cell culture	Less time-efficient and more costly as compared to direct PCR Can only be applied on viruses that propagate in cell culture

Adapted from (Fong *et al.* 2005).

14.3.2 Molecular assays (PCR and real-time PCR)

Nucleic acid amplification of viral genes is today the best method to detect and measure viral levels in the water environment (Girones *et al.* 2010). The most prominent methods used are reverse transcription (RT)-PCR, PCR and quantitative (real-time) PCR techniques, although other molecular methods exist such as nucleic acid sequence-based amplification (NASBA) and loop-mediated isothermal amplification (LAMP) (Lamhoujeb *et al.* 2008; Mattison *et al.* 2009; Mori *et al.* 2009). By amplifying the viral genomes exponentially, these detection methods become very sensitive. Real-time PCR is particularly useful since it enables a direct and convenient quantification of virus genes (Mackay *et al.* 2002; Nordgren *et al.* 2008; Nordgren *et al.* 2010). When applied on pure DNA/RNA samples free from inhibition, these assays can detect down to 5 viral genomes per PCR reaction (Nordgren *et al.* 2010). However, as stated before, the limitation of these assays is the inability to determine whether the viruses detected are infectious (Richards, 1999). Different water treatments, such as chlorination and UV irradiation can affect the viability of viruses, which the use of molecular techniques will not be able to determine since virus particles and/or virus genes can be present without being infectious. However, for many viruses, such as hepatitis A and norovirus, no good method to determine infectivity exists till today. Usually, having the concentrated water sample, the procedure for using molecular detection techniques would involve DNA/RNA extraction; reverse-transcription (for RNA viruses) and PCR/real-time PCR. It is important to consider the efficiency of virus recovery through each of these steps. Compounds and particles present in environmental waters can reduce the efficiency of DNA/RNA extraction, and inhibitors such as humic and fulvic acids, polysaccharides and metal ions present in the sample can hamper the efficiency of reverse transcription and/or PCR reactions considerably (Kermekchiev *et al.* 2009). Also, the use of beef extract and PEG in the elution buffer and/or flocculation/precipitation techniques for concentrating virus in water have been shown to be inhibitory for RT-PCR (Bosch *et al.* 2008). Several control mechanisms are frequently used, such as seeding the water samples with known amounts of internal control viruses or synthetic RNA, which are used to evaluate the percentage of virus recovery for each water sample as well as to evaluate the amount of inhibition in the RT and/or PCR reaction (Parshionikar *et al.* 2004; da Silva *et al.* 2007; Bosch *et al.* 2008). Dilution of the DNA/RNA (Nordgren *et al.* 2009), addition of inhibition-reducing substances into the PCR reaction (Kreader, 1996), or use of commercial kits (e.g. OneStep PCR Inhibitor Removal Kit, Zymo Research) can be used to reduce the amount PCR inhibition.

14.3.3 ICC-PCR and detection of viral mRNA

The integrated cell-culture PCR (ICC-PCR) is a combination approach that can be used to overcome some of the disadvantages of both the conventional cell culture and direct PCR assays. In the ICC-PCR assay, viruses are replicated in cell culture for a short period of time, followed by PCR detection (Mayer *et al.* 2010; Hamza *et al.* 2011). This approach significantly reduces the time necessary for cell culture assays, by being able to detect virus genes long before CPE/plaques are formed. Thus, using this approach, it is possible to assess virus viability in a shorter period of time, and also to quantify infectious viruses that do not produce CPE in cell-culture. (Rodriguez *et al.* 2009). Furthermore, this assay is less sensitive to environmental inhibitors as compared when using PCR directly. However, as for the cell-culture assays, this approach can only be used for viruses that can be propagated in cell-culture which limits its use. The use of ICC-PCR has been described for enteroviruses, hepatitis A, enteric adenovirus, and astrovirus (Rodriguez *et al.* 2009) .There are other alternative combination methods, such as detecting specific viral mRNA in cell culture. This approach has the same advantages and disadvantages as compared to ICC-PCR, but detecting viral mRNA in a cell-culture is a more solid verification of infectivity for a specific virus, since the presence of viral mRNA it is a clear indicator of virus replication within the cell (Rodriguez *et al.* 2009).

14.4 PERSPECTIVES

Technological improvements of detection methodologies have rapidly increased the knowledge of waterborne viruses. However there is still a scarcity of information, especially in developing countries, where the health implications of waterborne diseases are most acute. Many issues remain to be resolved. As of yet, there is no detection technique that complies with the criteria of low cost, fast-processing, and wide applicability. No international standard methods are validated for viruses in the water environment, and virus levels in waters are not routinely monitored. Many different methodologies are used, making results and risk assessments difficult to compare. There is yet no agreement of a reliable indicator for virus contamination, nor what levels of a particulate virus constitute an increased risk of contracting disease in a given water environment. The viruses further exhibit diverse seasonal pattern that varies from country to country, indicating that a multiplex system for detecting several viruses would be necessary and adapted to a particular geographical area. As for water treatment plants, reliable methods for evaluation of disinfections steps that can reduce or eliminate the virus contamination is needed. A simplification and standardization of methodology is required in order to alleviate these obstacles. By the use of standardized and widely applicable methods, it would further be possible to implement legislative measures and criteria regarding viruses for assessment of water quality, which would be greatly beneficial in reducing waterborne viral disease globally. The development of new as well as a thorough evaluation and improvement of current techniques will be essential in order to achieve this.

Acknowledgements

The authors are grateful to Fredrik Nyström and Magalí Martí Generó for valuable suggestions and critical reading of the chapter.

REFERENCES

Acharya S. K. and Panda S. K. (2011). Hepatitis E: water, water everywhere – now a global disease. *Journal of Hepatology*, **54**(1), 9–11.

Aslan A., Xagoraraki I., Simmons F. J., Rose J. B. and Dorevitch S. (2011). Occurrence of adenovirus and other enteric viruses in limited-contact freshwater recreational areas and bathing waters. *Journal of Applied Microbiology*, **111** (5), 1250–1261.

Begier E. M., Oberste M. S., Landry M. L., Brennan T., Mlynarski D., Mshar P. A., Frenette K., Rabatsky-Ehr T., Purviance K., Nepaul A., Nix W. A., Pallansch M. A., Ferguson D., Cartter M. L. and Hadler J. L. (2008). An outbreak of concurrent echovirus 30 and coxsackievirus A1 infections associated with sea swimming among a group of travelers to Mexico. *Clinical Infectious Diseases: An Official Publication of the Infectious Diseases Society of America*, **47**(5), 616–623.

Beller M., Ellis A., Lee S. H., Drebot M. A., Jenkerson S. A., Funk E., Sobsey M. D., Simmons O. D. 3rd Monroe S. S., Ando T., Noel J., Petric M., Middaugh J. P. and Spika J. S. (1997). Outbreak of viral gastroenteritis due to a contaminated well. International consequences. *JAMA: The Journal of the American Medical Association*, **278** (7), 563–568.

Bosch A. (1998). Human enteric viruses in the water environment: a minireview. *International Microbiology: The Official Journal of the Spanish Society for Microbiology*, **1**(3), 191–196.

Bosch A., Guix S., Sano D. and Pinto R. M. (2008). New tools for the study and direct surveillance of viral pathogens in water. *Current Opinion in Biotechnology*, **19**(3), 295–301.

Bucardo F., Lindgren P. E., Svensson L. and Nordgren J. (2011). Low Prevalence of rotavirus and high prevalence of norovirus in hospital and community wastewater after introduction of rotavirus vaccine in Nicaragua. *PLoS ONE*, **6**(10), e25962.

Calgua B., Barardi C. R., Bofill-Mas S., Rodriguez-Manzano J. and Girones R. (2011). Detection and quantitation of infectious human adenoviruses and JC polyomaviruses in water by immunofluorescence assay. *Journal of Virological Methods*, **171**(1), 1–7.

Cao J., Wang Y., Song H., Meng Q., Sheng L., Bian T., Mahemuti W., Yierhali A., Omata M. and Bi S. (2009). Hepatitis A outbreaks in China during 2006: application of molecular epidemiology. *Hepatology International*, **3**(2), 356–363.

Chapron C. D., Ballester N. A., Fontaine J. H., Frades C. N. and Margolin A. B. (2000). Detection of astroviruses, enteroviruses, and adenovirus types 40 and 41 in surface waters collected and evaluated by the information collection rule and an integrated cell culture-nested PCR procedure. *Applied and Environmental Microbiology*, **66**(6), 2520–2525.

Cheong S., Lee C., Song S. W., Choi W. C., Lee C. H. and Kim S. J. (2009). Enteric viruses in raw vegetables and groundwater used for irrigation in South Korea. *Applied and Environmental Microbiology*, **75**(24), 7745–7751.

Chobe L. P. and Arankalle V. A. (2009). Investigation of a hepatitis A outbreak from Shimla Himachal Pradesh. *The Indian Journal of Medical Research*, **130**(2), 179–184.

Chonmaitree T., Ford C., Sanders C. and Lucia H. L. (1988). Comparison of cell cultures for rapid isolation of enteroviruses. *Journal of Clinical Microbiology*, **26**(12), 2576–2580.

da Silva A. K., Le Saux J. C., Parnaudeau S., Pommepuy M., Elimelech M. and Le Guyader F. S. (2007). Evaluation of removal of noroviruses during wastewater treatment, using real-time reverse transcription-PCR: different behaviors of genogroups I and II. *Applied and Environmental Microbiology*, **73**(24), 7891–7897.

Decaro N., Cirone F., Mari V., Nava D., Tinelli A., Elia G., Di Sarno A., Martella V., Colaianni M. L., Aprea G., Tempesta M. and Buonavoglia C. (2010). Characterisation of bubaline coronavirus strains associated with gastroenteritis in water buffalo (Bubalus bubalis) calves. *Veterinary Microbiology*, **145**(3–4), 245–251.

Donia D., Bonanni E., Diaco L. and Divizia M. (2010). Statistical correlation between enterovirus genome copy numbers and infectious viral particles in wastewater samples. *Letters in Applied Microbiology*, **50**(2), 237–240.

Duizer E., Bijkerk P., Rockx B., De Groot A., Twisk F. and Koopmans M. (2004). Inactivation of caliciviruses. *Applied and Environmental Microbiology*, **70**(8), 4538–4543.

Ehlers M. M., Grabow W. O. and Pavlov D. N. (2005). Detection of enteroviruses in untreated and treated drinking water supplies in South Africa. *Water Research*, **39**(11), 2253–2258.

Enriquez C. E. and Gerba C. P. (1995). Concentration of enteric adenovirus 40 from tap, sea and waste water. *Water Research*, **29**(11), 2554–2560.

Estes M. K., Graham D. Y., Smith E. M. and Gerba C. P. (1979). Rotavirus stability and inactivation. *The Journal of General Virology*, **43**(2), 403–409.

Falkenhorst G., Krusell L., Lisby M., Madsen S. B., Bottiger B. and Molbak K. (2005). Imported frozen raspberries cause a series of norovirus outbreaks in Denmark, 2005. *Euro Surveillance: Bulletin Europeen sur les Maladies Transmissibles=European Communicable Disease Bulletin*, **10**(9), E0509222.

Fewtrell L., Kaufmann R. B., Kay D., Enanoria W., Haller L. and Colford J. M. Jr. (2005). Water, sanitation, and hygiene interventions to reduce diarrhoea in less developed countries: a systematic review and meta-analysis. *The Lancet Infectious Diseases*, **5**(1), 42–52.

FitzSimons D., Hendrickx G., Vorsters A. and Van Damme P. (2010). Hepatitis A and E: update on prevention and epidemiology. *Vaccine*, **28**(3), 583–588.

Fong T. T. and Lipp E. K. (2005). Enteric viruses of humans and animals in aquatic environments: health risks, detection, and potential water quality assessment tools. *Microbiology and Molecular Biology Reviews: MMBR*, **69**(2), 357–371.

Fumian T. M., Gagliardi Leite J. P., Rose T. L., Prado T. and Miagostovich M. P. (2011). One year environmental surveillance of rotavirus specie A (RVA) genotypes in circulation after the introduction of the Rotarix((R)) vaccine in Rio de Janeiro, Brazil. *Water Research*, **45**(17), 5755–5763.

Gantzer C., Senouci S., Maul A., Levi Y. and Schwartzbrod L. (1997). Enterovirus genomes in wastewater: concentration on glass wool and glass powder and detection by RT–PCR. *Journal of Virological Methods*, **65** (2), 265–271.

Gilgen M., Germann D., Luthy J. and Hubner P. (1997). Three-step isolation method for sensitive detection of enterovirus, rotavirus, hepatitis A virus, and small round structured viruses in water samples. *International Journal of Food Microbiology*, **37**(2–3), 189–199.

Girones R., Ferrus M. A., Alonso J. L., Rodriguez-Manzano J., Calgua B., Correa Ade A., Hundesa A., Carratala A. and Bofill-Mas S. (2010). Molecular detection of pathogens in water – the pros and cons of molecular techniques. *Water Research*, **44**(15), 4325–4339.

Grabow W. O., Botma K. L., de Villiers J. C., Clay C. G. and Erasmus B. (1999). Assessment of cell culture and polymerase chain reaction procedures for the detection of polioviruses in wastewater. *Bulletin of the World Health Organization*, **77**(12), 973–980.

Guthmann J. P., Klovstad H., Boccia D., Hamid N., Pinoges L., Nizou J. Y., Tatay M., Diaz F., Moren A., Grais R. F., Ciglenecki I., Nicand E. and Guerin P. J. (2006). A large outbreak of hepatitis E among a displaced population in Darfur, Sudan, 2004: the role of water treatment methods. *Clinical Infectious Diseases: An Official Publication of the Infectious Diseases Society of America*, **42**(12), 1685–1691.

Halliday M. L., Kang L. Y., Zhou T. K., Hu M. D., Pan Q. C., Fu T. Y., Huang Y. S. and Hu S. L. (1991). An epidemic of hepatitis A attributable to the ingestion of raw clams in Shanghai, China. *The Journal of Infectious Diseases*, **164** (5), 852–859.

Hamza I. A., Jurzik L., Uberla K. and Wilhelm M. (2011). Methods to detect infectious human enteric viruses in environmental water samples. *International Journal of Hygiene and Environmental Health*, **214**(6), 424–436.

Haramoto E., Katayama H., Oguma K., Yamashita H., Tajima A., Nakajima H. and Ohgaki S. (2006). Seasonal profiles of human noroviruses and indicator bacteria in a wastewater treatment plant in Tokyo, Japan. *Water Science and Technology*, **54**(11–12), 301–308.

Haramoto E., Katayama H. and Ohgaki S. (2004). Detection of noroviruses in tap water in Japan by means of a new method for concentrating enteric viruses in large volumes of freshwater. *Applied and Environmental Microbiology*, **70**(4), 2154–2160.

Haramoto E., Katayama H., Utagawa E. and Ohgaki S. (2009). Recovery of human norovirus from water by virus concentration methods. *Journal of Virological Methods*, **160**(1–2), 206–209.

Harley D., Harrower B., Lyon M. and Dick A. (2001). A primary school outbreak of pharyngoconjunctival fever caused by adenovirus type 3. *Communicable Diseases Intelligence*, **25**(1), 9–12.

Hauri A. M., Schimmelpfennig M., Walter–Domes M., Letz A., Diedrich S., Lopez-Pila J. and Schreier E. (2005). An outbreak of viral meningitis associated with a public swimming pond. *Epidemiology and Infection*, **133**(2), 291–298.

Hernandez-Morga J., Leon-Felix J., Peraza-Garay F., Gil-Salas B. G. and Chaidez C. (2009). Detection and characterization of hepatitis A virus and Norovirus in estuarine water samples using ultrafiltration – RT-PCR integrated methods. *Journal of Applied Microbiology*, **106**(5), 1579–1590.

Hewitt J., Bell D., Simmons G. C., Rivera-Aban M., Wolf S. and Greening G. E. (2007). Gastroenteritis outbreak caused by waterborne norovirus at a New Zealand ski resort. *Applied and Environmental Microbiology*, **73**(24), 7853–7857.

Hjertqvist M., Johansson A., Svensson N., Abom P. E., Magnusson C., Olsson M., Hedlund K. O. and Andersson Y. (2006). Four outbreaks of norovirus gastroenteritis after consuming raspberries, Sweden, June–August 2006. *Euro Surveillance: Bulletin Europeen sur les Maladies Transmissibles = European Communicable Disease Bulletin*, **11** (9), E060907 1.

Hoebe C. J., Vennema H., de Roda Husman A. M. and van Duynhoven Y. T. (2004). Norovirus outbreak among primary schoolchildren who had played in a recreational water fountain. *The Journal of Infectious Diseases*, **189**(4), 699–705.

Hundesa A., Bofill-Mas S., Maluquer de Motes C., Rodriguez-Manzano J., Bach A., Casas M. and Girones R. (2010). Development of a quantitative PCR assay for the quantitation of bovine polyomavirus as a microbial source-tracking tool. *Journal of Virological Methods*, **163**(2), 385–389.

Iwai M., Hasegawa S., Obara M., Nakamura K., Horimoto E., Takizawa T., Kurata T., Sogen S. and Shiraki K. (2009). Continuous presence of noroviruses and sapoviruses in raw sewage reflects infections among inhabitants of Toyama, Japan (2006 to 2008). *Applied and Environmental Microbiology*, **75**(5), 1264–1270.

Jiang S., Noble R. and Chu W. (2001). Human adenoviruses and coliphages in urban runoff-impacted coastal waters of Southern California. *Applied and Environmental Microbiology*, **67**(1), 179–184.

Karim M. R., Rhodes E. R., Brinkman N., Wymer L. and Fout G. S. (2009). New electropositive filter for concentrating enteroviruses and noroviruses from large volumes of water. *Applied and Environmental Microbiology*, **75**(8), 2393–2399.

Katayama H., Haramoto E., Oguma K., Yamashita H., Tajima A., Nakajima H. and Ohgaki S. (2008). One-year monthly quantitative survey of noroviruses, enteroviruses, and adenoviruses in wastewater collected from six plants in Japan. *Water Research*, **42**(6–7), 1441–1448.

Katayama H., Shimasaki A. and Ohgaki S. (2002). Development of a virus concentration method and its application to detection of enterovirus and norwalk virus from coastal seawater. *Applied and Environmental Microbiology*, **68**(3), 1033–1039.

Katzenelson E., Buium I. and Shuval H. I. (1976). Risk of communicable disease infection associated with wastewater irrigation in agricultural settlements. *Science*, **194**(4268), 944–946.

Katzenelson E., Fattal B. and Hostovesky T. (1976). Organic flocculation: an efficient second-step concentration method for the detection of viruses in tap water. *Applied and Environmental Microbiology*, **32**(4), 638–639.

Kermekchiev M. B., Kirilova L. I., Vail E. E. and Barnes W. M. (2009). Mutants of Taq DNA polymerase resistant to PCR inhibitors allow DNA amplification from whole blood and crude soil samples. *Nucleic Acids Research*, **37**(5), e40.

Kitajima M., Haramoto E., Phanuwan C. and Katayama H. (2011). Prevalence and genetic diversity of Aichi viruses in wastewater and river water in Japan. *Applied and Environmental Microbiology*, **77**(6), 2184–2187.

Koroglu M., Yakupogullari Y., Otlu B., Ozturk S., Ozden M., Ozer A., Sener K. and Durmaz R. (2011). A waterborne outbreak of epidemic diarrhea due to group A rotavirus in Malatya, Turkey. *The New Microbiologica*, **34**(1), 17–24.

Kreader C. A. (1996). Relief of amplification inhibition in PCR with bovine serum albumin or T4 gene 32 protein. *Applied and Environmental Microbiology*, **62**(3), 1102–1106.

Kremer J. R., Langlet J., Skraber S., Weicherding P., Weber B., Cauchie H. M., De Landtsheer S., Even J., Muller C. P., Hoffmann L. and Mossong J. (2011). Genetic diversity of noroviruses from outbreaks, sporadic cases and wastewater in Luxembourg 2008–2009. *Clinical Microbiology and Infection: The Official Publication of the European Society of Clinical Microbiology and Infectious Diseases*, **17**(8), 1173–1176.

Kukkula M., Arstila P., Klossner M. L., Maunula L., Bonsdorff C. H. and Jaatinen P. (1997). Waterborne outbreak of viral gastroenteritis. *Scandinavian Journal of Infectious Diseases*, **29**(4), 415–418.

Kukkula M., Maunula L., Silvennoinen E. and von Bonsdorff C. H. (1999). Outbreak of viral gastroenteritis due to drinking water contaminated by Norwalk–like viruses. *The Journal of Infectious Diseases*, **180**(6), 1771–1776.

Kuusi M., Nuorti J. P., Maunula L., Miettinen I., Pesonen H. and von Bonsdorff C. H. (2004). Internet use and epidemiologic investigation of gastroenteritis outbreak. *Emerging Infectious Diseases*, **10**(3), 447–450.

Lambertini E., Spencer S. K., Bertz P. D., Loge F. J., Kieke B. A. and Borchardt M. A. (2008). Concentration of enteroviruses, adenoviruses, and noroviruses from drinking water by use of glass wool filters. *Applied and Environmental Microbiology*, **74**(10), 2990–2996.

Lamhoujeb S., Fliss I., Ngazoa S. E. and Jean J. (2008). Evaluation of the persistence of infectious human noroviruses on food surfaces by using real-time nucleic acid sequence-based amplification. *Applied and Environmental Microbiology*, **74**(11), 3349–3355.

La Rosa G., Fontana S., Di Grazia A., Iaconelli M., Pourshaban M. and Muscillo M. (2007). Molecular identification and genetic analysis of Norovirus genogroups I and II in water environments: comparative analysis of different reverse transcription-PCR assays. *Applied and Environmental Microbiology*, **73**(13), 4152–4161.

Le Guyader F. S., Le Saux J. C., Ambert-Balay K., Krol J., Serais O., Parnaudeau S., Giraudon H., Delmas G., Pommepuy M., Pothier P. and Atmar R. L. (2008). Aichi virus, norovirus, astrovirus, enterovirus, and rotavirus involved in clinical cases from a French oyster-related gastroenteritis outbreak. *Journal of Clinical Microbiology*, **46**(12), 4011–4017.

Le Guyader F. S., Mittelholzer C., Haugarreau L., Hedlund K. O., Alsterlund R., Pommepuy M. and Svensson L. (2004). Detection of noroviruses in raspberries associated with a gastroenteritis outbreak. *International Journal of Food Microbiology*, **97**(2), 179–186.

Liu J., Wu Q. and Kou X. (2007). Development of a virus concentration method and its application for the detection of noroviruses in drinking water in China. *Journal of Microbiology*, **45**(1), 48–52.

Lodder W. J. and de Roda Husman A. M. (2005). Presence of noroviruses and other enteric viruses in sewage and surface waters in The Netherlands. *Applied and Environmental Microbiology*, **71**(3), 1453–1461.

Lukasik J., Scott T. M., Andryshak D. and Farrah S. R. (2000). Influence of salts on virus adsorption to microporous filters. *Applied and Environmental Microbiology*, **66**(7), 2914–2920.

Ma J. F., Naranjo J. and Gerba C. P. (1994). Evaluation of MK filters for recovery of enteroviruses from tap water. *Applied and Environmental Microbiology*, **60**(6), 1974–1977.

Mackay I. M., Arden K. E. and Nitsche A. (2002). Real-time PCR in virology. *Nucleic Acids Research*, **30**(6), 1292–1305.

Mahoney F. J., Farley T. A., Kelso K. Y., Wilson S. A., Horan J. M. and McFarland L. M. (1992). An outbreak of hepatitis A associated with swimming in a public pool. *The Journal of Infectious Diseases*, **165**(4), 613–618.

Mattison K. and Bidawid S. (2009). Analytical methods for food and environmental viruses. *Food and Environmental Virology*, **1**(3), 107–122.

Maunula L., Kalso S., Von Bonsdorff C. H. and Ponka A. (2004). Wading pool water contaminated with both noroviruses and astroviruses as the source of a gastroenteritis outbreak. *Epidemiology and Infection*, **132**(4), 737–743.

Maunula L., Miettinen I. T. and von Bonsdorff C. H. (2005). Norovirus outbreaks from drinking water. *Emerging Infectious Diseases*, **11**(11), 1716–1721.

Maunula L., Roivainen M., Keranen M., Makela S., Soderberg K., Summa M., von Bonsdorff C. H., Lappalainen M., Korhonen T., Kuusi M. and Niskanen T. (2009). Detection of human norovirus from frozen raspberries in a cluster of gastroenteritis outbreaks. *Euro Surveillance: Bulletin Europeen sur les Maladies Transmissibles = European Communicable Disease Bulletin*, **14**(49), 1–3.

Mayer B. K., Ryu H., Gerrity D. and Abbaszadegan M. (2010). Development and validation of an integrated cell culture-qRTPCR assay for simultaneous quantification of coxsackieviruses, echoviruses, and polioviruses in disinfection studies. *Water Science and Technology: A Journal of the International Association on Water Pollution Research*, **61**(2), 375–387.

Michen B. and Graule T. (2010). Isoelectric points of viruses. *Journal of Applied Microbiology*, **109**(2), 388–397.

Miyamura T. (2011). Hepatitis E virus infection in developed countries. *Virus Research*, **161**(1), 40–46.

Mori Y. and Notomi T. (2009). Loop-mediated isothermal amplification (LAMP): a rapid, accurate, and cost-effective diagnostic method for infectious diseases. *Journal of Infection and Chemotherapy: Official Journal of the Japan Society of Chemotherapy*, **15**(2), 62–69.

Naik S. R., Aggarwal R., Salunke P. N. and Mehrotra N. N. (1992). A large waterborne viral hepatitis E epidemic in Kanpur, India. *Bulletin of the World Health Organization*, **70**(5), 597–604.

Nakagawa-Okamoto R., Arita-Nishida T., Toda S., Kato H., Iwata H., Akiyama M., Nishio O., Kimura H., Noda M., Takeda N. and Oka T. (2009). Detection of multiple sapovirus genotypes and genogroups in oyster-associated outbreaks. *Japanese Journal of Infectious Diseases*, **62**(1), 63–66.

Nenonen N. P., Hannoun C., Horal P., Hernroth B. and Bergstrom T. (2008). Tracing of norovirus outbreak strains in mussels collected near sewage effluents. *Applied and Environmental Microbiology*, **74**(8), 2544–2549.

Nenonen N. P., Hannoun C., Olsson M. B. and Bergstrom T. (2009). Molecular analysis of an oyster-related norovirus outbreak. *Journal of Clinical Virology: The Official Publication of the Pan American Society for Clinical Virology*, **45**(2), 105–108.

Niu M. T., Polish L. B., Robertson B. H., Khanna B. K., Woodruff B. A., Shapiro C. N., Miller M. A., Smith J. D., Gedrose J. K. and Alter M. J. (1992). Multistate outbreak of hepatitis A associated with frozen strawberries. *The Journal of Infectious Diseases*, **166**(3), 518–524.

Nordgren J., Bucardo F., Dienus O., Svensson L. and Lindgren P. E. (2008). Novel light-upon-extension real-time PCR assays for detection and quantification of genogroup I and II noroviruses in clinical specimens. *Journal of Clinical Microbiology*, **46**(1), 164–170.

Nordgren J., Bucardo F., Svensson L. and Lindgren P. E. (2010). Novel light-upon-extension real-time PCR assay for simultaneous detection, quantification, and genogrouping of group A rotavirus. *Journal of Clinical Microbiology*, **48**(5), 1859–1865.

Nordgren J., Matussek A., Mattsson A., Svensson L. and Lindgren P. E. (2009). Prevalence of norovirus and factors influencing virus concentrations during one year in a full-scale wastewater treatment plant. *Water Research*, **43** (4), 1117–1125.

Nygard K., Torven M., Ancker C., Knauth S. B., Hedlund K. O., Giesecke J., Andersson Y. and Svensson L. (2003). Emerging genotype (GGIIb) of norovirus in drinking water, Sweden. *Emerging Infectious Diseases*, **9**(12), 1548–1552.

Okamoto H. (2011). Hepatitis E virus cell culture models. *Virus Research*, **161**(1), 65–77.

Parshionikar S. U., Cashdollar J. and Fout G. S. (2004). Development of homologous viral internal controls for use in RT-PCR assays of waterborne enteric viruses. *Journal of Virological Methods*, **121**(1), 39–48.

Patel M. M., Widdowson M. A., Glass R. I., Akazawa K., Vinje J. and Parashar U. D. (2008). Systematic literature review of role of noroviruses in sporadic gastroenteritis. *Emerging Infectious Diseases*, **14**(8), 1224–1231.

Paul J. H., Jiang S. C. and Rose J. B. (1991). Concentration of viruses and dissolved DNA from aquatic environments by vortex flow filtration. *Applied and Environmental Microbiology*, **57**(8), 2197–2204.

Pina S., Buti M., Cotrina M., Piella J. and Girones R. (2000). HEV identified in serum from humans with acute hepatitis and in sewage of animal origin in Spain. *Journal of Hepatology*, **33**(5), 826–833.

Polaczyk A. L., Narayanan J., Cromeans T. L., Hahn D., Roberts J. M., Amburgey J. E. and Hill V. R. (2008). Ultrafiltration–based techniques for rapid and simultaneous concentration of multiple microbe classes from 100-L tap water samples. *Journal of Microbiological Methods*, **73**(2), 92–99.

Polo D., Vilarino M. L., Manso C. F. and Romalde J. L. (2010). Imported mollusks and dissemination of human enteric viruses. *Emerging Infectious Diseases*, **16**(6), 1036–1038.

Ponka A., Maunula L., Von Bonsdorff C. H. and Lyytikainen O. (1999). Outbreak of calicivirus gastroenteritis associated with eating frozen raspberries. *Euro Surveillance: Bulletin Europeen sur les Maladies Transmissibles=European Communicable Disease Bulletin*, **4**(6), 66–69.

Puig M., Jofre J., Lucena F., Allard A., Wadell G. and Girones R. (1994). Detection of adenoviruses and enteroviruses in polluted waters by nested PCR amplification. *Applied and Environmental Microbiology*, **60**(8), 2963–2970.

Rhodes E. R., Hamilton D. W., See M. J. and Wymer L. (2011). Evaluation of hollow-fiber ultrafiltration primary concentration of pathogens and secondary concentration of viruses from water. *Journal of Virological Methods*, **176**(1–2), 38–45.

Richards G. P. (1999). Limitations of molecular biological techniques for assessing the virological safety of foods. *Journal of Food Protection*, **62**(6), 691–697.

Rodriguez R. A., Pepper I. L. and Gerba C. P. (2009). Application of PCR-based methods to assess the infectivity of enteric viruses in environmental samples. *Applied and Environmental Microbiology*, **75**(2), 297–307.

Roslev P. and Bukh A. S. (2011). State of the art molecular markers for fecal pollution source tracking in water. *Applied Microbiology and Biotechnology*, **89**(5), 1341–1355.

Rutjes S. A., Italiaander R., van den Berg H. H., Lodder W. J. and de Roda Husman A. M. (2005). Isolation and detection of enterovirus RNA from large–volume water samples by using the NucliSens miniMAG system and real-time nucleic acid sequence-based amplification. *Applied and Environmental Microbiology*, **71**(7), 3734–3740.

Sarguna P., Rao A. and Sudha Ramana K. N. (2007). Outbreak of acute viral hepatitis due to hepatitis E virus in Hyderabad. *Indian Journal of Medical Microbiology*, **25**(4), 378–382.

Sartorius B., Andersson Y., Velicko I., De Jong B., Lofdahl M., Hedlund K. O., Allestam G., Wangsell C., Bergstedt O., Horal P., Ulleryd P. and Soderstrom A. (2007). Outbreak of norovirus in Vastra Gotaland associated with recreational activities at two lakes during August 2004. *Scandinavian Journal of Infectious Diseases*, **39**(4), 323–331.

Scarcella C., Carasi S., Cadoria F., Macchi L., Pavan A., Salamana M., Alborali G. L., Losio M. M., Boni P., Lavazza A. and Seyler T. (2009). An outbreak of viral gastroenteritis linked to municipal water supply, Lombardy, Italy, June 2009. *Euro Surveillance: Bulletin Europeen sur les Maladies Transmissibles=European Communicable Disease Bulletin*, **14**(29), 1–3.

Schwab K. J., De Leon R. and Sobsey M. D. (1995). Concentration and purification of beef extract mock eluates from water samples for the detection of enteroviruses, hepatitis A virus, and Norwalk virus by reverse transcription–PCR. *Applied and Environmental Microbiology*, **61**(2), 531–537.

Seitz S. R., Leon J. S., Schwab K. J., Lyon G. M., Dowd M., McDaniels M., Abdulhafid G., Fernandez M. L., Lindesmith L. C., Baric R. and Moe C. L. (2011). Norovirus human infectivity and persistence in water. *Applied and Environmental Microbiology*, **77**(19), 6884–6888.

Serracca L., Verani M., Battistini R., Rossini I., Carducci A. and Ercolini C. (2010). Evaluation of Adenovirus and E. coli as indicators for human enteric viruses presence in mussels produced in La Spezia Gulf (Italy). *Letters in Applied Microbiology*, **50**(5), 462–467.

Shields P. A. and Farrah S. R. (1986). Concentration of viruses in beef extract by flocculation with ammonium sulfate. *Applied and Environmental Microbiology*, **51**(1), 211–213.

Sinclair R. G., Jones E. L. and Gerba C. P. (2009). Viruses in recreational water-borne disease outbreaks: a review. *Journal of Applied Microbiology*, **107**(6), 1769–1780.

Sobsey M. D. and Jones B. L. (1979). Concentration of poliovirus from tap water using positively charged microporous filters. *Applied and Environmental Microbiology*, **37**(3), 588–595.

Steyer A., Torkar K. G., Gutierrez-Aguirre I. and Poljsak-Prijatelj M. (2011). High prevalence of enteric viruses in untreated individual drinking water sources and surface water in Slovenia. *International Journal of Hygiene and Environmental Health*, **214**(5), 392–398.

Straub T. M., Bartholomew R. A., Valdez C. O., Valentine N. B., Dohnalkova A., Ozanich R. M., Bruckner-Lea C. J. and Call D. R. (2011). Human norovirus infection of caco-2 cells grown as a three-dimensional tissue structure. *Journal of Water and Health*, **9**(2), 225–240.

Straub T. M., Honer zu Bentrup K., Orosz-Coghlan P., Dohnalkova A., Mayer B. K., Bartholomew R. A., Valdez C. O., Bruckner-Lea C. J., Gerba C. P., Abbaszadegan M. and Nickerson C. A. (2007). *In vitro* cell culture infectivity assay for human noroviruses. *Emerging Infectious Diseases*, **13**(3), 396–403.

Swain S. K., Baral P., Hutin Y. J., Rao T. V., Murhekar M. and Gupte M. D. (2010). A hepatitis E outbreak caused by a temporary interruption in a municipal water treatment system, Baripada, Orissa, India, 2004. *Transactions of the Royal Society of Tropical Medicine and Hygiene*, **104**(1), 66–9.

Teunis P. F., Moe C. L., Liu P., Miller S. E., Lindesmith L., Baric R. S., Le Pendu J. and Calderon R. L. (2008). Norwalk virus: how infectious is it? *Journal of Medical Virology*, **80**(8), 1468–1476.

Tian P., Engelbrektson A. and Mandrell R. (2008). Two-log increase in sensitivity for detection of norovirus in complex samples by concentration with porcine gastric mucin conjugated to magnetic beads. *Applied and Environmental Microbiology*, **74**(14), 4271–4276.

van den Berg H., Lodder W., van der Poel W., Vennema H. and de Roda Husman A. M. (2005). Genetic diversity of noroviruses in raw and treated sewage water. *Research in Microbiology*, **156**(4), 532–540.

Villena C., El-Senousy W. M., Abad F. X., Pinto R. M. and Bosch A. (2003). Group A rotavirus in sewage samples from Barcelona and Cairo: emergence of unusual genotypes. *Applied and Environmental Microbiology*, **69**(7), 3919–3923.

Villena C., Gabrieli R., Pinto R. M., Guix S., Donia D., Buonomo E., Palombi L., Cenko F., Bino S., Bosch A. and Divizia M. (2003). A large infantile gastroenteritis outbreak in Albania caused by multiple emerging rotavirus genotypes. *Epidemiology and Infection*, **131**(3), 1105–1110.

Vivier J. C., Ehlers M. M. and Grabow W. O. (2004). Detection of enteroviruses in treated drinking water. *Water Research*, **38**(11), 2699–26705.

Wyn-Jones A. P., Carducci A., Cook N., D'Agostino M., Divizia M., Fleischer J., Gantzer C., Gawler A., Girones R., Holler C., de Roda Husman A. M., Kay D., Kozyra I., Lopez-Pila J., Muscillo M., Nascimento M. S., Papageorgiou G., Rutjes S., Sellwood J., Szewzyk R. and Wyer M. (2011). Surveillance of adenoviruses and noroviruses in European recreational waters. *Water Research*, **45**(3), 1025–1038.

Wyn-Jones A. P. and Sellwood J. (2001). Enteric viruses in the aquatic environment. *Journal of Applied Microbiology*, **91**(6), 945–962.

Wyn-Jones P. (2007). The detection of waterborne viruses. In: Human Viruses in Water, A. Bosch (ed.), Elsevier, Amsterdam, the Netherlands, pp. 177–204.

Chapter 15

Design of PCR primers for the detection of waterborne bacteria

Julien Gardès and Richard Christen

15.1 INTRODUCTION

Nucleic-acid-based technologies are progressively replacing culture methods for the identification of pathogenic micro-organisms, allowing a low detection limit, and even providing results within a few minutes (Belgrader *et al.* 1999). There are two main possibilities for the molecular identification of pathogens:

- A specific detection: the qualitative and/or quantitative identification of one or more pathogenic microorganism(s);
- A global detection: using one of the new deep-sequencing technology, for amplifying for example the ribosomal RNA (rRNA) genes using universal primers and then searching for sequences of any known or emerging pathogen.

The choice of the target gene(s) and PCR primers are the preliminary steps before setting up a detection protocol, since up to now amplification by PCR or cloning is necessary before sequencing (at least for a specific detection).

15.2 THE TARGET GENES

The choice of the genomic region chosen as target for identification should:

- Be specific, that is it should not be exactly found in two different species;
- Have a wide coverage, that is the primers selected for the PCR amplification step should amplify every known sequence of the gene.

As a result, a gene found only in the target species and under a strong evolutionary constraint could be the best possible target. However if none or only a single complete genome is known, the evolutionary constraint cannot be properly evaluated. Also if no closely related species has been sequenced, it is difficult to verify if a gene is specific of a single species. In order to simplify the detection, primers are often designed to amplify many species and detection is done at the sequence level, for example with a sequencing step. As a result, genetic markers used to identify species are often genes encoding either for proteins or rRNA genes. Finally, targeting RNA allows evaluating the state of activity of an organism and distinguishing alive and dead cells.

15.2.1 rRNA genes

The small subunit ribosomal RNA gene (16S–18S rRNA) is the most sequenced gene, with 2,851,421 sequences present in the public databases (DDBJ, GenBank and EMBL in June 2011). It is mandatory to provide this sequence for each publication of a new bacterial species, and it is almost always used for biodiversity studies of microbes (Hong *et al.* 2010; Ye & Zhang, 2011; Stecher *et al.* 2010; Hoffmann *et al.* 2009). As a result, any microorganism presenting an interest in health or in agriculture has been sequenced at least once. The advantage of the rRNA gene is also its composition of intermingled conserved and variable regions, allowing the use of amplifications by PCR and universal primers (Van de Peer *et al.* 1996; Mears *et al.* 2002). However, in some cases the phylogenetic signal it carries does not allow discrimination between related species. This problem is known for example for *Mycoplasma bovis* versus *Mycoplasma agalactiae* (Mattsson *et al.* 1994) and for *Bacillus anthracis/Bacillus cereus* versus *Bacillus thuringiensis* (Helgason *et al.* 2000). In this case, it is necessary to use housekeeping genes, which have a faster evolutionary rate.

15.2.2 Housekeeping genes

Several housekeeping genes are now known to satisfy the criteria described above (Lee and Lee, 2010), the most widely used are:

- *rpoB*, encoding the β subunit of bacterial RNA polymerase;
- *gyrB*, encoding the β subunit of DNA gyrase;
- *EF-Tu*, encoding the elongation factor Tu;
- *pgk*, encoding phosphoglycerate kinase;
- *dnaK*, encoding a heat shock protein.

Such housekeeping genes generally used in combination are now the state of the art for a molecular identification of prokaryotes (Gürtler *et al.* 2002). A dedicated database (http://pubmlst.org/) now hosts information for a precise identification using multilocus sequence typing (MLST). Sometimes however such analysis does not provide any or enough information about the pathogenicity of a bacterial strain. It is then necessary to directly target the pathogenicity genes.

15.2.3 Pathogenicity genes

Some bacterial species are known to be composed of commensal and pathogenic strains. This is for example the case of *Escherichia coli*. Using pathogenicity genes for the detection of an obligatory pathogen may also inform about the virulence and the toxicity of a given strain. Usually, genes encoding surface proteins are avoided, because they are constantly adapting in response to the immune system of the host, resulting in a high evolutionary rate.

15.2.4 Deep sequencing

When previous molecular techniques fail (for example when searching for an unknown of emergent pathogen, or in many cases for viruses), one may turn to a discovery tool that enables in-depth identification of individual pathogens, even in mixtures. Deep sequencing systems now provide fast and accurate detection of new agents, and enable high-quality *de novo* genome assembly. As the next generation sequencing technologies improve every year, they may rapidly become the best choice for a rapid characterization of both known and unknown pathogens (Rodriguez-Brito *et al.* 2010; Park *et al.* 2011; Simon & Daniel, 2011; Nakamura *et al.* 2009).

15.3 DESIGN OF PCR PRIMERS

The design of PCR primers is the first step in the elaboration of a molecular detection protocol. This step involves an *in silico* approach that aims to define and optimize the characteristics of the primers. In a multiplexing strategy (e.g. multiplex PCR), it is necessary to check the compatibility of the different couples [melting temperature (Tm), secondary structures and cross hybridizations]. In this section, we review the main steps of the design and most of the dedicated softwares freely available on the web.

15.3.1 The features of PCR primers

Thermodynamic features

Several thermodynamic properties are thought to determine good PCR primers, such as a size from 18 to 30 bp, a GC content from 40 to 60%, a melting temperature from 50 to 60°C and less than four repeats of the same nucleotide. In addition, secondary structures should be avoided (hairpins, intramolecular interactions and finally self or hetero dimerisations). Many softwares are now available to estimate the quality of known primers and even to design new primers.

Specificity and coverage

Sequences retrieved from the public databases are usually used to estimate these parameters. Two approaches allow retrieving every alternate sequence of a given gene: a search by sequence similarity or using annotations (search by keywords).

To perform an exhaustive search by keywords, every synonym used to describe a gene or a gene product should be used, since often there is no normalization. But the use of vague terms (e.g. gyrase to describe the DNA gyrase subunit a) or mis-annotations can result in a high number of false positives (e.g. *gyrb* instead of *gyra*). In addition, sequences without annotation or described with a rarely-used term can be missed. Thus the current state of annotation of a gene in the public databases does not allow an exhaustive and precise search by keywords. The most powerful tool for searching by keywords is ACNUC (Gouy & Delmotte, 2008). This database and its query system (Query_win) have several advantages compared to the popular Entrez of NCBI and SRS of EBI (Croce *et al.* 2006). It allows precise searches and extractions of sequences using combined keywords separated by spaces, the use of a text file containing a list of keywords, of sequences according to cellular location, and type of sequences (CDS, mRNA, rRNA…). Finally it automatically extracts subsequences from large genomic sequences. A keyword search is fast, easy to tune; but sequences poorly or not annotated are easily missed.

The most popular program to search sequences by similarity is BLAST. Among the different types of existing BLAST, tBLASTx is the most appropriate to perform an exhaustive search of protein coding sequences, as it searches a nucleotide sequence translated in all possible frames within the same translations of a database of DNA sequences (Altschul *et al.* 1990). tBLASTx thus disregards synonymous mutations which can saturate the similarity signal. The sensitivity of the search is hence increased while the selectivity is not diminished, since sequences of different genes are expected to be different at the amino acid level. However, the selection of the proper sequences from the BLAST results may be difficult for some gene families. Two scores inform on the similarity of a sequence: the percentage of similarity and the Expect value (e-value). The e-value represents the probability that a match would occur purely by chance and this value depends upon the size of the database. In addition, it is computed for each High-scoring Segment Pair (HSP) and can therefore be very misleading when only the e-value of the best HSP is used, as often done. An HSP consists of two sequence fragments of arbitrary length whose alignment is locally maximal and for which the alignment score exceeds a given threshold. As a result, a percentage of similarity cumulative on all non-overlapping HSPs should be preferred (as implemented for WU-Blast at EBI). BLAST [or other algorithms such as USEARCH (Edgar, 2010)] efficiently retrieves similar sequences, but it is difficult to identify a false positive, that is a sequence from a different but similar gene. Finally tuning every search parameter (open gap, extend gap, …) for a given gene is often difficult.

By combining these two approaches, it is possible to estimate the proportion of false positives or negatives of the two methods. Protein and gene names of sequences identified by the similarity method can for example be used to start a keyword approach. New sequences then found may be false negatives of the similarity method. Alternatively, it is possible to perform a keyword analysis from every annotation of the gene to verify that the results of the similarity method are exhaustive. Finally, a phylogenetic analysis of these sequences can be done to check the results of these searches.

15.3.2 The softwares for designing PCR primers

Many softwares are now available to help in designing primers. Year after year, more optimizations of more thermodynamic parameters have been included, as well as estimates of coverage and specificity. These softwares can be divided in four main groups:

- Design from a single target sequence;
- Design from an alignment of sequences;
- Design for specific tasks (e.g. SNP genotyping);
- Checking already-designed primers.

In this section, we will focus on freely available softwares (Table 15.1). Commercial softwares like Primer Premier (www.premierbiosoft.com) or Visual OMP™ (http://dnasoftware.com/VisualOMP/tabid/108/Default.aspx) are not reported, since they have never been used in our laboratory.

Primer design from a single target sequence

The most popular tool to design primers is perhaps Primer3 (Rozen and Skaletsky, 2000), available either stand-alone or as web servers. From one input sequence, Primer3 generates primers according to Tm, primer size, excluded region and product size. It also considers other factors such as GC content, 3′ stability, estimated secondary structure, dimer formation, DNA and salt concentrations. There are several other similar tools using one target sequence for designing PCR primers such as Primer3Plus (Untergasser *et al.* 2007), Pythia (Mann *et al.* 2009) or Gene2Oligo (Rouillard *et al.* 2004) (Table 15.1). The NCBI website now proposes Primer-BLAST (http://www.ncbi.nlm.nih.gov/tools/primer-blast/) which allows to check the specificity of newly designed primers, but does not take into account genetic variation present in a gene for a target group and does not recognize degenerated positions. In conclusion, the softwares cited above are not really relevant or easy to use when primers have to be designed in order to target every allele of a gene, and not a single sequence.

Primer design from a sequence alignment

Some programs propose to perform a BLAST analysis from one single input sequence to collect similar sequences (*e.g.* BatchPrimer3 (You *et al.* 2008)). Programs such as CODEHOP (Rose *et al.* 2003), PriFi (Fredslund & Lange, 2007), Primaclade (Gadberry *et al.* 2005) or PrimerHunter (Duitama *et al.* 2009)

Table 15.1 List of the main softwares available for the design of primers. Only freely available softwares are reported. "Speciality" describes if softwares can design primers for specific molecular methods and "Input" reports the several types of format compatible with each software. "Parameter settings" presents common parameters of softwares and the methods used to compute melting temperature (Tm). "Option" shows if a program supports degenerated positions (IUPAC code), checks specificity (BLAST of primers) and coverage (BLAST of input sequences) of primers, and allows using target sequences and non-target sequences.

		3PD	Amplicon	AutoDimer	Barcrawl	BatchPrimer3	BiSearch	ConservedPrimers	dCAPS Finder 2.0	dnaMATE	dPrimer	EC_oligos	Expeditor	FindGDPs	FractTM	Gene2Oligo	GeneFisher-P	GeneFisher2	GenoFrag	GenomePRIDE	iCODEHOP	Insilico.mutagenesis	LAVA	MELTING	MethPrimer	MFEPrimer	MPCRPs Designer	Mprime	MSPprimer	MultiPriDe	MuPlex	MutScreener	OHM	OligoCalc	OligoFaktory	OligoWiz	Optimus Primer
Type	*De novo* primer design	•	•	•	•	•	•	•	•			•	•	•		•	•	•	•	•	•	•	•		•		•	•	•	•	•	•			•	•	•
	Primer properties determination									•	•				•						•			•		•							•	•			
Speciality	Bisulfite-PCRs						•																		•												
	LAMP																						•														
	Long-Range PCRs	•										•							•	•																	
	Microarrays												•	•						•								•							•	•	
	Multiplex NGSs				•																								•								
	Multiplex PCRs																									•	•				•						
	NGSs				•	•														•								•	•			•					
	qPCRs																													•							
	Regular PCRs	•	•	•	•	•	•				•	•	•			•	•	•		•	•				•	•		•				•			•		
	RFLP	•																																			•
	SNP Genotyping / Mutagenesis					•		•	•				•									•									•	•					•
	SSR/Microsatellite Genotyping					•						•																									
Input	Ensembl "Exon View"												•																								
	ESTs or Transcript ID																																				
	GenBank ID																				•							•		•		•					•
	Gene names																											•	•								
	Multiple amino acid sequences																•	•			•																
	Multiple nucleic acid sequences		•	•	•	•		•		•	•	•					•	•	•				•			•	•	•			•	•	•		•	•	
	Reference SNP ID																																				
	Single amino acid sequence																•	•			•																
	Single nucleic acid sequence	•	•	•	•	•	•	•	•	•	•			•	•	•	•	•	•	•		•	•	•	•	•	•	•		•	•	•	•	•	•		•
Parameter settings	%GC	•	•	•		•	•	•					•	•			•	•	•							•	•	•	•		•	•		•	•		•
	Amplicon size	•	•			•	•	•				•	•	•			•	•	•				•		•		•	•	•		•	•			•		
	Primer size	•	•			•	•	•					•			•	•	•	•	•			•		•		•	•	•		•	•			•	•	•
	Secondary structure	•	•	•	•	•	•	•				•					•	•	•	•	•		•		•	•	•	•	•		•	•		•	•	•	•
	Substrate concentrations	•	•	•		•	•	•								•					•			•		•	•				•			•			•
	Basic method (Tm)		•							•																							•	•			
	Salt-adjusted method (Tm)		•							•											•						•						•	•			
	Nearest neighbor method (Tm)	•	•	•	•	•	•	•	•	•	•	•	•	•	•	•	•	•	•	•		•	•	•	•	•		•	•	•	•	•	•	•	•	•	•
Options	BLAST for input sequences					•	•	•					•				•				•								•		•	•					
	BLAST for primers																																				
	IUPAC code		•			•		•	•		•		•				•	•			•							•			•	•			•		
	Non-target sequences		•																•																		
	Region selection		•			•	•	•												•	•	•			•										•	•	

(*Continued*)

		Osprey	PCRTiler	PDA	PerlPrimer	PRIDE	PriFi	Prim-SNPing	Primer3	PrimeArray	PRIMEGENS-w3	Primer Prim'er	Primer-BLAST	PrimerHunter	PrimerIdent	Primers4clades	PrimerStation	PrimerZ	Primique	PRIMROSE	PRISE	PriSM	Pythia	QDD	QuantPrime	RExPrimer	RJPrimers	SciTools	SNP Cutter	SNPbox	SNPicker	ThermoPhyl	TROLL	UNAFold	UniPrime2	WASP
Type	**De novo** primer design	•	•	•	•	•	•	•	•	•	•	•	•	•	•	•	•	•	•	•	•	•	•	•	•	•	•	•	•	•	•	•	•		•	•
	Primer properties determination	•																								•		•						•		
Speciality	Bisulfite-PCRs				•																															
	LAMP																																			
	Long-Range PCRs										•												•													
	Microarrays	•								•																										
	Multiplex NGSs																																			
	Multiplex PCRs						•										•																			
	NGSs	•			•																	•		•		•		•								
	qPCRs				•																				•			•				•				
	Regular PCRs	•	•	•	•	•		•	•		•	•	•	•	•	•		•	•		•	•	•	•		•	•	•		•		•	•		•	
	RFLP							•																					•							
	SNP Genotyping / Mutagenesis							•																•		•			•	•	•					•
	SSR/Microsatellite Genotyping																													•			•			
Input	Ensembl "Exon View"																																			
	ESTs or Transcript ID									•																•										
	GenBank ID												•				•	•								•		•		•						
	Gene names																									•										•
	Multiple amino acid sequences																											•			•					
	Multiple nucleic acid sequences		•	•			•	•		•	•	•		•	•	•			•	•	•	•		•			•	•	•	•	•	•	•	•		•
	Reference SNP ID							•																		•			•	•						•
	Single amino acid sequence																											•			•					
	Single nucleic acid sequence	•	•	•	•	•	•	•	•	•	•	•	•	•	•	•		•	•	•	•	•	•				•	•	•	•	•	•	•	•	•	•
Parameter settings	%GC			•	•			•	•	•	•			•	•			•			•	•	•	•		•	•	•	•		•		•			•
	Amplicon size	•	•	•	•		•	•	•	•	•		•	•		•	•	•			•	•	•	•		•	•	•	•	•		•	•		•	•
	Primer size	•	•	•	•	•	•	•	•	•	•		•	•	•			•		•	•	•	•	•		•	•	•	•			•	•			•
	Secondary structure	•		•	•	•	•	•	•	•	•			•	•			•			•	•	•	•		•	•	•	•	•		•	•	•		•
	Substrate concentrations	•					•	•	•	•	•			•			•	•				•	•	•			•	•	•				•	•		
	Basic method (Tm)			•								•																•								
	Salt-adjusted method (Tm)											•																•								
	Nearest neighbor method (Tm)	•	•		•	•	•	•	•	•	•	•	•	•	•	•	•	•	•	•	•	•	•	•	•	•	•	•	•	•	•	•	•	•	•	•
Options	BLAST for input sequences																										•	•		•						•
	BLAST for primers		•					•					•															•								
	IUPAC code		•				•			•	•		•	•	•							•				•	•	•	•	•		•				•
	Non-target sequences													•					•		•							•				•				
	Region selection			•	•	•	•		•	•	•	•						•					•			•	•	•	•							

LAMP: loop-mediated isothermal amplification, NGS: Next-Generation Sequencing, qPCR: quantitative PCR, RFLP: Randow Fragment Length PCR, SNP: Single Nucleotide Polymorphism, SSR: Simple Sequence Repeat, EST: Expressed Sequence Tag.

can produce degenerated PCR primers, often required when gene sequences carry intrinsic variations like SNPs. With Amplicon (Jarman, 2004) or Primique (Fredslund & Lange, 2007), it is also possible to define a set of non-target sequences. This interesting option ensures a design of specific primers. Finally, the specificity of primers can be estimated by a BLAST analysis of primers, as it is proposed in PCRTiler (Gervais *et al.* 2010) or Prim-SNPing (Chang *et al.* 2009), but this can be difficult when a primer contains degenerated positions, which are not recognized by BLAST.

Primer design for specific tasks

There are many different types of molecular detection protocols, such as multiplex PCR, bisulfite-PCR, microarrays or SNP genotyping. Each method requires a particular design of primers, which respect to special features. For example, thermodynamic conditions of DNA hybridization are greatly modified during studies of DNA methylation by bisulfite-PCR. In next-generation sequencing, barcode and adaptor regions must be added to primers. Thus specific programs with complex pipelines have been proposed (see Table 15.1). Finally some softwares for the design of primer are focused on specific databases, such as 16S RNA gene [PRIMROSE (Ashelford *et al.* 2002)], eukaryote genomes [SOP3v2 (Ringquist *et al.* 2005)], or plant genomes [GeMprospector (Fredslund *et al.* 2006)].

Computing the features of primers

Some tools only check the thermodynamic properties of primers. NetPrimer (http://www.premierbiosoft.com/netprimer/index.html) or OligoCalc (Kibbe, 2007) can analyze one primer at a time, while dnaMATE (Panjkovich *et al.* 2005) or OHM (Croce *et al.* 2008) can assess a list of primers. OHM was specifically designed to compute Tm of primers against several target and non-target sequences. Its interesting feature is the ease of visualizing how primers bind to sequences, either as a picture (heat-map) or used with Treedyn (Chevenet *et al.* 2006) to annotate phylogenetic trees composed of target and non-target sequences. With a color code, the specificity and the sensitivity of each primer can be easily visually estimated. To our knowledge, only two softwares have the ability to assess the thermodynamic properties of degenerated primers: OligoAnalyzer (http://eu.idtdna.com/analyzer/Applications/OligoAnalyzer/) and dPrimer (Chen & Zhu, 1997).

In conclusion, one basic approach to get proper oligomers is a *de novo* design, using one of the numerous stand-alone softwares or web tools, which are best suited to the situation. These programs can deal with one or several target sequences and some are specialized for specific tasks like SNP genotyping or microarray. However, the most common practice when searching for primers is to first look for their description in scientific papers. To achieve this, one program has recently been made available. Oligomer-Extractor (http://patho-genes.org/Software/) is a stand-alone application able to find every oligomer present in a series of PDF files, and then select those present in a given set of DNA sequences.

15.4 DNA-BASED DETECTION TECHNOLOGIES

15.4.1 Specific detection

The advantages of a specific detection are that it is fast, reliable and easy to implement. If primers are extremely specific, the presence of a given pathogen can be simply deduced from the presence of an amplicon of the proper size, and quantification is obtained at the end of a quantitative PCR. For waterborne pathogenic bacteria, the most sequenced genes represent potential marker genes that are frequently used. A list of the ten most sequenced genes of the major water borne pathogenic bacteria is shown in Table 15.2, as well as a list of some published primers used to amplify these genes in Table 15.3.

Table 15.2 10 most sequenced genes found in major waterborne bacteria. Genes from transposons were not taken in account. Names of major waterborne bacteria are indicated in the first row. For each rank of most sequenced genes, gene name (Gene), protein name (Description) and number of entries found in the public databases (Sequences) are shown.

		Burkholderia pseudomallei	**Campylobacter coli**	**Campylobacter jejuni**	**Escherichia coli**	***Legionella* spp.**
1th	Gene	*hism*	*flaa/flab*	*flaa/flab*	*icd*	*rpob*
	Description	histidine transporter	flagellin a/b	flagellin a/b	isocitrate dehydrogenase	RNA polymerase subunit b
	Sequences	27	98	518	2360	687
2nd	Gene	*mucd*	*nadp*	*gyra*	*mdh*	*dota*
	Description	peptidase s1	alcohol dehygrogenase	DNA gyrase subunit a	malate dehydrogenase	organelle trafficking protein
	Sequences	27	85	173	1455	402
3rd	Gene	*flic*	*adk*	*pora*	*uida*	*mip*
	Description	phase 1 flagellin	adenylate kinase	major outer membrane protein	beta-D-glucuronidase	macrophage infect. potentiator
	Sequences	23	83	153	863	259
4th	Gene	*bipb*	*aroe*	*rplv*	*fimh*	*asd*
	Description	type III translocator	shikimate 5-dehydrogenase	ribosomal protein l22	type 1 fimbriae	aspartate dehydrogenase
	Sequences	21	83	145	835	177
5th	Gene	*cspd*	*carb*	*rpld*	*aspc*	*gyrb*
	Description	cold shock protein	carbomoyl-phosphate synthase	ribosomal protein l4	aspartate aminotransferase	DNA gyrase subunit b
	Sequences	21	83	143	821	156
6th	Gene	*sdhb*	*mdh*	*fspa2*	*lysp*	*flic*
	Description	succinate dehydrogenase b	malate dehydrogenase	flagella secreted protein	lysine transporter	phase 1 flagellin
	Sequences	21	83	109	808	139
7th	Gene	*bipc*	*pgi*	*cmer*	*fadd*	*omp28*
	Description	type iii secretion protein	glucose-6-P isomerase	transcriptional repressor	acyl-coA synthase	major outer membrane protein
	Sequences	20	83	104	804	135
8th	Gene	*amo*	*phes*	*htrb*	*clpx*	*pile3*
	Description	luciferase-like monooxygenase	phenylalanyl-tRNA synthetase	acyltransferase (lipid synthesis)	ATP-dependent clp protease	type IV pilin
	Sequences	18	83	86	715	129
9th	Gene	*irls*	*reca*	*cgta*	*eae*	*mlte*
	Description	sensor protein	recombinase a	acetylgalactosa minyltransferase	intimin	lytic murein transglycosylase
	Sequences	18	83	84	490	105
10th	Gene	*pena*	*trmu*	*glna*	*tar*	*mspa*
	Description	class a beta-lactamase	tRNA methyltransferase	glutamine synthetase type I	methyl-accepting chemotaxis	zinc metalloprotease
	Sequences	18	83	79	460	103

		Salmonella enterica	Salmonella typhi	Vibrio cholerae	Yersinia enterocolitica
1th	Gene	*flic*	*gyra*	*reca*	*lcrv*
	Description	phase 1 flagellin	DNA gyrase subunit a	recombinase protein a	v antigen
	Sequences	969	45	527	68
2nd	Gene	*fljb*	*gyrb*	*dnae*	*tufa*
	Description	phase 2 flagellin	DNA gyrase subunit b	DNA polymerase III sub. alpha	elongation factor tu a
	Sequences	424	29	386	58
3rd	Gene	*gyrb*	*parc*	*mdh*	*gyrb*
	Description	DNA gyrase subunit b	DNA topoisomerase subunit a	malate dehydrogenase	DNA gyrase subunit b
	Sequences	572	20	310	48
4th	Gene	*gyra*	*sty3695*	*ctxb*	*tufb*
	Description	DNA gyrase subunit a	DNA invertase	cholera toxin subunit b	elongation factor tu b
	Sequences	332	11	295	45
5th	Gene	*manb*	*dmsa*	*gyrb*	*cspb*
	Description	phosphomannomutase	dimethyl sulphoxide reductase	DNA gyrase subunit b	major cold shock protein
	Sequences	317	9	255	38
6th	Gene	*h1a*	*flic*	*icd*	*flic*
	Description	flagellin h1a	phase 1 flagellin	isocitrate dehydrogenase	phase 1 flagellin
	Sequences	311	9	210	35
7th	Gene	glna	sty4132	lap	ompf
	Description	glutamine synthetase	electron transport protein	leucine aminopeptidase	major outer membrane protein f
	Sequences	252	9	162	26
8th	Gene	*atpb*	*pduj*	*nagb*	*hsp60*
	Description	ATP synthase subunit beta	propanediol utilization protein	glucosamine-6-P deaminase	60 kDa heat shock protein
	Sequences	231	8	161	25
9th	Gene	*pduf*	*clya*	*pgm*	*glna*
	Description	propanediol utilization factor	cytolysin a	phophoglucomutase	glutamine synthetase
	Sequences	216	6	134	24
10th	Gene	*mdh*	*hisd*	*cat*	*ail*
	Description	malate dehydrogenase	histidinol dehydrogenase	catalase	attachment invasion protein
	Sequences	193	6	132	18

Table 15.3 Primers published for amplifying the 10 most sequenced genes found for major waterborne bacteria. Only primers that are specific and/or have a complete coverage are shown.

Species	Genes	Published Primers	Specificity	Coverage	References (PMID)
C. coli	*flaa/flab*	TTGCACAGCGTTACGTTGGCT	•		8098328
C. jejuni	*gyra*	AAACTGCTATATCTCCATGT	•		16048946
		AATTTCACTCATAGCCTCACG	•		17023497
		ACAGGACAAGGCAACTTTGG	•		17224260
		AGGCACACGCTTAATATAACCA	•		16339953
		AGTTGCCTTGTCCTGTAATA	•		12951353
		ATGCTCTTTGCAGTAACCAAAAAA	•		18086814
		ATTTTTAGCAAAGATTCTGAT	•		10488192
		CAAAGCATCATAAACTGC	•		21045366
		CATCGCAGCGGCACTATCAC	•		15673754
		CATGGAGATATAGCAGTTT	•		20224816
		CCACATGGAGATATAGCAGTTTATGATGC	•		11060054
		CCATAAATTATTCCACCTGT	•		10488192
		CCCTGTGCGATAAGCTTCTAT	•		17224260
		CGACTTACACGGCCGATTTC	•		18086814
		GCCTGACGCAAGAGATGGTT	•		15793099, 15673754
		GCTATGCAAAATGATGAGGC	•		8540709
		GCTCATGAGAAAGTTTACTC	•		11060054
		GGGTGCTGTTATAGGTCGTTATCA	•		20224816
		GGTGTGATTATGTTGGCAATTCAT	•		16339953
		GTTATTATAGGTCGTGCTTT	•		19041906
		TAAACTGCTATATCTCCA	•		14522100
		TAGAAGGTAAAACATCAGGTT	•		10488192
		TAGTGGGTGCTGTTATAGGTCGTTATC	•		15271386
		TCCGCGTTGTTATAGAGCT	•		9056011
		TGGGTGCTGTTATAGGTCGT	•		11060054
		TTATTATAGGTCGTGCTTTG	•		10488192
		TTTGCTTCAGTATAACGCATCGCAGC	•		11060054
		TTTTGCTTCAGTATAACG	•		15271386
		TTTTTAGCAAAGATTCTGAT	•		21045366, 10488192
	pora	AAGCACCTTCAAGTGTCC	•		15793099,15201231
		AGCTTCATGGCTGCAGAGC	•		10992471
		CAGCTATAGCAGCACTGAAG	•		10992471
		CTTCAGTGCTGCTATAGCTG	•		10992471
		GCTCTGCAGCCATGAAGCT	•		10992471
		TAATGCTGCTGATGGTGG	•		15793099, 15201231

	rplv	ATGAGTAAAGCATTAATTAAATTCATAAG		•	19014274, 16318913
	rpld	ATGAGTAAAGTAGTTGTTTTAAATGAT		•	16318913
	fspa2	AAGCCTCTAATGCGGGATTT	•		17517862
	cgta	CCTTTGAATCCTTGGGCTTT	•		20023103
		CGATGTGGATCATTACTACGATGC	•		11796612
		CGTTTCGGCGGTATTTTAAGGC	•		11796612
		GAATGTGGTTTTATCGATGGAG	•		20023103
		GGCCTATCATGCAAAAGAGC	•		20023103
		TATAAGCCACTCATCTTTTG	•		17631048
		TCCTTTATGCCCCTCCCCTA	•		17631048
		TGAAGAAAAACTTTGTTTTGTTGCT	•		20023103
		TGCTTATTCATCCCTTTTAG	•		17631048
		TTGAAAGAACCAGGAGATTG	•		17631048
L. spp.	*mip*	CTCGACAGTGACTGTATCCGATTT	•		18094136, 17878552
S. enterica	*flic*	AACGAAAGCGTAGCAGACAAG	•		15184437
		AAGTGACTTTTCCATCGGCTG	•		15184437
		AAGTTTCGCACTCTCGTTTTTGG	•		15184437
		ACTGCTAAAACCACTACT	•		16145086
		AGAAAGCGTATGATGTGAAA	•		18845003
		ATCAACGGTAACTTCATATTTG	•		18809549
S. enterica	*flic*	ATTCCGACTACGCAACCGAA	•		15969472
		ATTGTGGTTTTAGTTGCGCC	•		18845003
		CAACCTGGGCAATACCGTAAATAA	•		15969472
		CAAGTAATCAACACTAACAGT	•		15583311
		CAAGTAATCAACACTAACAGTC	•		9738029
		CACCGCCTGTTCTGAAGTTATGT	•		20110454
		CAGAAAGTTTCGCACTCTCG	•		10739340
		CAGCAGGCATTGTGGTCTTAG	•		20110454
		CATTAACATCCGTCGCGCTAG	•		18845003
		CCCGAAAGAAACTGCTGTAACCG	•		15184437
		CCGATACATCCAGTGTAGTA	•		11292720
		CCTGAACAGACAACTCACGCAC	•		20110454
		CCTGCTATTACTGGTGATC	•		18845003
		CGCCAGCCGCAAGGGTTACTGTAC	•		20110454
		CGTAACAGAGACAGCACGTTYTGYG	•		17208323
		CTGGTGATGACGGTAATGGT	•		10739340
		GAAAAGATCATGGCACAA	•		18834913
		GATCGGAATCAGAGTTAGTC	•		20110454
		GCACTGGCGTTACTCAATCTC	•		18845003

(*Continued*)

Table 15.3 Primers published for amplifying the 10 most sequenced genes found for major waterborne bacteria. Only primers that are specific and/or have a complete coverage are shown (*Continued*).

Species	Genes	Published Primers	Specificity	Coverage	References (PMID)
		GCATAGCCACCATCAATAACC	•		11825983
		GGCACAAGTAATCAACACTAACAGTCTGT	•		17208323
		GGCTAGTATTGTCCTTATCGG	•		19631095
		GGTACTACACTGGATGTATCGGG	•		19741087
		GGTGATCTGAAATCCAGCTTCAAG	•		15184437
		TAACGACGGCATTTCTATTG	•		20110454
		TAATGTAACCACTTATACTGATTC	•		20110454
		TACCGTCTACGCCACCAAGT	•		20110454
		TAGTGCTTAATGTAGCCGAAGG	•		11825983
		TCTGAAGTTGTTACTGCTAC	•		20110454
		TGACCAACTCAGCGCCATTA	•		16428399
		TGTTACTATTGGTGGCTTTACTGG	•		20110454
		TTAACGTAACAGAGACAGCAC	•		15583311, 9738029
		TTATCTGTATTAACCTCTTTAAGC	•		20110454
		TTTACCGTCTACGCCACCC	•		19741087
	gyra	CCAGGCAGCCGTTAATCACT	•		15215081
		CCATCAGTTCGTGGGCGATTTTCG	•		14576119
		CCGTACCGTCATAGTTATCC	•		12409384
		GAGACGGTGGATTTCGTGGAT	•		12019140
		GGTGCCATACACTGCGGAAT	•		11283069
		GGTGTCATACACTGCGAAAT	•		11283069
		GGTGTTATACACTGCGGAAT	•		11283069
		TACCGTCATAGTTATCCACGA	•		15145503
	mdh	TATCCAGCATAGCGTCCAGC	•		16081897
S. typhi	*gyra*	GAGACGGTGGATTTCGTGGAT		•	12019140
		TACCGTCATAGTTATCCACG		•	14576119
	gyrb	AAGCGCGATGGCAAAGAAG		•	21329318, 15070992
	clya	AAGTTTTGCTTATGGACAGC		•	19835951
		CCGCAGCAATAGAATAGGAAA		•	19835951
		GACCTTTGATGAAACCATAAAAGAG		•	18715828
		GCATCGATATCTTTATTCGCTTG		•	18715828
V. cholerae	*reca*	TGGACGAGAATAAACAGAAGGC	•		10024551, 15143042
	mdh	AAGCACTGCAAGGCGAATCT	•		10024551
		CCAGAACCACATCCGCACCTTC	•		21118541
		GTTTGACGGTCGGATACACC	•		10024551
		TGTATGATATTGCGCCTGTCACAC	•		21118541

	lap	CGGCAGCGTTGAGCGCCAAT	•		8890197
		GCATCATCATCCGCGCCGGG	•		8890197
V. cholerae	*cat*	CTAAGCCTTGTTCGTGCAGTTCAG	•		18790865
		GAAAATAGCCGTTACTCTGGTCAGC	•		18790865
Y. enterolitica	*lcrv*	CCATTAAGGACGACGAGTTG	•		10089162
		CTCATGTATATTTATGGTGCCAC	•		10089162
		GATTAGAGCCTACGAACAA		•	20438844
		TTGCACGAGGTTGGAGTTATAGCTGGGGC	•		10089162
	ompf	GGTTATGGTCAGTGGGAATAT	•		2626594, 17571800
	ail	ACCAAACTTATTACTGCCATAGAAGAAATCGTAT	•		18765735
		ACCGTGACCAAACTTATTACTGCCATAGAAG	•		18765735
		ACTCGATGATAACTGGGGAG	•		20646877, 21056771, 1401022, 9143098, 12788031
		ACTGGGGAGTAATAGGTTCG	•		18206260
		AGCTAGTTCTCTAATAGCCTG	•		17682042
		ATACAGATGACTTAACCTTTC	•		20646877
		ATGATAACTGGGGAGTAATAGGTTCG	•		18708521
		CAAGTAAGACGTCAATGGCATACGG	•	•	15784567
		CACTCGCAGCGTACACAT	•		1370953, 1688838
		CCCAGTAATCCATAAAGGCTAACATAT		•	18708521, 18206260
		CTATTGGTTATGCGCAAAGC	•		12270272
		CTTAACCTTTCCGTGAGCAGCAC	•		18765735
		GAACTCGATGATAACTGGG	•		1400967, 7811077
		GAATCGATACCCTGCACCAAGC	•		15784567
		GACATTACTAGCTAGTTCTC	•		15784567
		GGAGTATTCATATGAAGCGTC	•		11724866
		TAATGTGTACGCTGCGAG	•		12620874
		TCTATGGCAGTAATAAGTTTGGTCACGGTGATCT	•		18708521, 18206260
		TGACCAAACTTATTACTGCCATA	•	•	18708521, 18206260
		TGCTTATACCCATCAGGGATA	•		20646877
		TTAATGTGTACGCTGCGAGTG	•		11724866
		TTAATGTGTACGCTGCGAGTGAA	•		15784567
		TTTGGAAGTGGGTTGAATTGC	•		17682042

15.4.2 Global detection: the sequencing

For waterborne pathogenic bacteria, SSU (16S) rRNA gene sequences were retrieved using a list of 377 alternate keywords. Figure 15.1 shows the size distribution of these sequences, Table 15.4 shows the total number of sequences per organism.

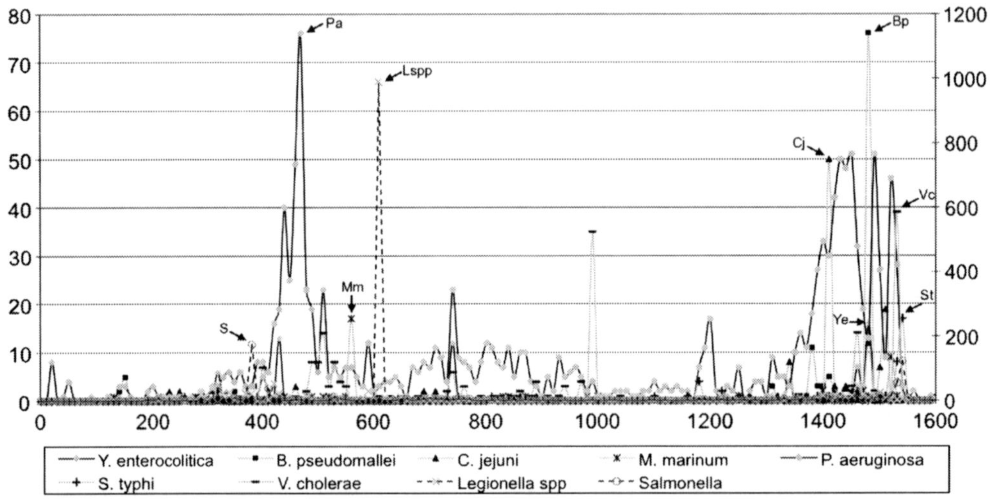

Figure 15.1 Number of 16S rRNA sequences for each class of waterborne pathogen. Axes used as described in Table 15.4: 1 for the left axis, 2 for the right axis.

Table 15.4 Number of 16S rRNA sequences per waterborne pathogen present in public databases.

Pathogen	Nbr seq	Axis
Burkholderia pseudomallei	123	1
Campylobacter jejuni	130	1
Legionella spp.	1711	2
ycobacterium marinum	48	1
Pseudomonas aeruginosa	1371	1
Salmonella typhi	43	1
Salmonella	723	2
Vibrio cholerae	194	1
Yersinia enterocolitica	105	1

Despite the fact that a specific detection of a given species is hard to achieve using 16S rRNA gene sequences as a target, it has sometimes been proposed. For example, Table 15.5 shows some of the primers that have been proposed for a specific detection of *Salmonella*.

Table 15.5 Analyses of primers extracted from 9 pdf files using Oligomer Extractor. In bold, primers found in more than 10,000 of the rRNA genes sequences contained in the Silva release 108 Reference sequences (which contain 726 *Salmonella* sequences); underlined, "universal" primers; in italics, *Salmonella* possible "specific" primers.

Primers	References (PMID)
AAGAGTTTGATCCTGGCTCAG	7542265
ACTCCTACGGGAGGCAGCAGT	15364475
AGAGTTTGATCATGGCTCAG	17289198, 12430773, 14572216
AGAGTTTGATCCTGGCTCAG	12430773, 14572216
AGTGTGGCTGGTCATCCTC	16487678, 16487678
ATTAGATACCCTGGTAGTCC	15364475
CACAAATCCATCTCTGGA	15364475
CATCGTTTACGGCGTGGACTACCA	15184165
CCAGCAGCCGCGGTAATACG	17083715
CCTGGCTCAGATTGAACGC	16487678
CGGGGAGGAAGGTGTTGTG	15184165
GAGCCCGGGGATTTCACATC	15184165
GGTTACCTTGTTACGACTT	7542265
GGTTACCTTGTTACGACTTC	15364475
GTGTGACGGGCGGTGTGTAC	17289198
TCCCGCATCTCTGCAGGA	17289198
TGCGGCTGGATCACCTCCTT	15582736
TGCTGCGGTTATTAACCAC	17289198
TGTTGTGGTTAATAACCGCA	15364475

Some of these primers are in fact not specific and often described as universal primers. In Trkov & Avgustin (2003), primers MINf: ACGGTAACAGGAAGMAG and MINr: TATTAACCACAACACCT have been described for the specific amplification of *Salmonella enterica*. Using the reference sequences of the Silva 108 release, it was in fact found that they would also amplify *Enterobacter sp.* (AM421983, GQ418089), *Haemophilus sp.* (FJ463822), *Citrobacter sp.* (FJ463782, EU644454, GU458277, GU458292), *Citrobacter koseri* (AF025366, AF025372, EF059858) and *Citrobacter farmeri* (DQ187383). They would indeed amplify 266 *S. enterica* rRNA sequences, but 275, 278 and 281 such sequences allowing respectively 1, 2 or 3 differences between primers and sequences. The forward primer was exactly present in 532 *Escherichia*, 88 *Shigella*, 85 *Enterobacter*, 37 *Citrobacter*, 18 *Cronobacter* and 10 *Enterobacteriaceae* sequences. The reverse primer was found in 218 *Enterobacter*, 132 *Cronobacter*, 64 *Pantoea* and 62 *Citrobacter* sequences. These data suggest that when using primers previously described as "specific", much care should be taken when analyzing the results, when the sequences of the amplicons are not used as further control.

15.5 CONCLUSIONS

In this chapter, we presented the different methods generally used for the detection of pathogens. Each section of this chapter shows the importance of performing *in silico* analyses before performing the real experiments. During detection of a pathogen, characterization of pathogenicity and toxicity levels may be important. Thus a well-studied virulence gene specific to the target species and with a moderate evolutionary rate will insure good specificity and coverage during the detection tests. For the global detection of pathogens, the current state of knowledge only allows the use of *16S rRNA* gene for the identification of bacterial pathogens, since it is the only gene allowing the use of "universal" primers.

Our recent analyses showed that two thirds of primers published as targeting the pathogenicity genes of *Vibrio cholerae* do not respect recommendations given above and cannot be used to detect every strain of this species. Moreover, the publication date and the number of citations of a primer are not factors allowing to estimate the quality of a published primer. Often one uses primers used in a recent work without performing even the simplest *in silico* analysis. For example, in an article describing the presence of *V. cholerae* in mussels following an outbreak in Denmark and Sweden (Collin & Rehnstam-Holm, 2011), *ctxA* genes of *V. cholerae* were not detected by PCR, while biochemical tests identified the presence of the gene product, due to an inappropriate reverse primer published in 1998 (Brasher *et al.* 1998).

A new website now groups information about published primers and gene sequences for major pathogenic bacteria (http://patho-genes.org). For each annotated gene, one has an easy access to every nucleic sequence, scientific article, published primer and link to other main public databases. Each published primer is analyzed, and its specificity, its coverage, its position within the gene and its melting temperatures are available. One can also check the coverage and the specificity of newly designed primers, if the target marker gene is an annotated coding sequence. It can be easily seen that no primer published for detecting *Bacillus anthracis* is a valid one. This is mainly due to a lack of specificity, with generally a binding to sequences of *Bacillus thuringiensis* or *Bacillus cereus. B. anthracis* is very difficult to distinguish from other species of its species group based on genotyping. Indeed, *B. anthracis*, *B. thuringiensis* and *B. cereus* are very closely related species and it has been shown that some *B. thuringiensis* and *B. cereus* strains carry virulence plasmids characteristic of *B. anthracis* (Helgason *et al.* 2000). Thus, even by using pathogenicity genes, it is sometimes difficult to perform a specific detection based on the genes presently used. The same situation might happen for *V. cholerae*, due to exchanges of genetic materials with closely related species such as *Vibrio mimicus* (Wang *et al.* 2011).

REFERENCES

Altschul S. F., Gish W., Miller W., Myers E. W. and Lipman D. J. (1990). Basic local alignment search tool. *J. Mol. Biol.*, **215**(3), 403–410.

Ashelford K. E., Weightman A. J. and Fry J. C. (2002). PRIMROSE: a computer program for generating and estimating the phylogenetic range of 16S rRNA oligonucleotide probes and primers in conjunction with the RDP-II database. *Nucleic Acids Res.*, **30**(15), 3481–3489.

Belgrader P., Benett W., Hadley D., Richards J., Stratton P., Mariella R. and Milanovich F. (1999). PCR detection of bacteria in seven minutes. *Science*, **284**(5413), 449–450.

Brasher C. W., DePaola A., Jones D. D. and Bej A. K. (1998). Detection of microbial pathogens in shellfish with multiplex PCR. *Curr. Microbiol.*, **37**(2), 101–107.

Chang H.-W., Chuang L.-Y., Cheng Y.-H., Hung Y.-C., Wen C.-H., Gu D.-L. and Yang C.-H. (2009). Prim-SNPing: a primer designer for cost-effective SNP genotyping. *BioTechniques*, **46**(6), 421–431.

Chen H. and Zhu G. (1997). Computer program for calculating the melting temperature of degenerate oligonucleotides used in PCR or hybridization. *BioTechniques*, **22**(6), 1158–1160.

Chevenet F., Brun C., Bañuls A.-L., Jacq B. and Christen R. (2006). TreeDyn: towards dynamic graphics and annotations for analyses of trees. *BMC Bioinformatics*, **7**, 439.

Collin B. and Rehnstam-Holm A.-S. (2011). Occurrence and potential pathogenesis of *Vibrio cholerae*, *Vibrio parahaemolyticus* and *Vibrio vulnificus* on the South Coast of Sweden. *FEMS Microbiol. Ecol.*, **78**(2), 306–313. doi: 10.1111/j.1574-6941.2011.01157.x. Epub 2011 Jul 18.

Croce O., Chevenet F. and Christen R. (2008). OligoHeatMap (OHM): an online tool to estimate and display hybridizations of oligonucleotides onto DNA sequences. *Nucleic Acids Res.*, **36**(Web Server issue), W154–W156.

Croce O., Lamarre M. and Christen R. (2006). Querying the public databases for sequences using complex keywords contained in the feature lines. *BMC Bioinformatics*, **7**, 45.

Duitama J., Kumar D. M., Hemphill E., Khan M., Mandoiu I. I. and Nelson C. E. (2009). PrimerHunter: a primer design tool for PCR-based virus subtype identification. *Nucleic Acids Res.*, **37**(8), 2483–2492.

Edgar R. C. (2010). Search and clustering orders of magnitude faster than BLAST. *Bioinformatics*, **26**(19), 2460–2461.

Fredslund J. and Lange M. (2007). Primique: automatic design of specific PCR primers for each sequence in a family. *BMC Bioinformatics*, **8**, 369.

Fredslund J., Madsen L. H., Hougaard B. K., Sandal N., Stougaard J., Bertioli D. and Schauser L. (2006). GeMprospector – online design of cross-species genetic marker candidates in legumes and grasses. *Nucleic Acids Res.*, **34**(Web Server issue), W670–W675.

Gadberry M. D., Malcomber S. T., Doust A. N. and Kellogg E. A. (2005). Primaclade – a flexible tool to find conserved PCR primers across multiple species. *Bioinformatics*, **21**(7), 1263–1264.

Gervais A. L., Marques M. and Gaudreau L. (2010). PCRTiler: automated design of tiled and specific PCR primer pairs. *Nucleic Acids Res.*, **38**(Web Server issue), W308–W312.

Gouy M. and Delmotte S. (2008). Remote access to ACNUC nucleotide and protein sequence databases at PBIL. *Biochimie*, **90**(4), 555–562.

Gürtler V., Barrie H. D. and Mayall B. C. (2002). Denaturing gradient gel electrophoretic multilocus sequence typing of *Staphylococcus aureus* isolates. *Electrophoresis*, **23**(19), 3310–3320.

Helgason E., Okstad O. A., Caugant D. A., Johansen H. A., Fouet A., Mock M., Hegna I. and Kolstø A. B. (2000). *Bacillus anthracis*, *Bacillus cereus*, and *Bacillus thuringiensis* – one species on the basis of genetic evidence. *Appl. Environ. Microbiol.*, **66**(6), 2627–2630.

Hoffmann C., Hill D. A., Minkah N., Kirn T., Troy A., Artis D. and Bushman F. (2009). Community-wide response of the gut microbiota to enteropathogenic *Citrobacter rodentium* infection revealed by deep sequencing. *Infect. Immun.*, **77**(10), 4668–4678.

Hong P.-Y., Hwang C., Ling F., Andersen G. L., LeChevallier M. W. and Liu W.-T. (2010). Pyrosequencing analysis of bacterial biofilm communities in water meters of a drinking water distribution system. *Appl. Environ. Microbiol.*, **76**(16), 5631–5635.

Jarman S. N. (2004). Amplicon: software for designing PCR primers on aligned DNA sequences. *Bioinformatics*, **20** (10), 1644–1645.

Kibbe W. A. (2007). OligoCalc: an online oligonucleotide properties calculator. *Nucleic Acids Res.*, **35**(Web Server issue), W43–W46.

Lee C. S. and Lee J. (2010). Evaluation of new *gyrB*-based real-time PCR system for the detection of *B. fragilis* as an indicator of human-specific fecal contamination. *J. Microbiol. Methods*, **82**(3), 311–318.

Mann T., Humbert R., Dorschner M., Stamatoyannopoulos J. and Noble W. S. (2009). A thermodynamic approach to PCR primer design. *Nucleic Acids Res.*, **37**(13), e95.

Mattsson J. G., Guss B. and Johansson K. E. (1994). The phylogeny of *Mycoplasma bovis* as determined by sequence analysis of the 16S rRNA gene. *FEMS Microbiol. Lett.*, **115**(2–3), 325–328.

Mears J. A., Cannone J. J., Stagg S. M., Gutell R. R., Agrawal R. K. and Harvey S. C. (2002). Modeling a minimal ribosome based on comparative sequence analysis. *J. Mol. Biol.*, **321**(2), 215–234.

Nakamura S., Yang C.-S., Sakon N., Ueda M., Tougan T., Yamashita A., Goto N., Takahashi K., Yasunaga T., Ikuta K., Mizutani T., Okamoto Y., Tagami M., Morita R., Maeda N., Kawai J., Hayashizaki Y., Nagai Y., Horii T., Iida T. and Nakaya T. (2009). Direct metagenomic detection of viral pathogens in nasal and fecal specimens using an unbiased high–throughput sequencing approach. *PloS One*, **4**(1), e4219.

Panjkovich A., Norambuena T. and Melo F. (2005). dnaMATE: a consensus melting temperature prediction server for short DNA sequences. *Nucleic Acids Res.*, **33**(Web Server issue), W570–W572.

Park E.-J., Kim K.-H., Abell G. C. J., Kim M.-S., Roh S. W. and Bae J.-W. (2011). Metagenomic analysis of the viral communities in fermented foods. *Appl. Environ. Microbiol.*, **77**(4), 1284–1291.

Ringquist S., Pecoraro C., Gilchrist C. M. S., Styche A., Rudert W. A., Benos P. V. and Trucco M. (2005). SOP3v2: web-based selection of oligonucleotide primer trios for genotyping of human and mouse polymorphisms. *Nucleic Acids Res.*, **33**(Web Server issue), W548–W552.

Rodriguez-Brito B., Li L., Wegley L., Furlan M., Angly F., Breitbart M., Buchanan J., Desnues C., Dinsdale E., Edwards R., Felts B., Haynes M., Liu H., Lipson D., Mahaffy J., Martin-Cuadrado A. B., Mira A., Nulton J., Pasić L., Rayhawk S., Rodriguez-Mueller J., Rodriguez-Valera F., Salamon P., Srinagesh S., Thingstad T. F., Tran T., Thurber R. V., Willner D., Youle M. and Rohwer F. (2010). Viral and microbial community dynamics in four aquatic environments. *ISME J.*, **4**(6), 739–751.

Rose T. M., Henikoff J. G. and Henikoff S. (2003). CODEHOP (COnsensus-DEgenerate Hybrid Oligonucleotide Primer) PCR primer design. *Nucleic Acids Res.*, **31**(13), 3763–3766.

Rouillard J.-M., Lee W., Truan G., Gao X., Zhou X. and Gulari E. (2004). Gene2Oligo: oligonucleotide design for in vitro gene synthesis. *Nucleic Acids Res.*, **32**(Web Server issue), W176–W180.

Rozen S. and Skaletsky H. (2000). Primer3 on the WWW for general users and for biologist programmers. *Methods Mol. Biol.*, **132**, 365–386.

Simon C. and Daniel R. (2011). Metagenomic analyses: past and future trends. *Appl. Environ. Microbiol.*, **77**(4), 1153–1161.

Stecher B., Chaffron S., Käppeli R., Hapfelmeier S., Freedrich S., Weber T. C., Kirundi J., Suar M., McCoy K. D., von Mering C., Macpherson A. J. and Hardt W.-D. (2010). Like will to like: abundances of closely related species can predict susceptibility to intestinal colonization by pathogenic and commensal bacteria. *PLoS Pathog.*, **6**(1), e1000711.

Trkov M. and Avgustin G. (2003). An improved 16S rRNA based PCR method for the specific detection of *Salmonella enterica. Int. J. Food Microbiol.*, **80**(1), 67–75.

Untergasser A., Nijveen H., Rao X., Bisseling T., Geurts R. and Leunissen J. A. M. (2007). Primer3Plus, an enhanced web interface to Primer3. *Nucleic Acids Res.*, **35**(Web Server issue), W71–74.

Van de Peer Y., Chapelle S. and De Wachter R. (1996). A quantitative map of nucleotide substitution rates in bacterial rRNA. *Nucleic Acids Res.*, **24**(17), 3381–3391.

Wang D., Wang H., Zhou Y., Zhang Q., Zhang F., Du P., Wang S., Chen C. and Kan B. (2011). Genome sequencing reveals unique mutations in characteristic metabolic pathways and the transfer of virulence genes between *V. mimicus* and *V. cholerae*. *PLoS ONE*, **6**(6), e21299.

Ye L. and Zhang T. (2011). Pathogenic bacteria in sewage treatment plants as revealed by 454 pyrosequencing. *Environ. Sci. Technol.*, **45**(17), 7173–7179.

You F. M., Huo N., Gu Y. Q., Luo M.-C., Ma Y., Hane D., Lazo G. R., Dvorak J. and Anderson O. D. (2008). BatchPrimer3: a high throughput web application for PCR and sequencing primer design. *BMC Bioinformatics*, **9**, 253.

Chapter 16

Fluid structure and boundary slippage in nanoscale liquid films

Nikolai V. Priezjev

16.1 ABSTRACT

During the last ten years, there has been enormous interest in understanding transport phenomena in micro and nanofluidic systems and, in particular, in accurate prediction of fluid flows with slip boundary conditions at liquid-solid interfaces. In this chapter, we discuss recent results obtained from molecular dynamics simulations of fluids that consist of monomers or linear polymer chains confined by flat crystalline surfaces. The effects of shear rate and wall lattice orientation on the slip behaviour are studied for a number of material parameters of the interface, such as fluid and wall densities, wall-fluid interaction energy, polymer chain length, and wall lattice type. A detailed analysis of the substrate-induced fluid structure and interfacial diffusion of fluid molecules is performed to identify slip flow regimes at low and high shear rates.

16.2 INTRODUCTION

The study of fluid transport through micro and nanochannels is important for biotechnological applications and energy conversion processes (Sparreboom *et al.* 2009). The precise control and manipulation of fluids in systems with large surface-to-volume ratios, however, require fundamental understanding of flow boundary conditions. Fluid velocity profiles can be significantly modified in the presence of slip at a solid surface. The degree of slip is quantified by the slip length, which is defined as a distance between locations of the real interface and imaginary plane where the extrapolated tangential velocity component vanishes. It was shown by numerous experimental studies that the main factors affecting slippage at the liquid-solid interface include surface roughness (Zhu & Granick, 2002; Sanchez-Reyes & Archer, 2003; Schmatko *et al.* 2006), surface wettability (Churaev *et al.* 1984; Baudry *et al.* 2001; Schmatko *et al.* 2005), fluid structure (Schmatko *et al.* 2005; McBride & Law, 2009; Baumchen *et al.* 2009), and shear rate (Zhu & Granick, 2001; Choi *et al.* 2003; Ulmanella & Ho, 2008). The slip length in the micron range is reported for complex flows near superhydrophobic surfaces (Rothstein, 2010) and flows of high molecular weight polymers (Baumchen *et al.* 2009), while the magnitude of the slip length in the range of a few tens of nanometers is typically measured for flows of water over smooth nonwetting surfaces (Bocquet & Charlaix, 2010).

In recent years, many molecular dynamics (MD) simulation studies have been performed in order to investigate the influence of structural properties of the interface between monatomic fluids and crystalline walls on the degree of slip (Heinbuch & Fischer, 1989; Thompson & Robbins, 1990; Bocquet & Barrat, 1994; Barrat & Bocquet, 1999; Travis & Gubbins, 2000; Sokhan *et al.* 2001; Galea & Attard, 2004; Priezjev, 2007; Priezjev, 2007a; Thompson & Troian, 1997; Asproulis & Drikakis, 2010; Liu & Li, 2009; Pahlavan & Freund, 2011; Yong & Zhang, 2010) and references therein. Most notably, it was demonstrated that the slip length is strongly correlated with the intensity of structure induced within the first fluid layer by the periodic surface potential (Thompson & Robbins, 1990). In general, the slippage is suppressed by the strong wall-fluid attraction and/or due to the formation of commensurate structures between solid wall and adjacent fluid layer. A theoretical estimate of the slip length at low shear rates can be obtained via the Green-Kubo relation between the friction coefficient at the interface and the time integral of the autocorrelation function of the lateral force that acts on the adjacent fluid from the solid wall (Barrat & Bocquet, 1999). In most of the MD studies, the solid walls are modelled as an array of atoms arranged on sites of a periodic lattice. Two types of walls are usually considered, solid and thermal, where the wall atoms are either fixed at the lattice sites or allowed to oscillate under the harmonic potential. It was recently found that the slip length weakly depends on the value of the spring stiffness coefficient for sufficiently strong harmonic bonds (Priezjev, 2007; Asproulis & Drikakis, 2010). In addition, the slope of the shear rate dependence of the slip length is not significantly affected by stiff springs (Priezjev, 2007), except at very high shear rates (Martini *et al.* 2008).

At the interface between simple fluids and atomically smooth, weakly attractive surfaces, the slip length is constant only at relatively low shear rates and it increases nonlinearly at high shear rates, as originally reported by Thompson and Troian (1997) and later confirmed by several studies (Priezjev, 2007; Asproulis & Drikakis, 2010; Yang & Fang, 2005; Niavarani & Priezjev, 2010). For sufficiently strong wall-fluid interaction and incommensurate structures of the liquid and solid phases at the interface, the slip length varies almost linearly with shear rate (Priezjev, 2007; Priezjev, 2007a). It should be emphasized, however, that if the slip length at low shear rates is less than about a molecular diameter then the boundary conditions for dense monatomic fluids remain independent of shear rate (Thompson & Robbins, 1990; Priezjev, 2007; Thompson & Troian, 1997). Also, it was shown that molecular-scale surface roughness reduces the magnitude of the slip length and the slope of its rate dependence (Bocquet & Barrat, 1994; Priezjev, 2007; Niavarani & Priezjev, 2010; Priezjev & Troian, 2006; Sofos *et al.* 2009).

It was recently demonstrated that the effective slip length for flows over anisotropic surfaces with two-component texture of different wettability is largest (smallest) for parallel (perpendicular) orientation of stripes with respect to the mean flow (Feuillebois *et al.* 2009). These conclusions hold when the stripe width is comparable to the molecular diameter (Priezjev *et al.* 2005). For the transverse orientation of the flow relative to the stripes, the slip is reduced because of the molecular scale corrugation of the composed surface potential, while for the parallel orientation, the fluid molecules are transported along homogeneous stripes with either no-slip or partial slip conditions; and, therefore, the effective slip length is enhanced (Priezjev *et al.* 2005; Priezjev, 2011). More recently, it was observed that the slip length also depends on the crystal lattice plane in contact with the fluid and on the lattice orientation with respect to the flow direction (Soong *et al.* 2007). In this chapter, we will show that at sufficiently high shear rates, the slip flow is anisotropic for atomically flat crystalline surfaces; and, in particular, the slip length is enhanced when the shear flow is oriented along the crystallographic axis of the wall lattice.

Recent studies of friction between adsorbed monolayers and smooth crystalline surfaces are relevant to the analysis of flow boundary conditions (Smith *et al.* 1996; Tomassone *et al.* 1997). It was found that the slip time, which represents the transfer of momentum between the adsorbed monolayer and the substrate, is proportional to the phonon lifetime divided by the normalized peak value of the structure factor computed in the monolayer at the main reciprocal lattice vector (Smith *et al.* 1996). Also, the simulation results have shown that the slip time is independent of the sliding direction if the slip velocity of the monolayer is much smaller than the speed of sound (Smith *et al.* 1996). In the linear regime between friction force and sliding velocity and in the range of film coverages from submonolayer to bilayer, the slip times were computed directly from the decay of the film velocity and from the decay of the velocity correlation function at equilibrium (Tomassone *et al.* 1997).

During the last two decades, a number of MD studies have examined slip boundary conditions at the interface between polymeric fluids and flat crystalline surfaces (Thompson *et al.* 1995; Manias *et al.* 1996; Khare *et al.* 1996; Stevens *et al.* 1997; Koike & Yoneya, 1998; Jabbarzadeh *et al.* 1999; Niavarani & Priezjev, 2008; Priezjev & Troian, 2004; Servantie & Muller, 2008; Priezjev, 2009; Priezjev, 2010). The velocity profiles with stick boundary conditions were observed when a highly viscous interfacial layer was formed because of the strong wall-fluid interaction (Manias *et al.* 1996; Jabbarzadeh *et al.* 1999; Servantie & Muller, 2008), high fluid density and pressure (Thompson *et al.* 1995; Manias *et al.* 1996; Priezjev, 2009), or chemical structure of chain molecules (Kong *et al.* 2010). The variation of the slip length as a function of shear rate was reported for flat polymer-solid interfaces with weak wall-fluid interactions (Martini *et al.* 2008; Koike & Yoneya, 1998; Jabbarzadeh *et al.* 1999; Priezjev & Troian, 2004; Niavarani & Priezjev, 2008; Priezjev, 2009; Dhondi *et al.* 2009; Niavarani & Priezjev, 2008). In our previous studies (Niavarani & Priezjev, 2008; Priezjev, 2010), it was shown that the rate dependence of the slip length acquires a local minimum at low shear rates followed by a rapid growth at higher shear rates. More recently, a correlation between the shear rate threshold for the boundary slip in dense unentangled polymer films and a chain relaxation dynamics in the interfacial region was reported (Priezjev, 2009).

In this chapter, we investigate slip boundary conditions at the interface between fluids and crystalline surfaces using molecular dynamics simulations. We find that at sufficiently high shear rates, the slip flow over atomically flat crystalline surfaces is anisotropic. The nonlinear shear rate dependence of the slip length is analyzed in terms of the friction coefficient at the liquid-solid interface and slip velocity. The simulation results indicate that the friction coefficient in the linear slip regime is a function of a single variable that is a product of the height of the normalized main peak in the structure factor and the contact density of the first fluid layer near the solid wall. We will show that the onset of the nonlinear regime between the wall shear stress and slip velocity is determined by the diffusion of fluid monomers within the first layer.

The rest of this chapter is organized as follows. In the next section, we describe the details of molecular dynamics simulations, equilibration procedure, and parameter values for twenty liquid-on-solid systems. The fluid density, velocity and temperature profiles for various flow conditions are discussed in Section IIIA. The dynamic response of shear viscosity and slip length is examined in Section IIIB. The numerical analysis of the friction coefficient and induced fluid structure is presented in Section IIIC. The conclusions are given in the last section.

16.3 MOLECULAR DYNAMICS SIMULATION MODEL

The geometry of the computational domain and the steady flow profile are shown schematically in Figure 16.1. The fluid undergoes planar shear flow between two atomically flat walls. For all simulations in this study, the fluid phase consists of $N_f = 9600$ monomers. The interaction between any two fluid monomers is modelled via the truncated Lennard-Jones (LJ) potential

$$V_{LJ}(r) = 4\varepsilon\left[\left(\frac{\sigma}{r}\right)^{12} - \left(\frac{\sigma}{r}\right)^{6}\right] \quad \text{for } r \leq r_c = 2.5\sigma \qquad (16.1)$$

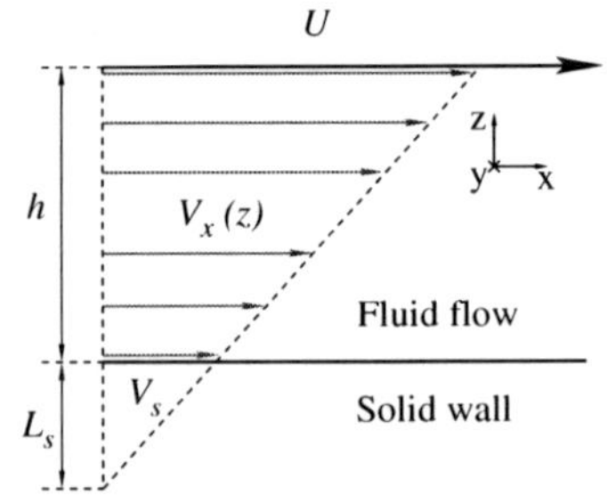

Figure 16.1 A schematic representation of the flow with slip boundary conditions. The steady shear flow is induced by the upper wall moving with a constant speed U in the $\hat{x}$ direction. The slip velocity and slip length L_s are related via $V_s = \dot{\gamma} L_s$, where $\dot{\gamma}$ is the shear rate computed from the slope of the velocity profile.

where ε and σ are the energy and length scales of the fluid phase. The interaction between wall atoms and fluid monomers is also modelled by the LJ potential with parameters ε_{wf} (the values ε_{wf} for each system are listed in Table 16.1) and $\varepsilon_{wf} = \sigma$. The wall atoms do not interact with each other.

Table 16.1 The fluid monomer density $\rho = N_f/A_{xy}(h-\sigma)$, number of monomers per chain N, distance between the wall lattice planes in contact with fluid h, wall area in the xy plane, fluid pressure at equilibrium (i.e., $U = 0$), wall density ρ_w, lattice type, Miller indices for the xy plane, lattice orientation along the shear flow direction ($\hat{x}$ direction), the $\hat{x}$ and $\hat{y}$ components of the first reciprocal lattice vector $\mathbf{G}_1(k_x, k_y)$, wall-fluid interaction energy, and the spring stiffness coefficient for thermal walls.

No.	$\rho\sigma^3$	N	h/σ	A_{xy}/σ^2	$P/\varepsilon\sigma^{-3}$	$\rho_w\sigma^3$	Type	(ijk)	$\hat{x}$	$(k_x\sigma, k_y\sigma)$	$\varepsilon_{wf}/\varepsilon$	$\kappa/\varepsilon\sigma^{-2}$
1	0.91	20	22.02	502.28	1.0	1.40	fcc	(111)	$[11\bar{2}]$	(7.23,0)	0.9	Fixed
2	0.91	20	22.02	502.28	1.0	1.40	fcc	(111)	$[1\bar{1}0]$	(6.26,3.62)	0.9	Fixed
3	0.88	20	19.46	589.79	0.5	1.10	fcc	(111)	$[11\bar{2}]$	(6.67,0)	0.8	1200
4	0.88	20	19.46	589.79	0.5	1.10	fcc	(111)	$[1\bar{1}0]$	(5.78,3.34)	0.8	1200
5	0.89	20	26.44	424.73	0.5	1.80	fcc	(111)	$[11\bar{2}]$	(7.86,0)	1.0	Fixed
6	0.89	20	26.44	424.73	0.5	1.80	fcc	(111)	$[1\bar{1}0]$	(6.81,3.93)	1.0	Fixed
7	0.83	10	23.93	502.28	0.0	1.40	fcc	(111)	$[11\bar{2}]$	(7.23,0)	0.7	1200
8	0.83	10	23.93	502.28	0.0	1.40	fcc	(111)	$[1\bar{1}0]$	(6.26,3.62)	0.7	1200
9	0.88	20	24.72	459.42	0.5	1.60	fcc	(111)	$[11\bar{2}]$	(7.56,0)	0.8	1200
10	0.88	20	24.72	459.42	0.5	1.60	fcc	(111)	$[1\bar{1}0]$	(6.55,3.78)	0.8	1200
11	0.89	20	19.12	595.87	0.5	1.90	bcc	(001)	[100]	(6.18,0)	0.4	Fixed
12	0.89	20	19.12	595.87	0.5	1.90	bcc	(001)	[100]	(6.18,0)	0.5	Fixed
13	0.89	20	19.12	595.87	0.5	1.90	bcc	(001)	[100]	(6.18,0)	0.6	Fixed
14	0.85	10	19.98	595.87	0.5	1.90	bcc	(001)	[100]	(6.18,0)	0.4	1200
15	0.85	10	19.98	595.87	0.5	1.90	bcc	(001)	[100]	(6.18,0)	0.5	1200
16	0.85	10	19.98	595.87	0.5	1.90	bcc	(001)	[100]	(6.18,0)	0.6	1200
17	0.81	1	34.86	350.61	2.36	2.40	fcc	(111)	$[11\bar{2}]$	(8.65,0)	0.4	Fixed
18	0.81	1	34.86	350.61	2.36	2.40	fcc	(111)	$[1\bar{1}0]$	(7.49,4.33)	0.4	Fixed
19	0.81	1	34.86	350.61	2.36	2.40	fcc	(111)	$[11\bar{2}]$	(8.65,0)	0.3	Fixed
20	0.81	1	34.86	350.61	2.36	2.40	fcc	(111)	$[1\bar{1}0]$	(7.49,4.33)	0.3	Fixed

Adapted from (Priezjev, 2010).

Three types of fluid were considered in the present study, that is, monatomic (or simple) fluid and polymer melts with the number of monomers per chain $N = 10$ and $N = 20$. In the case of polymers, the nearest-neighbour monomers in a chain interact through the finitely extensible nonlinear elastic (FENE) potential (Bird *et al.* 1987)

$$V_{\mathrm{FENE}}(r) = -\frac{k_s}{2} r_0^2 \ln\left[1 - \frac{r^2}{r_0^2}\right] \tag{16.2}$$

with the standard choice of parameters $k_s = 30\varepsilon\sigma^{-2}$ and $r_0 = 1.5\sigma$ (Kremer & Grest, 1990). As an example, a snapshot of an unentangled polymer melt with linear flexible chains $N = 20$ confined between solid walls is presented in Figure 16.2.

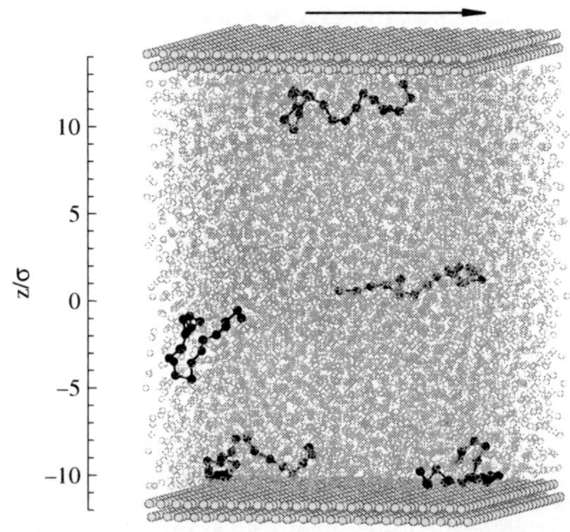

Figure 16.2 A snapshot of fluid monomers (open circles) and wall atoms (filled circles). positions. Five polymer chains are marked by solid lines and filled black circles. The black arrow indicates the upper wall velocity $U = 0.5\sigma/\tau$ in the $\hat{x}$ direction. The fluid monomer density is $\rho = 0.89\ \sigma^{-3}$ and the wall density is $\rho_w = 1.80\sigma^{-3}$. The rest of parameters for the system 5 are given in Table 16.1. Reprinted from (Priezjev, 2010).

The heat exchange between the fluid phase and the external heat bath was regulated via a Langevin thermostat (Grest & Kremer, 1986), which was applied only to the direction of motion perpendicular to the plane of shear (Thompson & Robbins, 1990). The equations of motion for fluid monomers in all three directions are given as follows:

$$m\ddot{x}_i = -\sum_{i \neq j} \frac{\partial V_{ij}}{\partial x_i} \tag{16.3}$$

$$m\ddot{y}_i + m\Gamma\dot{y}_i = -\sum_{i \neq j} \frac{\partial V_{ij}}{\partial y_i} + f_i \tag{16.4}$$

$$m\ddot{z}_i = -\sum_{i \neq j} \frac{\partial V_{ij}}{\partial z_i} \tag{16.5}$$

where the summation is performed over the fluid monomers and wall atoms within the cutoff radius $r_c = 2.5\sigma$, $\Gamma = 1.0\tau^{-1}$ is the friction coefficient, and f_i is a random force with zero mean and variance $\langle f_i(0) f_j(t)\rangle = 2mk_B\tau\Gamma\delta(t)\delta_{ij}$ determined from the fluctuation-dissipation theorem. The Langevin thermostat temperature is $T = 1.1\varepsilon/k_B$, where k_B is the Boltzmann constant. The equations of motion were integrated using the fifth-order Gear predictor-corrector algorithm (Allen & Tildesley, 1989) with a time step $\Delta t = 0.002\tau$, where $\tau = \sqrt{m\sigma^2/\varepsilon}$ is the characteristic time of the LJ potential. The small time step $\Delta t = 0.002\tau$ was used in our previous studies (Priezjev, 2007; Priezjev, 2009; Priezjev, 2010; Niavarani & Priezjev, 2008) for similar MD setups in order to compute accurately the trajectories of fluid molecules and wall atoms near interfaces. Typical values for liquid argon are $\sigma = 0.34$ nm, $\varepsilon/k_B =$ 120K and $\tau = 2.16 \times 10^{-12}$ s (Allen & Tildesley, 1989).

Each confining wall is composed of 1152 atoms arranged in two layers of either face-centered cubic (fcc) or body centered cubic (bcc) lattices. The wall density, lattice type, its orientation with respect to the shear flow direction, and wall-fluid interaction energy are listed in Table 16.1. The wall atoms are either fixed at the lattice sites or allowed to oscillate about their equilibrium lattice positions under the harmonic potential $V_{sp} = \frac{1}{2}k(r - r_{eq})^2$ with the spring stiffness coefficient $k = 1200\varepsilon/\sigma^2$. It was previously shown that this value of the stiffness coefficient does not significantly affect the shear rate dependence of the slip length (Priezjev, 2007). In case of thermal walls, the Langevin thermostat was applied to the $\hat{x}$, $\hat{y}$, and $\hat{z}$ components of the wall atom equations of motion. For example, the $\hat{x}$ component of the equation of motion is given by

$$m_w\ddot{x}_i + m_w\Gamma\dot{x}_i = -\sum_{i\neq j}\frac{\partial V_{ij}}{\partial x_i} - \frac{\partial V_{sp}}{\partial x_i} + f_i \tag{16.6}$$

where $m_w = 10$m, the friction coefficient is $\Gamma = 1.0\tau^{-1}$ and the sum is taken over the neighbouring fluid monomers within the cutoff radius $r_c = 2.5\sigma$. With these parameters, the oscillation time $2\pi\sqrt{m_w/\kappa} \approx 0.6\tau$ of wall atoms is much larger than the integration time step $\Delta t = 0.002\tau$. Periodic boundary conditions were imposed along the $\hat{x}$ and $\hat{y}$ directions parallel to the confining walls.

Initially, the fluid was equilibrated at a constant normal pressure applied on the upper wall (shown in Table 16.1) for about 5×10^4 τ while the lower wall was at rest. Then, the channel height was fixed and the system was additionally equilibrated for 5×10^4 τ at a constant density ensemble while both walls were at rest. The steady flow was generated by moving the upper wall with a constant speed U in the $\hat{x}$ direction parallel to the immobile lower wall (see Figure 16.1). The lowest speed of the upper wall is $U = 0.05\sigma/\tau$. Both fluid velocity and density profiles were computed within horizontal bins of thickness $\Delta z = 0.01\sigma$ for time intervals up to 6×10^5 τ. An estimate of the Reynolds number at the highest shear rates considered in the present study is O(10), which is indicative of laminar flow conditions.

16.4 RESULTS

16.4.1 Fluid density, velocity, and temperature profiles

In this section we present typical fluid density, velocity, and temperature profiles for the polymer system $N = 20$ (system 5 in Table 16.1). First, the averaged fluid density and velocity profiles are displayed in Figure 16.3 for the upper wall speeds $U = 0.5\sigma/\tau$ and $U = 4.0\sigma/\tau$. Notice that the density profiles exhibit a typical layered structure which extends for about $5\sigma - 6\sigma$ away from the solid walls. The amplitude of the first peak in the density profile determines the contact density ρ_c, an important parameter, which will be used in the subsequent analysis. We emphasize that the thickness of the averaging bins $\Delta z = 0.01\sigma$ is small enough so that the value of the contact density does not depend on the bin thickness and bin location relative to the walls. On the other hand, the shape of the density profiles will remain unchanged if thinner bins are used; however, the averaging will require additional computational resources. As evident from Figure 16.3 (b), the contact density is reduced at the higher upper wall speed.

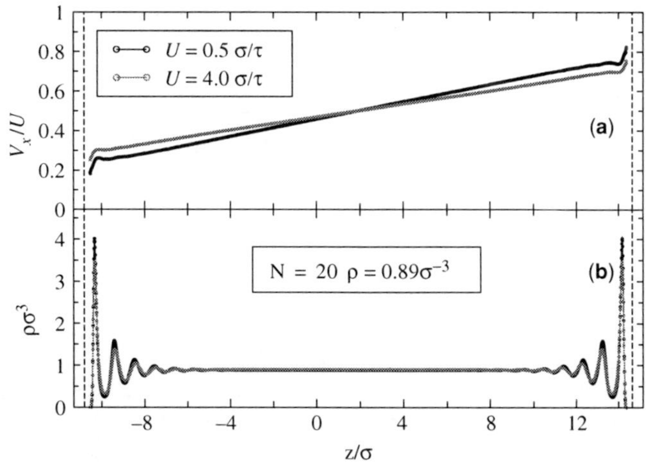

Figure 16.3 Averaged normalized velocity (a) and density (b) profiles across the channel for the upper wall speeds $U = 0.5\sigma/\tau$ and $U = 4.0\sigma/\tau$. The uniform monomer density of the polymer melt $N = 20$ away from the walls is $\rho = 0.89\ \sigma^{-3}$ (system 5 in Table 16.1). The vertical axes indicate the location of the fcc lattice planes (at $z/\sigma = -11.30$ and 15.14) in contact with the fluid. The dashed lines at $z/\sigma = -10.80$ and 14.64 denote reference planes for computing the slip length. Reprinted from (Priezjev, 2010).

The representative velocity profiles normalized by the upper wall speed are shown in Figure 16.3 (a) for the same flow conditions as in Figure 16.3 (b). It is apparent that the slip velocity is larger at the higher upper wall speed. It should be mentioned that simulations were performed at sufficiently low shear rates so that the fluid slip velocities remained the same at the lower and upper walls. In our study, the location of the liquid-solid interface (marked by the dashed vertical lines in Figure 16.3) is defined at the distance 0.5σ away from the wall lattice planes to take into account the excluded volume due to wall atoms. The slip length was computed from the linear fit to the velocity profiles excluding regions of about 2σ from the solid walls, where a slight nonlinearity appears due to fluid layering. At all shear rates examined in the present study, the velocity profiles are linear across the channel and the slip length is larger than about 3σ.

In steady flow, the fluid temperature was estimated from the local kinetic energy as follows:

$$k_B T = \frac{m}{3N} \sum_{i=1}^{N} [\dot{r}_i - \mathbf{v}(r_i)]^2 \tag{16.7}$$

where $\dot{r}_i$ is the instantaneous velocity of the fluid monomer and $\mathbf{v}(r_i)$ is the local flow velocity averaged inside a narrow bin. Averaged temperature profiles across the channel are shown in Figure 16.4 for selected values of the upper wall speed. At relatively low shear rates, $\dot{\gamma} \lesssim 0.01\tau^{-1}$, the fluid temperature is equal to its equilibrium value of $T = 1.1\varepsilon/k_B$ set by the Langevin thermostat. With a further increase of the shear rate, the fluid heats up and the temperature profile in steady state becomes non-uniform across the channel. As seen in Figure 16.4, the fluid temperature is higher near the interfaces because of the large slip velocity, which becomes comparable to the thermal velocity, $v_T^2 = k_B T/m$, at high shear rates. We also found that at high shear rates, the temperature in the $\hat{y}$ direction, in which the Langevin thermostat is applied, is slightly smaller than its value in the $\hat{x}$ and $\hat{z}$ directions (shown in the inset of

Figure 16.4). This difference implies that the kinetic energy in the $\hat{y}$ direction dissipates faster than the energy transfer from the other directions. Similar temperature profiles for a system of linear polymer chains $N = 20$, fluid density $\rho = 0.91\ \sigma^{-3}$, and thermal walls with density $\rho_w = 1.40\sigma^{-3}$ were reported in the previous MD study (Niavarani & Priezjev, 2008). Further discussion on the relation between temperature profiles and slip velocity in thin polymer films can be found in (Khare *et al.* 2006).

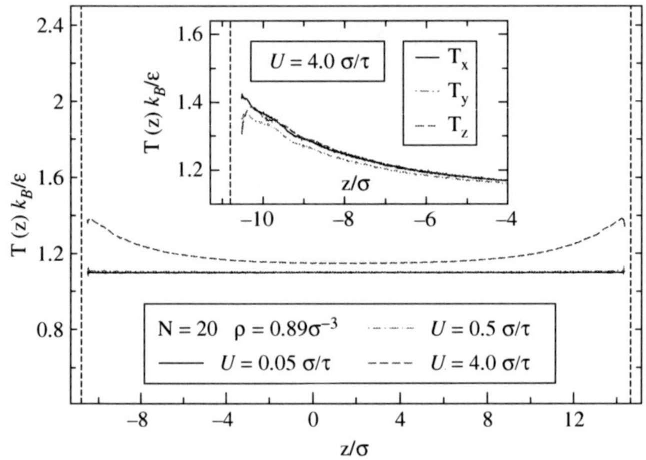

Figure 16.4 Temperature profiles across the channel for the indicated upper wall velocities (system 5 in Table 16.1). The vertical dashed lines at $z/\sigma = -10.80$ and 14.64 indicate the location of liquid-solid interfaces. The inset shows the $\hat{x}$, $\hat{y}$, and $\hat{z}$ components of the temperature profile near the stationary lower wall when the upper wall velocity is $U = 4.0\sigma/\tau$.

16.4.2 Shear viscosity and slip length

The fluid viscosity was estimated from the relation between shear rate and shear stress, which was computed using the Kirkwood formula (Irving & Kirkwood, 1950). The variation of viscosity as a function of shear rate is presented in Figure 16.5 for selected systems listed in Table 16.1. In agreement with previous studies with a similar setup (Priezjev, 2007; Thompson & Troian, 1997), the viscosity of monatomic fluids is independent of shear rate and equals $\mu = (2.2 \pm 0.2)\varepsilon\tau\sigma^{-3}$ when the fluid density is $\rho = 0.81\sigma^{-3}$. As expected, the shear viscosity of polymer melts with chains $N = 10$ and $N = 20$ is higher than the viscosity of simple monatomic fluids. For similar flow conditions, the transition from a Newtonian to a shear-thinning flow regime occurs at lower shear rates for polymers with longer chains $N = 20$ because of their slower relaxation dynamics. The slope of the shear-thinning region -0.37, which is shown in Figure 16.5 by the dashed line, is consistent with the results reported in earlier studies for polymer melts $N = 20$ at different fluid densities (Niavarani & Priezjev, 2008; Priezjev, 2009). As usual, the errors arising from averaging over thermal fluctuations are greater at lower shear rates.

The nonlinear rate dependence of the slip length is shown in Figure 16.6 for polymer melts with chains $N = 10$ and $N = 20$. The shear flow direction is oriented along the crystallographic axis of the (111) plane of the fcc wall lattice for systems 6 and 8 (see Table 16.1). In contrast, the fcc lattice plane is rotated by 90° with respect to the flow direction for systems 5 and 7 as indicated by open circles and the blue vertical arrow in the inset of Figure 16.6. At low shear rates $\dot{\gamma}\tau \lesssim 0.02$, the slip length is independent of the wall lattice orientation relative to the shear flow direction; while at higher shear rates, the slip length is greater when the shear flow is parallel to the crystallographic axis of the triangular lattice. The same trend for the slip length is observed for monatomic fluids (not shown). These results demonstrate that at sufficiently high shear rates the slip flow is anisotropic even for atomically flat crystalline surfaces.

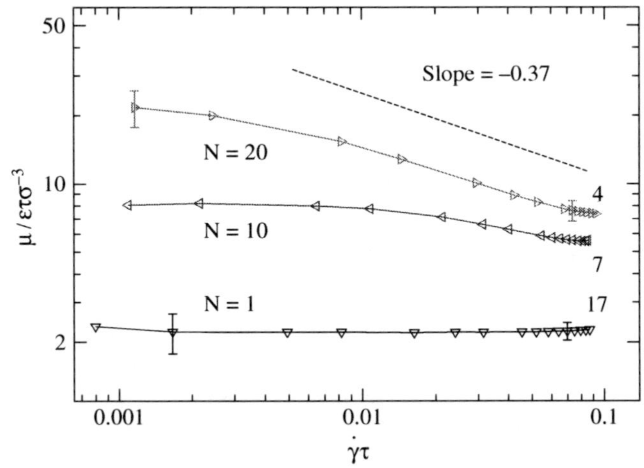

Figure 16.5 Fluid viscosity μ (in units $\varepsilon\tau\sigma^{-3}$) as a function of shear rate for the indicated systems listed in Table 16.1. The dashed line with a slope -0.37 is shown for reference. Solid curves are a guide for the eye. Adapted from (Priezjev, 2010).

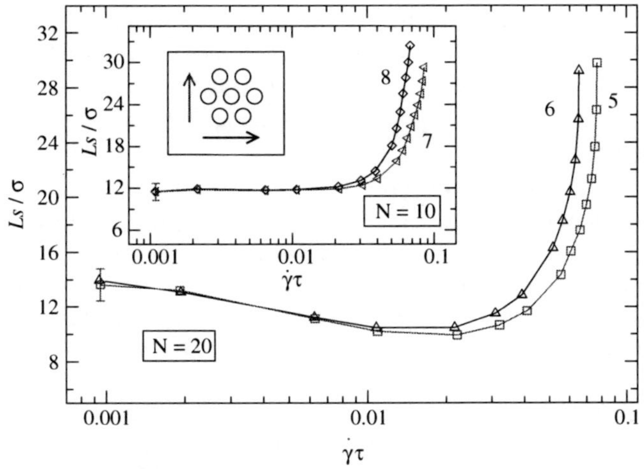

Figure 16.6 Slip length L_s/σ as a function of shear rate for polymer melts with chains $N = 20$ and $N = 10$ (see inset). The system's parameters are listed in Table 16.1. Open circles in the inset represent the (111) face of the fcc lattice atoms in contact with the fluid. The vertical arrow indicates the shear flow direction with respect to the $[11\bar{2}]$ fcc lattice orientation (systems 5 and 7). The horizontal arrow shows the flow direction along the $[1\bar{1}0]$ orientation (systems 6 and 8). Reprinted from (Priezjev, 2010).

The appearance of the local minimum in the shear rate dependence of the slip length reported in Figure 16.6 for polymer melts with chains $N = 20$ can be explained using simple physical arguments. The initial decay of the slip length at low shear rates is associated with a slight decrease of the polymer melt viscosity. In this regime, the friction coefficient at the liquid-solid interface (defined as $k = \mu\ /L_s$)

remains constant and independent of shear rate (or slip velocity). With increasing shear rate, both the friction coefficient and polymer viscosity decrease; however, their ratio, the slip length, $L_s = \mu/k$ grows rapidly because of the strong dependence of the friction coefficient on the slip velocity (see also next section). Similar behaviour of the slip length for polymer chains $N = 20$ was reported in recent MD studies (Niavarani & Priezjev, 2008; Niavarani & Priezjev, 2008a). Furthermore, the transition to the shear-thinning regime occurs at higher shear rates for polymer melts with shorter chains $N = 10$, and, therefore, the slip length remains nearly constant at low shear rates and then increases rapidly at higher rates (see inset in Figure 16.6). These results agree well with the previous simulation results for slip flow of polymers with chain lengths $N \leq 16$ and lower fluid density (Priezjev & Troian, 2004).

16.4.3 Friction coefficient versus slip velocity

It is difficult to make further progress in the analysis of the shear rate dependent slip length without taking into account shear-thinning effects explicitly. Instead, it is advantageous to reformulate the boundary conditions in terms of the friction coefficient at the liquid-solid interface and slip velocity. In steady shear flow, the shear stress in the bulk of the film ($\dot{\gamma}\mu$) is equal to the wall shear stress (kV_s). In addition, if velocity profiles are linear across the channel, then by definition $V_s = \dot{\gamma} L_s$ and the friction coefficient is given by $k = \mu/L_s$. In what follows, the MD data for the slip length and shear viscosity will be used to compute the friction coefficient, and its dependence on the slip velocity will be investigated.

We next briefly review the results from previous MD studies on slip flows of polymer melts with chains $N = 20$ confined by atomically flat walls with weak surface energy (Niavarani & Priezjev, 2008; Priezjev, 2009). It was found that in the range of fluid densities $0.86 \leq \rho\sigma^3 \leq 1.02$, the velocity profiles across the channel are linear at all shear rated examined (Niavarani & Priezjev, 2008). Therefore, the friction coefficient was computed from the relation $k = \mu/L_s$ and studied as a function of the slip velocity. The data for different fluid densities could be well fitted by the following equation:

$$k/k^* = [1 + (V_s/V_s^*)^2]^{-0.35} \tag{16.8}$$

where k^* is the friction coefficient at small slip velocities when $V_s \ll V_s^*$ and V_s^* is the characteristic slip velocity that marks the onset of the nonlinear regime (Niavarani & Priezjev, 2008). In the subsequent study (Priezjev, 2009), the simulations were performed at higher polymer densities, $1.04 \leq \rho\sigma^3 \leq 1.11$, while the rest of the system parameters were kept unchanged. Due to the formation of a highly viscous interfacial layer, the velocity profiles at low shear rates acquired a pronounced curvature near solid walls; and, as a result, the definition $k = \mu/L_s$ could not be applied (Priezjev, 2009). Therefore, the friction coefficient was computed directly from the ratio of the wall shear stress and slip velocity of the first fluid layer. Interestingly, for polymer melt densities $\rho = 1.04\sigma^{-3}$ and $1.06\sigma^{-3}$, the data were again well described by Eq. (16.8), while at higher melt densities ($1.08 \leq \rho\sigma^3 \leq 1.11$) only the nonlinear regime ($k \sim V_s^{-0.7}$) was observed (Priezjev, 2009). We finally note that when the functional form of the slip length versus shear rate reported for monatomic fluids (Thompson & Troian, 1997) is expressed in terms of k and V_s, the friction coefficient is also well fitted by Eq. (16.8) in the range $k/k^* \gtrsim 0.3$ (Niavarani & Priezjev, 2008).

We extend the analysis of the friction coefficient at liquid-solid interfaces which are described by the parameters listed in Table 16.1. Figure 16.7 shows the friction coefficient as a function of the slip velocity normalized by the parameters k^* and V_s^* respectively. Remarkably, the data for all twenty systems in Table 16.1 are well fitted by Eq. (16.8) over about three orders of magnitude. These results further support our previous conclusions that the boundary condition Eq. (16.8) describes slip flows of both monatomic and polymeric fluids over smooth solid walls. We also noticed the inverse correlation between the friction coefficient k^* and the characteristic slip velocity V_s^*, which is summarized in Figure 16.8.

Note that for every two systems, where the only difference is the orientation of the fcc wall lattice, the values of k^* are nearly the same, but the slip velocity V_s^* is slightly smaller when the shear flow direction is parallel to the crystallographic axis. This is consistent with the dynamic response of the slip length reported in Figure 16.6 for two different orientations of the fcc lattice; namely, the onset of rapid growth of the slip length occurs at smaller shear rates when the the flow is oriented along the crystallographic axis.

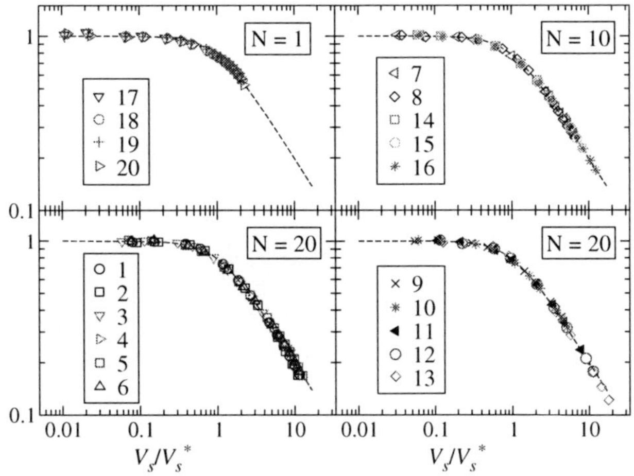

Figure 16.7 Log-log plot of the friction coefficient $k=\mu/L_s$ (in units $\varepsilon\tau\sigma^{-4}$) as a function of the slip velocity $V_s = L_s\dot{\gamma}$ (in units σ/τ) for systems listed in Table 16.1. The values of the normalization parameters V_s^* and k^* are presented in Figure 16.8. The dashed curves $y=(1+x^2)^{-0.35}$ are the best fit to the data. Reprinted from (Priezjev, 2010).

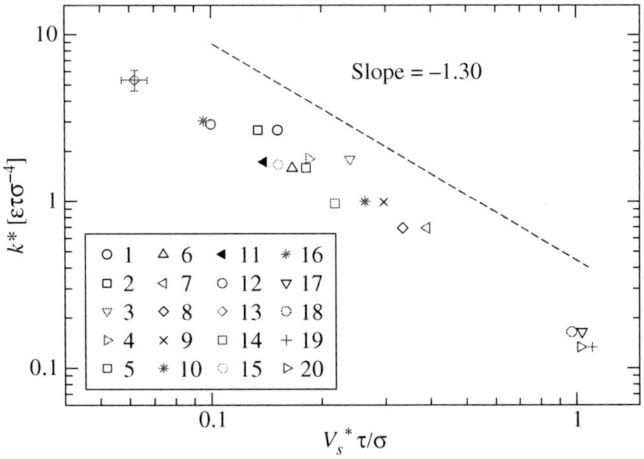

Figure 16.8 The normalization parameters V_s^* (in units σ/τ) and k^* (in units $\varepsilon\tau\sigma^{-4}$) used to fit the data in Figure 16.7 to Eq. (16.8). The inset shows system indices listed in Table 16.1. The dashed line with a slope -1.30 is shown for reference. Reprinted from (Priezjev, 2010).

We next argue that the onset of the nonlinear regime in Eq. (16.8) is determined by the diffusion time of fluid monomers over the distance between the nearest minima of the periodic surface potential. In the absence of shear flow, the typical trajectories of fluid monomers in the first layer near the solid wall are presented in Figure 16.9. It can be observed that the diffusive motion of fluid monomers in contact with the lower wall atoms is strongly influenced by the corrugation of the surface potential; that is, most of the time the monomer resides near the local minima of the surface potential with occasional jumps between the minima. Therefore, the elementary relaxation time within the first fluid layer can be estimated from the diffusion of monomers over the distance between the nearest minima.

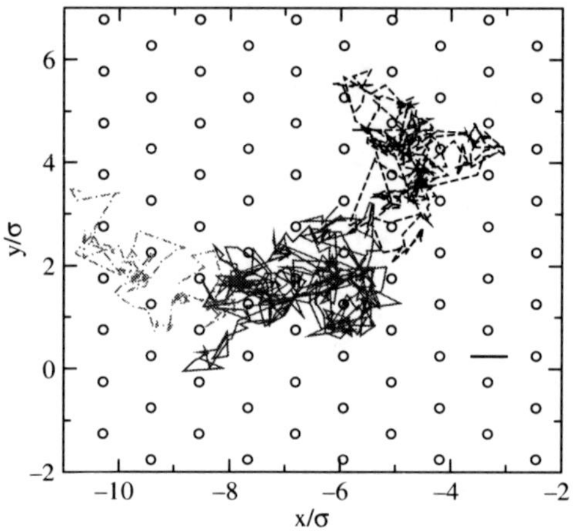

Figure 16.9 Typical trajectories, projected onto the xy plane, of monomers in the first fluid layer near the lower wall for the system 1 (see Table 16.1) at equilibrium (i.e. when both walls are at rest). The fcc lattice sites of the (111) plane are indicated by open circles. The horizontal line segment denotes the distance between nearest minima of the periodic surface potential.

In Figure 16.10 we plot the mean square displacement curves for fluid monomers within the first layer for selected systems in Table 16.1 at equilibrium (i.e. when both walls are at rest). The displacement vector as a function of time was computed along the trajectory of a fluid monomer only if it remained in the first layer during the time interval between successive measurements of the monomer position. Note that for monatomic fluids, there is a linear dependence between the mean square displacement and time, and, consequently, the diffusion coefficient is well defined by the Einstein relation. In contrast, fluid monomers that belong to a polymer chain diffuse slower than monomers in simple fluids since their dynamics is bounded by diffusion of the center of mass of the polymer chain. It should be mentioned that the evaluation of the mean square displacement curves at large times requires significant computational resources because of the low probability that a fluid monomer will remain within the first layer for a long time interval. The slope of the subdiffusive regime at intermediate times is indicated by the straight dashed line in Figure 16.10.

Finally, the comparison of the characteristic slip time of the first fluid layer in steady shear flow and the diffusion time of fluid monomers between nearest minima of the surface potential at equilibrium is presented in Figure 16.11. The diffusion time was estimated from the mean square displacement of fluid monomers within the first layer at the distance between nearest minima of the periodic surface potential. The same distance divided by the slip velocity V_s^* defines the characteristic slip time of the first fluid layer. In the case when the shear flow direction is parallel to the $[1\bar{1}0]$ fcc lattice orientation (indicated by the vertical blue arrow in the inset of Figure 16.6), the slipping distance of the first layer was computed by projecting the vector, which connects the nearest minima of the surface potential, onto the direction of flow. As seen in Figure 16.11, there is a strong correlation between the characteristic slip time of the first fluid layer and the diffusion time of fluid monomers in that layer at equilibrium. These results indicate that the linear relation between the wall shear stress and slip velocity in Eq. (16.8) holds when the slip velocity is smaller than the diffusion velocity of fluid monomers in contact with crystalline surfaces.

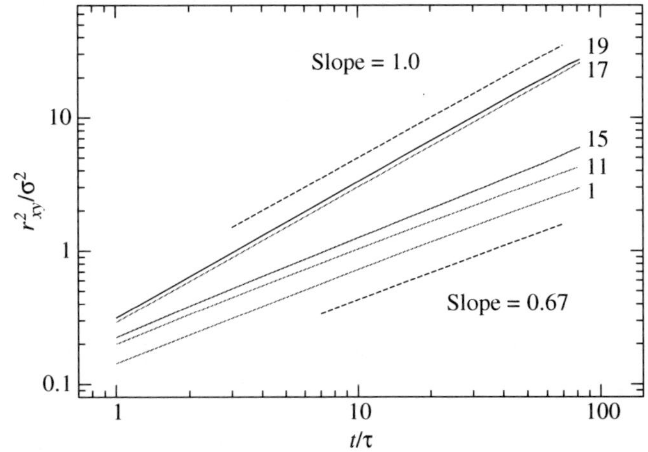

Figure 16.10 The mean square displacement of monomers in the first fluid layer at equilibrium (i.e. when both walls are at rest) as a function of time for five systems listed in Table 16.1. The dashed lines indicate slopes of 1.0 and 0.67. Adapted from (Priezjev, 2010).

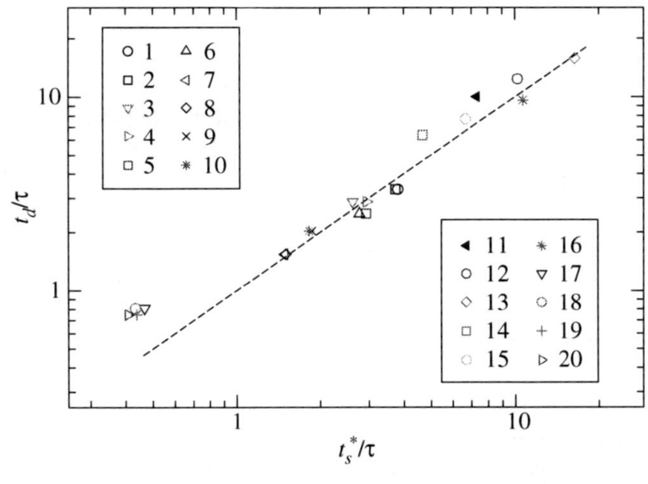

Figure 16.11 A correlation between the characteristic slip time t_s^* of the first fluid layer and the diffusion time t_d of fluid monomers between nearest minima of the periodic surface potential. The error bars are about the symbol size. The system parameters are listed in Table 16.1. The dashed line $y = x$ is shown as a reference. Adapted from (Priezjev, 2010).

16.4.4 Friction coefficient and induced fluid structure

The fluid structure near flat solid walls is characterized by density layering perpendicular to the surface and ordering of fluid monomers within the layers (Kaplan & Kauffmann, 2006). Examples of oscillatory density profiles in a polymer melt near confining walls were presented in Figure 16.3. It is intuitively expected that enhanced fluid density layering normal to the surface (obtained, for example, by increasing fluid pressure or wall-fluid interaction energy) would correspond to a larger friction coefficient at the liquid-solid interface. However, this correlation does not always hold; for example, the amplitude of fluid density oscillations near flat structureless walls might be large, but the friction coefficient is zero. As emphasized in the original paper by Thompson and Robbins (1990), the surface-induced fluid ordering within the first layer of monomers correlates well with the degree of slip at the liquid-solid interface. The measure of the induced order in the adjacent fluid layer is the static structure factor, which is defined as follows:

$$S(\mathbf{k}) = \frac{1}{N_\ell}\left|\sum_{j=1}^{N_\ell} e^{i\mathbf{k}\cdot\mathbf{r}_j}\right|^2 \tag{16.9}$$

where $\mathbf{k}$ is a two-dimensional wave vector, $\mathbf{r}_j = (x_j, y_j)$ is the position vector of the jth monomer, and $N_\ell = S(0)$ is the number of monomers within the layer (Thompson & Robbins, 1990). The probability of finding fluid monomers is greater near the minima of the periodic surface potential; and, therefore, the structure factor typically contains a set of sharp peaks at the reciprocal lattice vectors. It is well established that the magnitude of the largest peak at the first reciprocal lattice vector is one of the main factors that determines the value of the slip length at the interface between flat crystalline surfaces and monatomic fluids (Thompson & Robbins, 1990; Barrat & Bocquet, 1999; Priezjev, 2007; Priezjev, 2007a) or polymer melts (Thompson *et al.* 1995; Niavarani & Priezjev, 2008; Priezjev, 2009; Priezjev & Troian, 2004).

Next, we discuss the influence of the wall-fluid interaction energy, wall lattice type and orientation, and slip velocity on the structure factor computed in the first fluid layer. First, the effect of the wall-fluid interaction energy is illustrated in Figure 16.12 for monatomic fluids in contact with the (111) plane of the fcc wall lattice. The height of the surface-induced peaks in the structure factor is slightly larger at higher surface energy. The magnitude of the peak in the shear flow direction is $S(8.65\sigma^{-1}, 0) = 0.98$ for $\varepsilon_{wf} = 0.3\varepsilon$ and $S(8.65\sigma^{-1}, 0) = 1.06$ for $\varepsilon_{wf} = 0.4\varepsilon$. Notice that the height of the circular ridge, which is characteristic of short range order of fluid monomers, is larger than the amplitude of the induced peaks at the main reciprocal lattice vectors. A similar trend in the height of the peaks in the structure factor was observed previously for monatomic fluids near fcc walls with higher density $\rho_w = 2.73\sigma^{-3}$ (Priezjev, 2007).

Figure 16.13 shows the structure factor computed in the first fluid layer for polymer melts with chains $N = 20$ in contact with the (111) plane of the fcc wall lattice. As indicated by the horizontal arrows, the shear flow is oriented along the $[11\bar{2}]$ direction in Figure 16.13 (a) and parallel to the $[1\bar{1}0]$ direction in Figure 16.13 (b). Due to the hexagonal symmetry of the (111) lattice plane, the structure factor exhibits six peaks at the shortest reciprocal lattice vectors. Note that only two main peaks are present in the first quadrant. The magnitude of the peaks is the same at small slip velocities. The lattice orientation with respect to the shear flow direction determines the location of the main peaks.

Lastly, the effect of slip velocity on the magnitude of the substrate-induced peaks in the structure factor is presented in Figure 16.14 for polymer chains in contact with the (001) plane of the bcc wall lattice. In the case of small slip velocity $V_s = 0.012\sigma/\tau$, the magnitude of the main peaks at the main reciprocal lattice vectors $\mathbf{G}_1 = (6.18\sigma^{-1}, 0)$ and $\mathbf{G}_2 = (0, 6.18\sigma^{-1})$ is the same [see Figure 16.14(a)]. With increasing slip velocity, the height of the induced peak along the shear flow direction decreases significantly, whereas the amplitude of the peak in the perpendicular direction is less affected by the slip velocity [see Figure 16.14(b)]. More quantitatively, the velocity dependence of the main peaks in the structure factor along with the contact density and temperature of the first layer are presented in Figure 16.15. This behaviour is consistent with the density and temperature profiles reported in Section IIIA. Similar conclusions were obtained in the previous study on slip flows in dense polymer films (Priezjev, 2009).

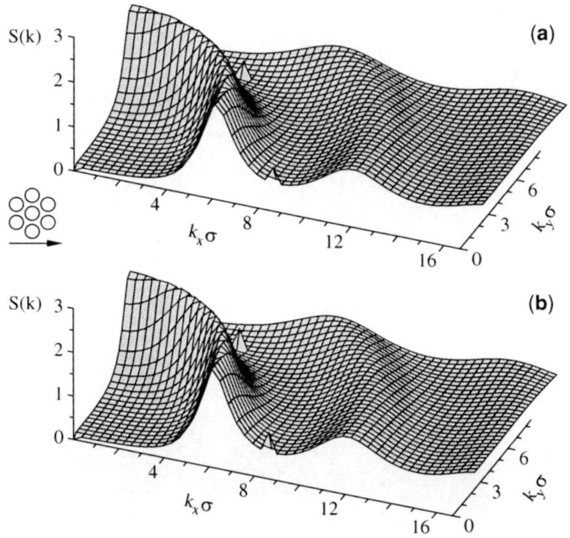

Figure 16.12 Two-dimensional structure factor $S(k_x, k_y)$ computed in the first fluid layer for $N = 1$ and $U = 0.05\sigma/\tau$ [systems (a) 19 and (b) 17 in Table 16.1]. The wall-fluid interaction energy is (a) $\varepsilon_{wf} = 0.3\varepsilon$ and (b) $\varepsilon_{wf} = 0.4\varepsilon$. The shear flow direction (denoted by the horizontal arrow) is parallel to the $[11\bar{2}]$ orientation of the (111) face of the fcc wall lattice (open circles). Reprinted from (Priezjev, 2010).

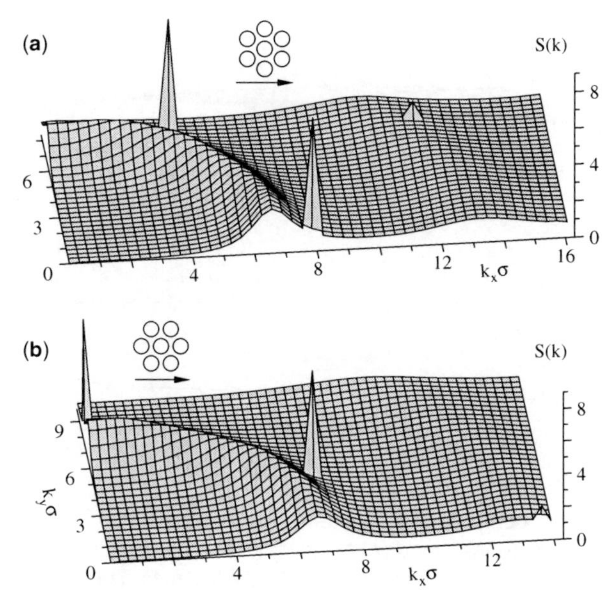

Figure 16.13 Structure factor $S(k_x, k_y)$ averaged in the first fluid layer for $N = 20$ polymer systems (a) 5 and (b) 6 (see parameters in Table 16.1). The sharp peaks are located at (a) $(7.86\sigma^{-1}, 0)$ and $(3.93\sigma^{-1}, 6.81\sigma^{-1})$, and (b) $(6.81\sigma^{-1}, 3.93\sigma^{-1})$ and $(0, 7.86\sigma^{-1})$. In each case, horizontal arrows indicate the shear flow direction with respect to the orientation of the (111) plane of the fcc wall lattice (denoted by open circles). The upper wall speed is $U = 0.05\sigma/\tau$ in both cases. Adapted from (Priezjev, 2010).

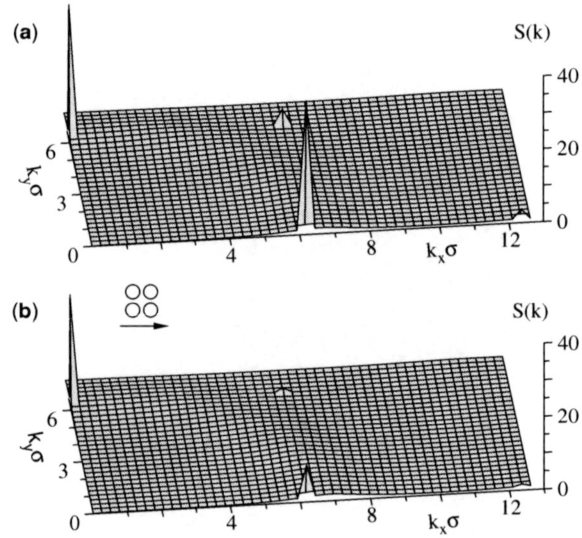

Figure 16.14 Structure factor $S(k_x, k_y)$ computed in the first fluid layer for $N = 20$ polymer system 12 (see Table 16.1). The upper wall speed and slip velocity are (a) $U = 0.05\sigma/\tau$ and $V_s = 0.012\sigma/\tau$, and (b) $U = 2.0\sigma/\tau$ and $V_s = 0.51\sigma/\tau$ respectively. The location of the main induced peak in the shear direction is $(6.18\sigma^{-1}, 0)$. The horizontal arrows denote the shear flow direction with respect to the orientation of the (001) face of the bcc wall lattice (open circles). Reprinted from (Priezjev, 2010).

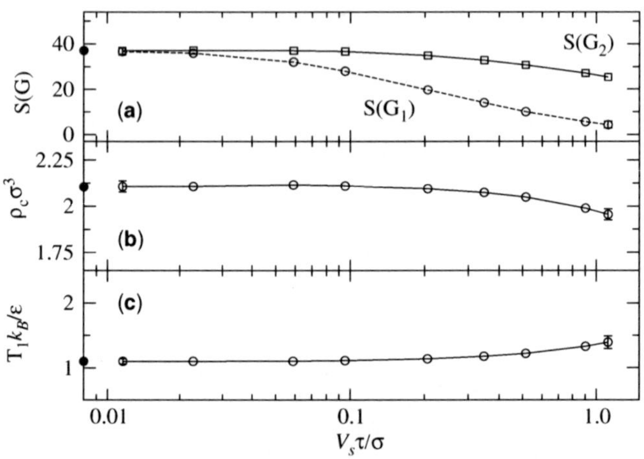

Figure 16.15 (a) Structure factor computed at the main reciprocal vectors of the bcc lattice $\mathbf{G}_1 = (6.18\sigma^{-1}, 0)$ and $\mathbf{G}_2 = (0, 6.18\sigma^{-1})$, (b) contact density, and (c) temperature of the first fluid layer as a function of the slip velocity $V_s = L_s\dot{\gamma}$ for polymer chains $N = 20$ (system 12 in Table 16.1). The same parameters at equilibrium (i.e. when $V_s = 0$) are denoted by closed symbols.

The correlation between surface-induced structure in the first fluid layer and the friction coefficient was investigated previously for polymer melts with chains $N = 20$ confined by atomically flat walls (Niavarani & Priezjev, 2008; Priezjev, 2009). The simulations were performed at fluid densities $0.86 \le \rho\sigma^3 \le 1.11$ and the wall density $\rho_w = 1.40\sigma^{-3}$. It was found that the data for the friction coefficient at different shear rates and fluid densities collapsed onto a master curve when plotted as a function of a variable $S(0)/[S(\mathbf{G}_1)\rho_c]$, where $\mathbf{G}_1$ is the first reciprocal lattice vector in the shear flow direction (Niavarani & Priezjev, 2008; Priezjev, 2009). The collapse of the data holds at relatively small values of the friction coefficient $k \lesssim 4\varepsilon\tau\sigma^{-4}$ and for slip lengths larger than approximately 5σ. Although these results are promising, the simulations were limited to a single wall density and the $[11\bar{2}]$ orientation of the (111) plane of the fcc wall lattice. Therefore, it is desirable to perform additional simulations in a more extended parameter range.

In the present study, a number of parameters that affect slippage at the liquid-solid interface have been examined, that is, fluid and wall densities, polymer chain length, wall lattice type and orientation, wall-fluid interaction energy, thermal and solid walls (see Table 16.1). We first consider the linear-response regime where the friction coefficient weakly depends on the slip velocity ($k/k^* \gtrsim 0.8$ in Figure 16.7). Figure 16.16 shows the ratio L_s/μ (an inverse friction coefficient) as a function of the variable $S(0)/[S(\mathbf{G}_1)\rho_c]$ computed in the first fluid layer for twenty systems listed in Table 16.1. In the case when the shear flow direction is parallel to the $[1\bar{1}0]$ orientation of the fcc lattice [e.g., see Figure 16.13 (b)], the structure factor was computed at the shortest reciprocal lattice vector $\mathbf{G}_1$ aligned at an angle of 30° with respect to the $\hat{x}$ axis. The data in Figure 16.16 are well described by a power-law fit with the slope 1.13. Remember also that the amplitude of the main peaks in the structure factor and the contact density of the first fluid layer are the same at equilibrium and small slip velocities (e.g. see Figure 16.15). Therefore, the results in Figure 16.16 suggest that at the interface between simple or polymer fluids and crystalline surfaces, the ratio of the slip length and fluid viscosity at low shear rates can be predicted from equilibrium measurements of the structure factor and contact density of the first fluid layer. In other words, the value of parameter k^* in Eq. (16.8) is determined by the induced fluid structure in the first layer at equilibrium.

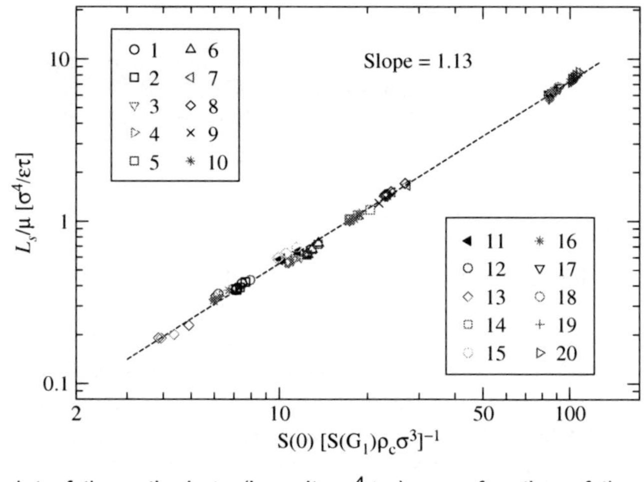

Figure 16.16 Log-log plot of the ratio L_s/μ (in units $\sigma^4/\varepsilon\tau$) as a function of the variable $S(0)/[S(\mathbf{G}_1)\rho_c]$ computed in the first fluid layer at low shear rates. The system parameters are listed in Table 16.1. The dashed line $y = 0.041x^{1.13}$ is the best fit to the data. Adapted from (Priezjev, 2010).

In Figure 16.17, we report the dependence of the friction coefficient ($k = \mu/L_s$) on the structure factor and contact density of the first fluid layer at all shear rates examined in this study. These results call for several comments. Note that at higher shear rates the derivative of L_s/μ with respect to the variable $S(0)/[S(\mathbf{G}_1)\rho_c]$ for several systems listed in Table 16.1 deviates significantly from the slope 1.13, which is indicated by a straight line in Figure 16.17 (a). In addition, for any two systems with the same ρ_w and ε_{wf}, the ratio L_s/μ as a function of $S(0)/[S(\mathbf{G}_1)\rho_c]$ depends on the orientation of the fcc wall lattice with respect to the shear flow direction. Although the data in Figure 16.17(b) are somewhat scattered, the results show the same trend; namely, the friction coefficient decreases when the magnitude of the normalized peak in the structure factor is reduced. The collapse of the data for L_s versus $S(\mathbf{G}_1)/S(0)$ was reported by Thompson & Robbins (1990) for monatomic fluids and crystalline walls when $L_s \lesssim 3.5\sigma$ and the boundary conditions are rate independent. In the present study, the slip lengths are greater than about 5σ except for the systems 13 and 16 where $L_s \approx 3\sigma$ at low shear rates.

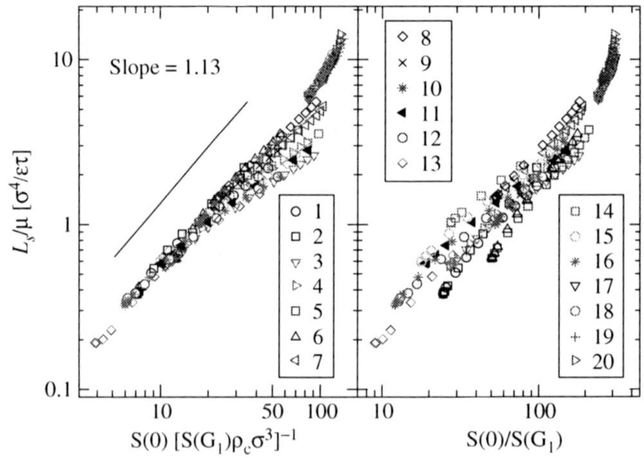

Figure 16.17 Log-log plot of the ratio L_s/μ (in units of $\sigma^4/\varepsilon\tau$) as a function of variables (a) $S(0)/\ [S(\mathbf{G}_1)\rho_c]$ and (b) $S(0)/S(\mathbf{G}_1)$ computed in the first fluid layer at all shear rates examined. The system parameters are given in Table 16.1. The same data as in Figure 16.7. The black line with a slope 1.13 is shown for reference. Reprinted from (Priezjev, 2010).

16.5 CONCLUSIONS

In summary, we investigated the dynamic behaviour of the slip length at interfaces between polymeric or monatomic fluids and dense crystalline surfaces using molecular dynamics simulations. The polymer melt was modelled as a collection of bead-spring linear flexible chains well below the entanglement length. We considered shear flow conditions at relatively low fluid densities (pressures) and weak wall-fluid interaction energies so that fluid velocity profiles are linear across the channel and slip lengths are larger than several molecular diameters at all shear rates examined. The simulations were performed for different wall and fluid densities, chain lengths, surface energies, lattice types, thermal or solid walls. It was found that the slip length does not depend on the wall lattice orientation with respect to the flow direction only at low shear rates, whereas the slip is enhanced at high shear rates when the flow direction is parallel to the crystallographic axis of the substrate.

We reformulated the boundary conditions in terms of the friction coefficient at the liquid-solid interface and slip velocity. It was shown that in the steady shear flow of either monatomic fluids or polymer melts, the friction coefficient undergoes a transition from a constant value to the power-law decay as a function of the slip velocity. The numerical results indicate that the characteristic velocity of the transition is determined by the diffusion time of fluid monomers over the distance between nearest minima of the substrate potential. It was demonstrated that the friction coefficient at small slip velocities is a function of the magnitude of the surface-induced peak in the structure factor and the contact density of the first fluid layer. These results suggest that the value of the slip length in the low-shear-rate limit can be predicted from equilibrium measurements of the induced structure in the first fluid layer. Future studies will show how general these conclusions are and whether they hold for more realistic potentials or different thermostating procedures.

Acknowledgments

Financial support from the National Science Foundation (CBET-1033662) and the Petroleum Research Fund of the American Chemical Society (48631-G9) is gratefully acknowledged. Computational work in support of this research was performed at Michigan State University's High Performance Computer Facility.

REFERENCES

Allen M. P. and Tildesley D. J. (1989). Computer Simulation of Liquids. Oxford University Press, New York, USA.

Asproulis N. and Drikakis D. (2010). Boundary slip dependency on surface stiffness. *Physical Review E*, **81**(6), 061503.

Barrat J.-L. and Bocquet L. (1999). Influence of wetting properties on hydrodynamic boundary conditions at a fluid/solid interface. *Faraday Discuss*, **112**, 119–128.

Baudry J., Charlaix E., Tonck A. and Mazuyer D. (2001). Experimental evidence for a large slip effect at a nonwetting fluid-solid interface. *Langmuir*, **17**, 5232–5236.

Baumchen O., Fetzer R. and Jacobs K. (2009). Reduced interfacial entanglement density affects the boundary conditions of polymer flow. *Physical Review Letters*, **103**, 247801.

Baumchen O. and Jacobs K. (2010). Slip effects in polymer thin films. *Journal of Physics Condensed Matter*, **22**, 033102.

Bird R. B., Armstrong R. C. and Hassager O. (1987). Dynamics of Polymeric Liquids. John Wiley and Sons.

Bocquet L. and Barrat J. (1994). Hydrodynamic boundary conditions, correlation functions, and Kubo relations for confined fluids. *Physical Review E*, **49**, 3079–3092.

Bocquet L. and Charlaix E. (2010). Nanofluidics, from bulk to interfaces. *Chemical Society Reviews*, **39**(3), 1073–1095.

Choi C., Westin K. J. and Breuer K. S. (2003). Apparent slip flows in hydrophilic and hydrophobic microchannels. *Physics of Fluids*, **15**, 2897–2902.

Churaev N. V., Sobolev V. D. and Somov A. N. (1984). Slippage of liquids over lyophobic solid surfaces. *Journal of Colloid and Interface Science*, **97**, 574–581.

Dhondi S., Pereira G. G. and Hendy S. C. (2009). Molecular dynamics simulations of polymeric fluids in narrow channels: methods to enhance mixing. *Physical Review E*, **80**(3), 036309.

Feuillebois F., Bazant M. Z. and Vinogradova O. I. (2009). Effective slip over superhydrophobic surfaces in thin channels. *Physical Review Letters*, **102**(2), 026001.

Galea T. M. and Attard P. (2004). Molecular dynamics study of the effect of atomic roughness on the slip length at the fluid–solid boundary during shear flow. *Langmuir*, **20**, 3477–3482.

Grest G. S. and Kremer K. (1986). Molecular dynamics simulation for polymers in the presence of a heat bath. *Physical Review A*, **33**(5), 3628–3631.

Heinbuch U. and Fischer J. (1989). Liquid flow in pores: slip, no-slip, or multilayer sticking. *Physical Review A*, **40**, 1144–1146.

Irving J. H. and Kirkwood J. G. 1950. The statistical mechanical theory of transport processes. IV. The equations of hydrodynamics. *The Journal of Chemical Physics*, **18**, 817–829.

Jabbarzadeh A., Atkinson J. D. and Tanner R. I. 1999. Wall slip in the molecular dynamics simulation of thin films of hexadecane. *The Journal of Chemical Physics*, **110**, 2612–2620.

Kaplan W. D. and Kauffmann Y. (2006). Structural order in liquids induced by interfaces with crystals. *Annual Review of Materials Research*, **36**, 1–48.

Khare R., de Pablo J. J. and Yethiraj A. (1996). Rheology of confined polymer melts. *Macromolecules*, **29**, 7910–7918.

Khare R., Keblinski P. and Yethiraj A. (2006). Molecular dynamics simulations of heat and momentum transfer at a solid–fluid interface: Relationship between thermal and velocity slip. *International Journal of Heat and Mass Transfer*, **49**, 3401–3407.

Kim B. H., Beskok A. and Cagin T. (2008). Thermal interactions in nanoscale fluid flow: molecular dynamics simulations with solid-liquid interfaces. *Microfluidics and Nanofluidics*, **5**, 551–559.

Koike A. and Yoneya M. (1998). Chain length effects on frictional behavior of confined ultrathin films of linear alkanes under shear. *The Journal of Physical Chemistry B*, **102**, 3669–3675.

Kong L.-T., Denniston C. and Müser M. H. (2010). The crucial role of chemical detail for slip-boundary conditions: molecular dynamics simulations of linear oligomers between sliding aluminum surfaces. *Modelling and Simulation in Materials Science and Engineering*, **18**, 034004.

Kremer K. and Grest G. S. (1990). Dynamics of entangled linear polymer melts: a molecular–dynamics simulation. *The Journal of Chemical Physics*, **92**, 5057–5086.

Liu C. and Li Z. (2009). Flow regimes and parameter dependence in nanochannel flows. *Physical Review E*, **80**(3), 036302.

Manias E., Hadziioannou G. and ten Brinke G. (1996). Inhomogeneities in sheared ultrathin lubricating films. *Langmuir*, **12**, 4587–4593.

Martini A., Hsu H.-Y., Patankar N. A. and Lichter S. (2008). Slip at high shear rates. *Physical Review Letters*, **100**(20), 206001.

McBride S. P. and Law B. M. (2009). Viscosity-dependent liquid slip at molecularly smooth hydrophobic surfaces. *Physical Review E – Statistical, Nonlinear, and Soft Matter Physics*, **80**, 060601(R).

Niavarani A. and Priezjev N. V. (2008). Rheological study of polymer flow past rough surfaces with slip boundary conditions. *The Journal of Chemical Physics*, **129**, 144902.

Niavarani A. and Priezjev N. V. (2008). Slip boundary conditions for shear flow of polymer melts past atomically flat surfaces. *Physical Review E – Statistical, Nonlinear, and Soft Matter Physics*, **77**, 041606.

Niavarani A. and Priezjev N. V. (2010). Modeling the combined effect of surface roughness and shear rate on slip flow of simple fluids. *Physical Review E*, **81**(1), 011606.

Pahlavan A. A. and Freund J. B. (2011). Effect of solid properties on slip at a fluid–solid interface. *Physical Review E*, **83** (2), 021602.

Priezjev N. V. and Troian S. M. (2006). Influence of periodic wall roughness on the slip behaviour at liquid/solid interfaces: molecular-scale simulations versus continuum predictions. *Journal of Fluid Mechanics*, **554**, 25–46.

Priezjev N. V. (2007). Effect of surface roughness on rate-dependent slip in simple fluids. *The Journal of Chemical Physics*, **127**, 144708.

Priezjev N. V. (2007). Rate-dependent slip boundary conditions for simple fluids. *Physical Review E*, **75**(5), 051605.

Priezjev N. V. (2009). Shear rate threshold for the boundary slip in dense polymer films. *Physical Review E*, **80**(3), 031608.

Priezjev N. V. (2010). Relationship between induced fluid structure and boundary slip in nanoscale polymer films. *Physical Review E*, **82**(5), 051603.

Priezjev N. V. (2011). Molecular diffusion and slip boundary conditions at smooth surfaces with periodic and random nanoscale textures. *The Journal of Chemical Physics*, **135**, 204704.

Priezjev N. V., Darhuber A. A. and Troian S. M. (2005). Slip behavior in liquid films on surfaces of patterned wettability: comparison between continuum and molecular dynamics simulations. *Physical Review E*, **71**(4), 041608.

Priezjev N. V. and Troian S. M. (2004). Molecular origin and dynamic behavior of slip in sheared polymer films. *Physical Review Letters*, **92**(1), 018302.

Rothstein J. P. (2010). Slip on superhydrophobic surfaces. *Annual Review of Fluid Mechanics*, **42**, 89–109.

Sanchez-Reyes J. and Archer L. A. (2003). Interfacial slip violations in polymer solutions: role of microscale surface roughness. *Langmuir*, **19**, 3304–3312.

Schmatko T., Hervet H. and Leger L. (2005). Friction and slip at simple fluid-solid interfaces: the roles of the molecular shape and the solid-liquid interaction. *Physical Review Letters*, **94**, 244501.

Schmatko T., Hervet H. and Leger L. (2006). Effect of nanometric-scale roughness on slip at the wall of simple fluids. *Langmuir*, **22**, 6843–6850.

Servantie J. and Muller M. (2008). Temperature dependence of the slip length in polymer melts at attractive surfaces. *Physical Review Letters*, **101**(2), 026101.

Smith E. D., Robbins M. O. and Cieplak M. (1996). Friction on adsorbed monolayers. *Physical Review B*, **54**(11), 8252–8260.

Sofos F. D., Karakasidis T. E. and Liakopoulos A. (2009). Effects of wall roughness on flow in nanochannels. *Physical Review E*, **79**(2), 026305.

Sokhan V. P., Nicholson D. and Quirke N. (2001). Fluid flow in nanopores: an examination of hydrodynamic boundary conditions. *The Journal of Chemical Physics*, **115**, 3878–3887.

Soong C. Y., Yen T. H. and Tzeng P. Y. (2007). Molecular dynamics simulation of nanochannel flows with effects of wall lattice-fluid interactions. *Physical Review E*, **76**(3), 036303.

Sparreboom W., Van Den Berg A. and Eijkel J. C. (2009). Principles and applications of nanofluidic transport. *Nature Nanotechnology*, **4**, 713–720.

Stevens M. J., Mondello M., Grest G. S., Cui S. T., Cochran H. D. and Cummings P. T. (1997). Comparison of shear flow of hexadecane in a confined geometry and in bulk. *The Journal of Chemical Physics*, **106**, 7303–7314.

Thompson P. A. and Robbins M. O. (1990). Shear flow near solids: Epitaxial order and flow boundary conditions. *Physical Review A*, **41**, 6830–6837.

Thompson P. A., Robbins M. O. and Grest G. S. (1995). Structure and shear response in nanometer thick films. *Israel Journal of Chemistry*, **35**, 93.

Thompson P. A. and Troian S. M. (1997). A general boundary condition for liquid flow at solid surfaces. *Nature*, **389**, 360–362.

Tomassone M. S., Sokoloff J. B., Widom A. and Krim J. (1997). Dominance of phonon friction for a xenon film on a silver (111) surface. *Physical Review Letters*, **79**(24), 4798–4801.

Travis K. P. and Gubbins K. E. (2000). Poiseuille flow of Lennard-Jones fluids in narrow slit pores. *The Journal of Chemical Physics*, **112**, 1984–1994.

Ulmanella U. and Ho C.-M. (2008). Molecular effects on boundary condition in micro/nanoliquid flows. *Physics of Fluids*, **20**, 101512.

Yang S. C. and Fang L. B. (2005). Effect of surface roughness on slip flows in hydrophobic and hydrophilic microchannels by molecular dynamics simulation. *Molecular Simulation*, **31**, 971–977.

Yong X. and Zhang L. T. (2010). Investigating liquid-solid interfacial phenomena in a Couette flow at nanoscale. *Physical Review E*, **82**(5), 056313.

Zhu Y. and Granick S. (2001). Rate-dependent slip of Newtonian liquid at smooth surfaces. *Physical Review Letters*, **87**, 961051–961054.

Zhu Y. and Granick S. (2002). Limits of the hydrodynamic no-slip boundary condition. *Physical Review Letters*, **88**, 1061021–1061024.

Chapter 17

Understanding slip at the nanoscale in fluid flows using atomistic simulations

T. E. Karakasidis and A. Liakopoulos

17.1 INTRODUCTION – DEFINITION OF SLIP

17.1.1 Continuum theory and slip

In classical fluid mechanics it is generally accepted that the relative velocity between a solid wall and the adjacent moving fluid, at a solid/fluid interface, is zero. This condition is known as the no-slip boundary condition in viscous flow theory. If this is not the case, a non-zero tangential velocity component at the wall is termed slip velocity.

The no-slip boundary condition at a fluid/solid interface is of fundamental importance in classical fluid dynamics. The celebrated boundary-layer theory, originated by L. Prandtl, lies firmly on this assumption, in the sense, that if one accepts the no-slip boundary condition, the concept of a thin boundary layer existing near a solid wall for high Reynold number flows is a natural consequence. In the continuum theory of laminar viscous flow, the no-slip condition coupled with the constant-density, constant-viscosity Navier-Stokes equations determine analytically the velocity distribution in three simple geometries: a) Linear velocity profile in "shear-driven" flow between two infinite parallel plates (planar Couette flow, b) Parabolic velocity profile for pressure- driven fully-developed flow between two infinite parallel plates (planar Poiseuille flow), and c) Paraboloid of revolution profile in the case of pressure-driven fully-developed flow in a straight pipe of circular cross section (axisymmetric Poiseuille flow).

The no-slip condition assumption has been disputed in earlier stages of the development of viscous flow theory (see e.g. Goldstein 1938; Goldstein 1969). Recently, the question of the validity of the no-slip condition, especially in the case of hydrophobic surfaces, has come back to the forefront of research activity with the new developments in micro-fluid mechanics. Thompson and Troian (1997) in their study placed the appropriate research framework on this subject. Recent literature reviews on the subject are those by Lauga *et al.* (2005), Neto *et al.* (2005) and Maali and Bhushan (2008). From both experimental measurements and atomic scale simulations it has been argued that wall slip is possible and, under certain conditions, this is the case. When wall velocity is not zero, we often reduce the problem to the point on which, based on the behavior of the fitting curve, we would have had zero velocity and the difference in length is called slip length. Alternatively, slip length b is defined as the ratio of relative velocity to the slope of the velocity normal to the surface, that is

$$u - u_s = b\frac{\partial u}{\partial n} \tag{17.1}$$

where u_s is the velocity at the solid surface (see Figure 17.1).

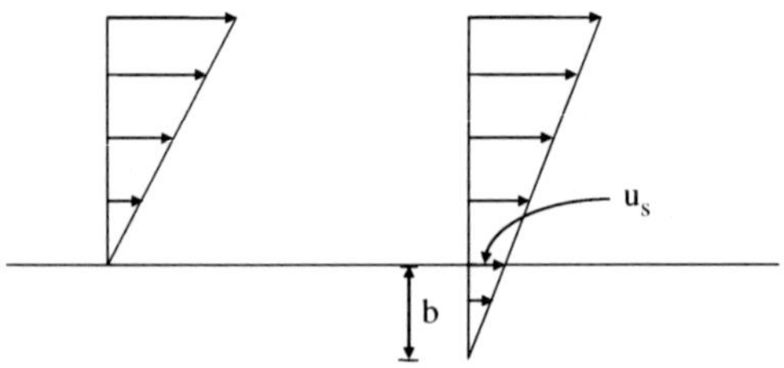

Figure 17.1 Definition sketch of slip length, b, and slip velocity, u_s.

Since measuring the flow velocity very close to the wall is technically difficult, the velocity profile at some distance from the wall is often measured and the wall velocity is extracted through fitting an appropriate curve, such as a linear curve for the case of Couette flow, or a parabolic curve for Poiseuille flow, and so on. based mainly on the continuum theory results mentioned above.

Nowadays, it is generally accepted that slip is observed on solid hydrophobic surfaces, while the scientific community has not yet reached a consensus regarding hydrophilic surfaces: while many researchers claim that they have not observed slip, others claim exactly the opposite. It should be noted here that there is a strong correlation between surface wettability and observed fluid slip, while the surface's crystallographic structure seems to also affect the above phenomena (Voronov *et al.* 2006).

17.1.2 Incorporating velocity slip in continuum models

In flows of fluids like poly melts or crude oil compressibility may play an important role in cases of high pressure (Hatzikiriakos & Dealy 1992). The stick-slip polymer extrusion instability is attributed to the combined effect of compressibility with non-monotonic slip laws relating the wall shear stress to slip velocity (Hatzikiriakos & Dealy 1992) as confirmed by phenomenological models (Debbeldam & Molenaar, 2003) and computer simulations (Taliadourou *et al.* 2007). A mathematical model describing time-dependent pressure driven flow of compressible polymeric fluids subject to pressure dependent slip was developed by Tang and Kalyon (2008a, b). Poya *et al.* (2011) present a model for planar and axisymmetric steady, laminar Poiseuille flows of a weakly compressible Newtonian fluid assuming that slip occurs along the wall following Navier's slip equation and that the density obeys a linear equation of state. Slip models for polymers are also reviewed in (Hatzikiriakos, 2011).

17.2 IMPORTANCE OF SLIP

Continuous flow in microconduits and nanoconduits requires large pressure differences (pressure drop) due to the large hydrodynamic resistance. In case of partial slippage, the situation is expected to improve significantly, especially if the slip length is of the same order of magnitude as the channel radius, or even longer (Bocquet & Barrat, 2007). Slippage on the fluid-solid interface reduces the drag force of fluid flow, something that is particularly important in some applications, including biosensors of micro- and nano-scale (Bhushan, 2007). Large slip lengths are required in many practical applications (Reyes *et al.* 2002, Whitesides, 2006) such as membranes used for waste water treatment or water desalination (Holt *et al.* 2006).

In this area, super-hydrophobic surfaces play an important role (Shirtcliffe *et al.* 2010). Super-hydrophobicity is the phenomenon in which roughness and hydrophobicity combine so as to produce unusually hydrophobic surfaces. Other than their theoretical interest, super-hydrophobic surfaces can constitute a revolutionary suggestion to micro-flow systems, reducing viscous drag forces and enhancing transport phenomena and transverse flows. The significant fluid movement on such super-hydrophobic surfaces is translated into their "self-cleaning" and, as a result, fluid drops rotate (instead of slipping) under the influence of gravity and bounce (instead of spreading) after collision. A combination of materials with low surface energy and surface structure of specific roughness can produce a super-hydrophobic surface.

The application of super-hydrophobic surfaces is not simply limited to a theoretical or laboratory level, since the textile industry has already adopted the results of relevant research and has advanced to massive production (Xue *et al.* 2008). Another possible application is the area of agriculture and environment, since the self-cleaning property can be specifically used for pollutant removal. Furthermore, several phenomena, such as snow sticking on the streets, conductivity of electricity conductors, surface oxidation, textiles and padding, micro-fluidic systems and non-wetting liquid transfer can be studied in light of the new developments that the design of super-hydrophobic surfaces brings about (Feng *et al.* 2002).

It is becoming clear that proposals for the construction of super-hydrophobic surfaces with appropriate characteristics and properties are really significant. Slip length and velocity, friction force, pressure, viscosity, diffusion and thermal conductivity of such surfaces present an intense interest and can provide answers to basic scientific questions, as well as to subjects of technological interest that are focused on practical matters of surface construction with desirable properties.

The literature that refers to micro-system design, construction and applications is voluminous and cannot be adequately covered in the framework of this chapter. We indicatively refer to the most important applications of micro-fluid mechanical formations in the areas of Environment-Sensors, Water Technology and Biomedicine. Below we mention a few more details on the applications. In environmental applications detecting dangerous species is of particularly great importance in various cases, such as in special workplaces. The potential of microdevices to be able to perform chemical analysis for the detection of many chemicals makes them very important, as they can be embodied on devices and so on. Slipping length increase means the reduction of power required for fluid movement. In water technology, researchers are interested in nano-membranes for the purification/treatment of water as well as other nanotechnology applications on seawater desalination (Holt *et al.* 2006). The elements of the most innovative desalination devices contain micro-conduits and nano-conduits, mainly carbon nano-conduits with walls presenting important hydrophobicity thus having smaller pressure drops, similar to the conduits that are proposed to be studied research activity section described next. In biomedical research, a micro-fluid mechanic setup is defined by the fact that it has channels that are smaller than 1 mm. Usually, such setups are used for manipulation and discharge of liquid phase biological samples (blood, cells) and bio-molecular solutions (proteins, DNA, biochemical). Given the fact that, in biological samples, the size of cells can be several tens of μm, the relevant biomedical setups usually have channels in the order of 100 nm. Such setups are already used in cellmetrics, in capillary electrophoresis, DNA analysis, cell separation and other clinical diagnostic applications. They are also used on the fast developing field of biosensors for the interaction of a biological sample with the natural transducer in a way that reading electronics is not affected, continuous sample flow is ensured, and analysis of many parameters per sample at the same time is achieved. Other than the above-mentioned setup, there exist, in an experimental stage, other setups with channels smaller than 1 μm for research of cell behavior on external stimulations and reactive reagents.

17.3 EXPERIMENTAL MEASUREMENT OF SLIP

Solid-fluid interface phenomena are particularly important in flows inside small channels and confined media. Unfortunately, the exact measurement of velocity near a solid wall and of wall shear stress is beyond the potential of even the most advanced experimental methods. Optical methods such as microPIV (micro Particle Image Velocimetry) are also limited by their resolution capabilities that reaches several μm from the flow walls (Sinton, 2003). Early research showed that fluids can exhibit slippage when in contact with non-wetted surfaces (Schnell 1956; Churaev *et al.* 1984). Further experiments with Surface Force Apparatus (SFA) or Atomic Force Microscopy (AFM) (Vinogradova & Yakubov, 2003) and Particle Image Velocimetry (Tretheway & Meinhart, 2002) reported slip length on hydrophobic surfaces while there was no mention of slip on hydrophilic surfaces (Baudry *et al.* 2001; Cottin-Bizone *et al.* 2005; Vinogradova & Yakubov, 2003; Tretheway & Meinhart, 2002; Honig & Ducker 2007; Maali *et al.* 2008). Slip on hydrophilic surfaces has been reported in other experiments SFA (Zhu & S. Granick, 2001), AFM (Craig *et al.* 2001; Bonaccurso *et al.* 2002), PIV (Joseph & Tabeling, 2005), as well as other methods (Pit *et al.* 2000; Schmatko *et al.* 2005). These unexpected results rather pertain to sample preparation and measurement analysis. For example, there is electrostatic double-layer force and Stoke's friction on experimental results that need to be removed.

Friction resistance reduction has been measured by PIV methods and by methods of direct pressure measurement in turbulent flows over super-hydrophobic surfaces. Significant friction reduction in non-turbulent flows was not observed. Friction reduction comes about as a result of the existence of a shear-free interface in super-hydrophobic surfaces. Slip velocities and friction reduction appeared to increase with Reynolds number. These results can provide a new mechanism for friction reduction in floatable media (Daniello *et al.* 2009).

Using a special experimental method (Internal Reflection-Fluorescence Recovery), the local velocity was measured (average in 50 nm from the solid wall) for two different fluids (squalane, hexadecane) over flat surfaces with similar roughness but with gradually reducing solid-fluid interaction (Schmatko *et al.* 2005). Measurements show that it is not only the solid-fluid interactions that play a role, but also the molecular shape (simple molecule or more complex molecules), which affects friction and the degree of slippage on the wall. The study of the effect of crystallographic orientation of channel wall surface with molecular dynamics showed that there is an important effect on slip, velocity profile and flowrate (Soong *et al.* 2007). In this study, slip was found to reach 50 μm, if the nanostructured surface has contact angle greater than 160° (Li *et al.* 2007).

17.4 ATOMISTIC SIMULATIONS

Understanding slip necessitates the simulation of fluids at a scale smaller than that described by continuum theory especially for cases where the characteristic length of the system ranges from several nm to some μm. Molecular dynamics (MD) is very well suited since it describes fluids at the atomic level providing a description of the dynamical behavior of the system under study.

17.4.1 Methodological issues

In this section we provide a brief description of MD. The reader can consult specialized references such as (Allen & Tildesley 1987; Rappaport 1995).

Let us consider a system of N interacting particles in a volume V. In the absence of external forces acting on the system, the total energy of the system, E, is conserved in time:

$$E = \sum_{i=1}^{N} \frac{1}{2} m_i v_i^2 + U(\vec{r}_1, \ldots, \vec{r}_N) = const \qquad (17.2)$$

where $\vec{r}_i$ denotes the position vector of the ith atom and $\vec{v}_i$ its velocity. The first term in Eq. (17.1) represents the kinetic energy of the system while U denotes the potential energy of the system.

We explore the system's properties in the microcanonical (NVE) ensemble (constant number of particles N, constant volume V and constant energy E). At every instant the position of an atom i is determined by the equation of motion:

$$\vec{F}_i = m_i \frac{d^2 \vec{r}_i}{dt^2} = -\frac{\partial U(\vec{r}_1, \ldots, \vec{r}_N)}{\partial \vec{r}_i} \quad i = 1, \ldots, N \qquad (17.3)$$

As we can see, for a system in three dimensions, we must solve 3N coupled second order differential equations. This way we can explore the phase space of the system and calculate average temporal quantities, which are comparable to macroscopic measurable quantities through statistical mechanics relations.

Several methods are used in MD simulation to solve the equations of motion (for a review see (Berendsen 1985). In the present study we used Verlet's algorithm (Verlet 1967), which is accurate and robust. In this scheme, provided we know two previous successive positions of the atom i at times t and $t - dt$, we can calculate its new position at time $t + dt$ using equation:

$$\vec{r}_i(t + dt) = 2\vec{r}_i(t) - \vec{r}_i(t - dt) + \frac{\vec{F}_i(t)}{m_i} dt^2 + O(dt^4) \qquad i = 1, 2, \ldots, N \qquad (17.4)$$

The choice of the timestep dt is a compromise between accuracy, stability and efficiency. In fact a very small timestep will give high precision solutions of Eq (17.4) but it will not permit to simulate long times, while a too large time step will lead to divergence of the trajectory. In general, one chooses $dt \ll 1/f_{max}$ where f_{max} is the maximum phonon frequency of the simulated system. In general dt is of the order of few femtoseconds (10^{-14} s). The total time usually accessible by this type of simulations is of the order of nanosecond (10^{-9} s). So one has to limit the study to phenomena having characteristic times less than this limit.

In order to describe interactions between atoms a model of interactions must be chosen. We must select analytical form function depending on geometrical quantities such as distance, angles or atom coordination. In general one derives several properties of the material as a function of the potential parameters and then fits them to experimental values. Also we can derive several properties from ab-initio calculations and fit to them. In constructing a potential function, we must reproduce properties as good as possible, the potential function should be applicable to situations for which the potential function was not fit and must be efficient as possible because force calculations are the most time consuming part of MD simulations.

The simplest choice for U is to write it as sum of pairwise interactions:

$$U(r_1, \ldots, r_N) = \frac{1}{2}\sum_i \sum_{j \neq i} \phi(|r_i - r_j|) = \sum_i \sum_{j>i} \phi(|r_i - r_j|) = \sum_i \sum_{j>i} \phi(r_{ij}) \tag{17.5}$$

A well known example that has been even used in the early applications of MD is the Lennard-Jones potential which has the following analytic form:

$$\phi(r_{ij}) = 4\varepsilon\left[\left(\frac{\sigma}{r_{ij}}\right)^{12} - \left(\frac{\sigma}{r_{ij}}\right)^{6}\right] \tag{17.6}$$

where r_{ij} denotes the distance between atoms i and j and σ, ε are parameters empirically derived by fitting calculated quantities to experimental data. For argon, the values of the parameters are $\sigma = 3.408$ Å and $\varepsilon/k_B = 119.8$ K, k_B being the Boltzmann's constant. The cut-off distance for the interactions is usually set to 2.5σ.

This model is good for rare gases but not for metals or semiconductors. However, even nowadays constitutes an extremely important and commonly employed model system. There exist also other forms of interaction potentials as the many body potentials for semiconductors with main representatives the Stillinger-Weber potential which is very good for describing situations with tetrahedral coordinates, and does not work well in other cases. In this model:

$$U(r_N) = \frac{1}{2}\sum_i^N \sum_{j \neq i}^{Nn} \phi(r_{ij}) + \sum_{ijk}^N g(r_{ij})g(r_{ik})\left[\cos(\theta_{ijk}) + \frac{1}{3}\right]^2 \tag{17.7}$$

where θ_{ijk} is the angle between bonds r_{ij} and r_{ik}.

Another model employed in semiconductors is the Tersoff Potential; Tersoff potentials work well in a broader spectrum than the Stillinger-Weber potentials. The strength of a bond between two atoms is not constant but depends on the local environment, bond ij is weakened by the presence of other bonds ik involving the atom i:

$$V(r_N) = \frac{1}{2}\sum_{ij}^N \Phi_R(r_{ij}) + \frac{1}{2}\sum_{ij}^N B_{ij}\Phi_A(r_{ij}) \quad \text{with } B_{ij} = B(G_{ij}),$$
$$G_{ij} = \sum_k f_c(r_{ik})g(\theta_{jik})f(r_{ij} - r_{ik}) \tag{17.8}$$

where Φ_R and Φ_A are appropriate repulsive and attractive terms.

Potentials for polymers and biological molecules have been developed in the frame of models employed also in CHARMM, AMBER, GROMOS which are the most commonly used biomolecular force fields (for details refer to Mackerell *et al.* 2004). For example the form of the potential energy function in CHARMM has the following form Brooks *et al.* (2009).

$$U(r_1, \ldots, r_N) = \sum_{\text{bonds}} K_b(b-b_o)^2 + \sum_{\text{angles}} K_\theta(\theta-\theta_o)^2 + \sum_{\substack{\text{Urey-Bradley}\\ \text{(UB) terms}}} K_{UB}(S-S_o)^2$$
$$+ \sum_{\text{dihedrals}} K_\phi[1+\cos(n\phi-\delta)] + \sum_{\substack{\text{improper}\\ \text{angles}}} K_\omega(\omega-\omega_o)^2$$
$$+ \sum_{\substack{\text{non-bonded}\\ \text{pairs}}} \left\{ \varepsilon_{ij}\left[\left(\frac{\sigma}{r_{ij}}\right)^{12} - 2\left(\frac{\sigma}{r_{ij}}\right)^{6}\right] + \frac{q_i q_j}{4\pi\varepsilon_0\varepsilon_1 r_{ij}} \right\} \tag{17.9}$$

where there are terms representing bonded and non-bonded contributions as a function of atomic positions. The first four terms on the right-hand side of Eq. (17.9) correspond to the bonded contributions and depend, respectively, on the covalent bonds (b), valence-angles (θ), distances (S) between the extreme atoms A, C of a valence angle A-B-C [Urey–Bradley (UB) terms], dihedral angles (ϕ) and improper angles (ω). The parameters K_b, K_θ, K_{UB}, K_ϕ, and K_ω are the respective force constants, and the variables with the subscript 0 are the respective equilibrium values. In the case of proteins and peptides a correction term is applied for the protein residues (in the case of CHARMM code see Brooks *et al.* (2009) and references therein).

In many simulations water is also implicated. There are several models employed in order to model interaction in the water molecule, the choice is based on the desired accuracy and the computational efficiency. The simplest interaction models consider the water molecule as rigid and are based on non-bonded interactions. The electrostatic interaction is modeled using Coulomb's law and the dispersion and repulsion forces using the Lennard-Jones potential. The potential for models such as TIP3P and TIP4P is represented by:

$$E_{ab} = \sum_i^{\text{on } a} \sum_j^{\text{on } b} \frac{k_C q_i q_j}{r_{ij}} + \frac{A}{r_{OO}^{12}} - \frac{B}{r_{OO}^{6}} \tag{17.10}$$

where k_C is the electrostatic constant, q_i are the partial charges relative to the charge of the electron; r_{ij} is the distance between two atoms or charged sites; and A and B are Lennard-Jones parameters. The charged sites may be on the atoms or on dummy sites depending on the model employed. The majority of models include Lennard-Jones terms only for oxygen atoms' interactions. Three-site models (TIP3P) have three interaction sites, corresponding to the three atoms of the water molecule (Jorgensen *et al.* 1983). Each atom is attributed a point charge, and the oxygen atom also has the Lennard-Jones parameters. Such models are widely used in MD simulations due to their simplicity and computational efficiency. Most models use a rigid geometry matching the known geometry of the water molecule. A modified TIP3P model is implemented in the CHARMM force field is a slightly modified version of the original. The difference lies in the Lennard-Jones parameters where Lennard-Jones parameters are employed for hydrogen atoms too. The 4-site models (TIP4P) place the negative charge on a dummy atom located next to the oxygen along the bisector of the HOH angle, a fact that results in better electrostatic distribution around the water molecule. There are several parameterizations of the TIP4P model for specific uses (Horn *et al.* 2004, Jorgensen *et al.* 1983).

In many cases periodic boundary conditions are applied in the three directions x, y, z or in appropriate directions. In order to start the simulation, atoms are given velocities in order to reach the appropriate temperature conditions.

17.4.2 Property calculations

During MD simulations several quantities are calculated and their corresponding time series are recorded. Here we discuss the most often employed ones.

Temperature, which is defined by:

$$T = \frac{1}{3Nk_B}\sum_{i=1}^{N} m_i v_i^2 \tag{17.11}$$

potential energy:

$$U = \sum_{i=1}^{N}\sum_{j>i}^{N} \phi(r_{ij}) \tag{17.12}$$

and pressure which is computed by:

$$P = \frac{N}{V}k_B T - \frac{1}{3V}\sum_{i=1}^{N}\sum_{j>i}^{N} r_{ij}\frac{\partial \phi(r_{ij})}{\partial r_{ij}} \tag{17.13}$$

where, as before, r_{ij} stands for the distance between two particles i and j, and v_i denotes the velocity of particle i.

17.4.3 Transport properties

The diffusion coefficient can be obtained using either the Einstein's relation:

$$D = \lim_{t\to\infty}\frac{1}{2dNt}\left\langle \sum_{j=1}^{N}\left[\mathbf{r}_j(t) - \mathbf{r}_j(0)\right]^2 \right\rangle \tag{17.14}$$

where $\mathbf{r}_j$ is the position vector of the *jth* atom and d is the dimensionality of the system ($d = 1$ for diffusivity calculation in one direction, $d = 2$ in two directions and $d = 3$ in three directions) or Green-Kubo's relation:

$$D = \frac{1}{3N}\int_0^{\infty}\left\langle \sum_{j=1}^{N} \mathbf{v}_j(0)\cdot \mathbf{v}_j(t) \right\rangle \tag{17.15}$$

where $\mathbf{v}_j$ is the velocity vector of the *jth* atom.

The two relations are equivalent and provide the same results (Rappaport 1995; Karniadakis *et al.* 2005). Equations (14) and (15) are derived for systems in equilibrium, but they can be used for non-equilibrium systems as well, provided one excludes the drift contribution from the flow (Karniadakis *et al.* 2005). In the present work we have used Einstein's relation for the calculation of the diffusion coefficient. The computation is carried out in two steps. In the first step the Mean Square Displacement (MSD) is obtained from the definition:

$$MSD(t) = \frac{1}{N}\left\langle \sum_{j=1}^{N}\left[\mathbf{r}_j(t) - \mathbf{r}_j(0)\right]^2 \right\rangle \tag{17.16}$$

and subsequently D is evaluated based on Eq. (17.2) which can be written as:

$$D = \lim_{t\to\infty} \frac{1}{2dt} MSD(t) \tag{17.17}$$

Shear viscosity and thermal conductivity for systems in equilibrium can be calculated using the Green-Kubo formalism, as described in (Evans 1986). These relations can be used without modification in NEMD as long as we stay in the linear regime close to equilibrium (Murad *et al.* 1993). When the system is not in equilibrium, due to the existence of an external force, the magnitude of the external force should be small enough for linearization to hold (Binder *et al.* 2005, Liu *et al.* 2006).

Shear viscosity η_s for a pure fluid is computed by the relation:

$$\eta_s = \frac{1}{Vk_B T} \int_0^\infty dt \left\langle J_p^{xy}(t) \cdot J_p^{xy}(0) \right\rangle \tag{17.18}$$

where J_p^{xy} is the off-diagonal component of the microscopic stress tensor:

$$J_p^{xy} = \sum_{i=1}^{N} m_i v_i^x v_i^y - \sum_{i=1}^{N} \sum_{j>1}^{N} r_{ij}^x \frac{\partial u(\mathbf{r}_{ij})}{\partial r_{ij}^y} \tag{17.19}$$

$u(\mathbf{r}_{ij})$ is the LJ potential of atom i interacting with atom j, $\mathbf{r}_{ij}$ is the distance vector between atoms i and j, and v_i^j is the j-component (j= x, y or z) of the velocity of atom i.

On the other hand, thermal conductivity λ can be calculated by the integration of the time-autocorrelation function of the microscopic heat flow J_q^x, that is,

$$\lambda = \frac{1}{Vk_B T^2} \int_0^\infty dt \left\langle J_q^x(t) \cdot J_q^x(0) \right\rangle \tag{17.20}$$

where the microscopic heat flow J_q^x is given by

$$J_q^x = \frac{1}{2} \sum_{i=1}^{N} m_i (v_i)^2 v_i^x - \sum_{i=1}^{N} \sum_{j>1}^{N} \left[r_{ij}^x : \frac{\partial \phi(\mathbf{r}_{ij})}{\partial r_{ij}^x} - \mathbf{I} \cdot \phi(\mathbf{r}_{ij}) \right] \cdot v_i^x \tag{17.21}$$

where v_i is the speed velocity magnitude of atom i and $\mathbf{I}$ is the unitary matrix.

Understanding the behaviour of transport properties such as D, η, λ very close to the wall influence greatly the interpretation as well as the error estimates of μPIV and nano PIV results. As several MD results have shown especially in nanochannel the existence of an important variation these properties as a function of the distance from the solid walls (see e.g. Sofos *et al.* 2009a, b, 2010). Moreover there are research results that correlate the variation of slip length with some transport properties (see e.g. Priezjev, 2011).

17.4.4 Slip velocity/length calculation

When it comes to the calculation of the slip velocity and slip length there are several practical questions that arise in order to calculate them from simulations which are not clear in general in the calculations presented in the literature. If ones extracts velocities from a velocity profile, then when trying to extract the velocity at the wall there are several questions that arise. Where exactly the wall is located? As it was found by many researchers the distribution of atoms close to the wall differs considerably from that of the bulk region of the liquid (see e.g. Figure 17.2). Is the wall located at the position of the geometric centres of the wall atoms or at slightly larger distance since atoms in reality have a finite dimension. The other question that arises is what profile can one fit. In the case of Poiseuille flow one can fit a parabolic profile using all velocity values. But this refers to a flat wall channel. What occurs when the wall is not flat as is the case containing protrusions and so on. An option is to perform linear extrapolations using values close to the defined wall. Another possibility would be to calculate the velocity as an average over particles located in a fine layer close to the wall and define this velocity as slip velocity and use it in subsequent calculations. When a surface presents an heterogeneous behaviour due to its composition or/and variation of its geometrical characteristics then we have in spatial variations in velocity profiles close to the wall resulting in variations of the slip at a local scale. In such cases in general flow profiles are averaged at length scales larger than the typical length of the spatial variation of the surface properties and can be used in order to define an effective slip length (L_s).

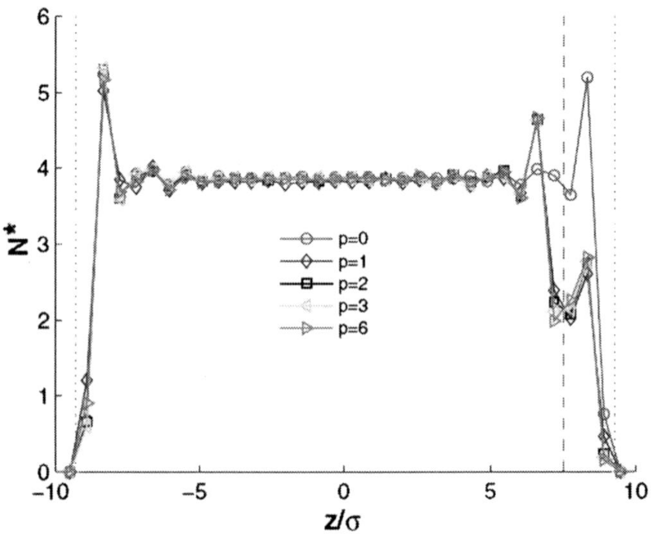

Figure 17.2 Average number density profiles in a channel consisting of an atomically flat wall and an upper wall with periodic orthogonal protrusions. Solid lines are a guide for the eye. Dashed lines denote the protrusion limits and dotted lines the cavities and lower wall limits. For the interpretation of the parameter p see Figure 17.5 below (Reproduced with permission from Sofos *et al.* 2009a).

17.5 ATOMISTIC SIMULATIONS RESULTS ABOUT SLIP

Early simulations were focused on the atomistic investigation of Poiseuile and Couette flows. The initial effort was mainly concentrated on the study of velocity profiles. In the case of pressure driven or body force driven Poiseuille flow between two parallel plates, the velocity profile is expected to follow a parabolic behaviour and if there is no slip the fluid velocity to be zero at the wall boundary. An atomistic model of flat surfaces usually employed in MD simulation is schematically represented in Figure 17.3. The type of fluids studied range from simple fluids (like argon using Lennard-Jones potentials) to more complex fluids (like chains of Lennard-Jones atoms, hexadecane and so on). Simple fluids are studied as model systems in a twofold aim: on the one hand the simplicity of interaction potentials permits for long simulation runs that are necessary for the calculations of several transport properties with statistical significance and on the other hand one can explore tendencies of the fluid behavior in relatively simple cases before attacking more complex situations.

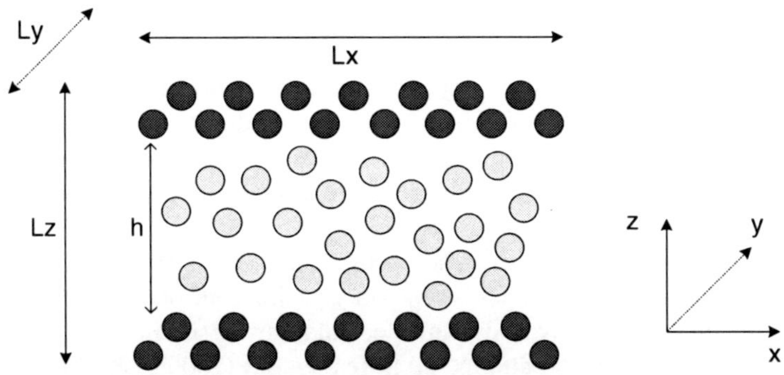

Figure 17.3 Schematic view of a typical parallel plate system employed in molecular dynamics simulations. Dark circles represent wall atoms while grey circles represent fluid atoms. Periodic boundary conditions are usually imposed in the x- and y-directions.

Using a molecular model for the fluid and the walls, Ceiplak *et al.* (1999) studied flows in high and medium Knudsen number values. Results showed that velocity profiles are different than those anticipated by the theory of continuous hydrodynamics and present an unexpected dependency on the system's density and dimension, as we move from low to high values of Knudsen number (Cieplak *et al.* 1999). For fluids, density increase results in viscosity increase and, thus, flow deceleration. An opposite trend is noted in the gas phase due to molecular interaction with the walls and among themselves. Likewise, studies with molecular dynamic in Poisseuile flows indicate that Navier-Stokes equation no longer apply in small channels (Travis *et al.* 1997) while, at the same time, the constant shear viscosity assumption is wrong. Furthermore, study of molecular dynamics in flows showed that the heat flow profile does not agree with the cubic profile that is predicted by classical continuum equations and presents important oscillations in a distance of one molecular diameter from the walls (Travis & Gubbins, 2000).

Early simulations had shown that there is no slip close to contact lines (Koplik & Banavar 1988; Thompson & Robbins 1989). More recent studies have reported that molecular slip increases as the following quantities are reduced: solid-fluid interactions (Barrat & Bocquet 1999; Cieplak *et al.* 2001; Nagayama & Cheng, 2004), fluid density (Koplik *et al.* 1989; Thompson & Robbins 1990) and wall density (Thompson & Troian 1997), while it is reduced with pressure (Barrat & Bocquet 1999). The model used for the solid wall and molecular roughness such as the wall density, the size and mass of wall atoms, was found to significantly affect slipping (Barrat & Bocquet 1999; Bocquet & Barrat 1993; Galea & P. Attard, 2004; Jabbarzadeh *et al.* 2000; Sun & Ebner 1992; Thompson & Troian 1997).

The effect of wettability on slip was studied with molecular dynamics simulations (Nagayama & Cheng 2004). Such studies take into account the hydrophilic or hydrophobic character of the wall-fluid interactions that are ignored in the continuum equations. In the sense hydrophilic or hydrophobic one treat surfaces that tend to attract and repel the liquid (not necessarily water). The molecular dynamics study showed that there exists a void between fluid and surface and, as a result, frictionless resistance and slipping are not constant but depend on external force. Likewise, the non-uniform profiles of temperature and pressure close to the wall are due to the effect of surface wetness (Nagayama & Cheng 2004).

A study of flow between flat plates has also been presented by Sofos *et al.* (2009b) focused on the effect of temperature, driving force, solid/fluid interactions and fluid density on density, as well as velocity and temperature distribution along the channel for channels from 0.9 nm-17.1 nm. Results show that the effect becomes important, as typical dimensions get smaller and we deviate from the continuum theory. Velocity profiles have a parabolic form in specific ranges of channel width and moving force.

The study of hydrophobic surfaces and specifically of super-hydrophobic surfaces is of great interest. Super-hydrophobic surfaces have gathered a great interest since, if the fluid is repelled by the surface, one can imagine that a thin gas layer is formed that could have a lubrication effect (Biben & Joly 2008).

Simulation results in agreement with experimental results report slip length on hydrophobic surfaces, while there was no mention of slip on hydrophilic surfaces (Wang *et al.* 2009a, b) on which, slip of 30 atomic diameters was measured on a surface with contact angle of 140° and no slipping on hydrophilic surfaces.

Slip has been studied by Lattice-Boltzmann mesoscopic approaches (Sbragaglia *et al.* 2006; J. Harting *et al.* 2006) and the potential of designing smart surfaces with controlled slipping has been proposed. Slip length has been estimated with hydrodynamic calculations in Philip and Angew (1972) and was calculated for surfaces with periodic patterns in Sofos *et al.* (2009a, 2012) and for surfaces with roughness in combination with controlled hydrophobicity in Lauga and Stone (2003). Experimental results for flows in super-hydrophobic surfaces are reported in Ou *et al.* (2004, 2005).

More recently Kannam *et al.* (2011) using MD simulations investigated the hydrodynamic boundary condition for simple nanofluidic system such as argon and methane flowing in graphene nanochannels. They calculated the fluid-graphene interfacial friction coefficient from which they predicted the slip length and the slip velocity. The results show that the non-slip boundary condition is violated due to the atomic smoothness and lyophobicity of grapheme surfaces. They mention that the friction coefficient is an order of magnitude smaller than the one between the same fluids and molecular Lennard-Jones crystal walls generally employed in computer simulations, while slip was found to be independent of the flow type, indicating this way its intrinsic nature. The flow rates are found to be an order of magnitude higher compared to classical hydrodynamic no-slip boundary condition predictions.

17.5.1 Wall roughness effects

Roughness exists at the atomic level even for macroscopically flat surfaces since surfaces are made from atoms and thus they are not smooth. Moreover, atom thermal vibrations enhance such roughness. Of course the latter is affected by several parameters such as the wall stiffness, atom mass, and so on.

The effect of surface roughness on nanoflow and on simple fluid slipping on hydrophobic and hydrophilic nanochannels was examined with the method of molecular dynamics in a number of studies. A very good literature review is the one by Lauga *et al.* (2005), in which references are made on experimental and simulation methods. Simulation results have been reported for various fluids (Barrat & Bocquet 1999; Cieplak *et al.* 2001; Cottin-Bizonne *et al.* 2004; Cottin-Bizonne *et al.* 2003; Fan *et al.* 2002; Galea & Attard 2004; Heinbuch & Fischer, 1989; Jabbarzadeh *et al.* 2000; Koplik *et al.* 1988; Koplik *et al.* 1989; Nagayama & Cheng 2004; Sokhan *et al.* 2002; Sun Ebner 1992; Thompson & Robbins, 1989; Thompson & Robbins, 1990; Thompson & Troian 1997).

Gallea and Attard (2004) studied the effect of solid roughness on the slip boundary condition in a Couette flow. They modeled atomic roughness through variation of the size and spacing between solid atoms at constant packing ratio while keeping the interaction parameters and thermodynamic parameters of the fluid constant. Fluid slip length was found to follow nonmonotonic behavior as the solid structure is varied from smooth to rough. Slip occurs for both smooth and rough surfaces and stick occurs only for surfaces commensurate with the liquid. They claim that this is due to the potential felt by the fluid is important in determining the slip or no slip condition. In fact Sofos *et al.* (2009a) in a case of periodic wall pattern formation found a variation of potential energy close a hydrophobic and hydrophilic wall (see Figure 17.4) and found that in the second case there are regions were fluid atoms can be trapped for some time. This trapping seems to be correlated with the lower slip observed in the latter case.

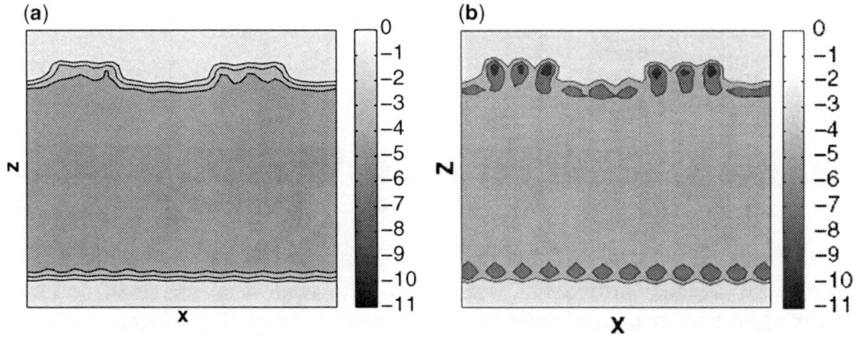

Figure 17.4 Potential energy contour plots for p = 2 (channel with periodic patterns) in the case of (a) hydrophobic and (b) hydrophilic wall-fluid interactions (Sofos 2009a). [The darker the regions the more attractive the interactions between fluid particles and the wall (Reproduced with permission from Sofos *et al.* 2009a)].

Asproulis and Drikakis (2010a, b, 2011) have investigated the effects of wall stiffness and particle mass on slip length. The results show that the nanoscale corrugation is primarily determined by the stiffness and wall temperature, with softer surfaces leading to enhanced roughness. The effect of wall-mass results in a weak influence on the vibration of solid wall atoms. The results showed that slip length variation as a function of the oscillating frequency of wall atoms can follow a master curve of the form:

$$\frac{L_S}{L_{S,\max}} = a + b\frac{\omega}{\omega_{\max}} + \cdots + f\left(\frac{\omega}{\omega_{\max}}\right)^5 \qquad (17.22)$$

where L_S is the slip length and ω the wall-atom vibrating frequency, the values of the coefficients $a, b, \ldots, f$ have been obtained through data fitting and are $a = -0.55$, $b = 4.27$, $c = -4.46$, $d = 2.21$, $e = -0.53$, and $f = -0.05$.

The effect of the size of the molecules was also examined b Cieplak *et al.* (2001) who studied slip length for monatomic fluid and a fluid of chains (both consisting of LJ atoms) and where they varied the strength of interactions between the fluid and the wall. It was observed that the slip length was related to the fluid organization close to the channel walls. For repulsive wall interaction and low fluid densities slip length was attaining its maximum value. On the other hand, the slip length behavior for chain molecules was more sensitive to the variation of the potential interactions.

17.5.2 Effect of periodic wall patterns

In addition to the atomic and thermal roughness, periodic modification of the solid surface in contact with the fluid presents interest. The study of periodic wall perturbation is mainly focused on orthogonal protrusions like the ones in Figure 17.5.

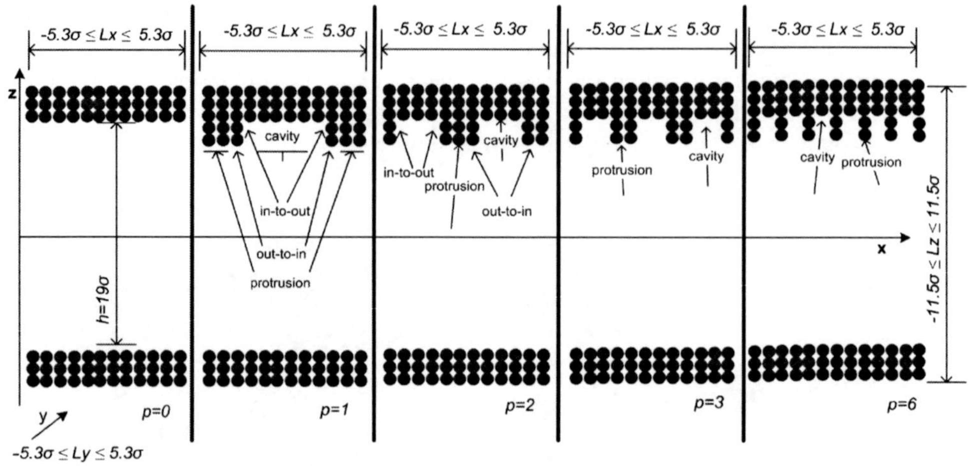

Figure 17.5 Walls with periodic orthogonal protrusions. The system is assumed periodic in x- and y-directions. Each module shown corresponds to a computational cell (Reproduced with permission from Sofos *et al.* 2012).

Slippage of the fluid over hydrophobic surfaces with orthogonal protrusions was observed in many cases. The existence of nanobubbles on hydrophobic surfaces is a possible explanation of this behavior (Ishida *et al.* 2000; Lu *et al.* 2000; Tyrrell & Attard 2002; Holmberg *et al.* 2003; Simonsen *et al.* 2004; Agrawal *et al.* 2005; Bhushan *et al.* 2008; Wang *et al.* 2009). Nanobubbles can be preserved for several hours (Wang *et al.* 2009; Yang *et al.* 2003) and are constant in both ambient conditions and in conditions of reduced water pressure as low as -6 Mpa (Borkent *et al.* 2007). Two factors are considered important for slippage: surface wetness and surface roughness. When a fluid flows over a completely wettable surface, roughness increases (Richardson 1973; Schmatko *et al.* 2006). However, if the fluid wets the surface in part, roughness favors the formation of gas pockets which, in turn, results in a large slipping length (Cottin-Bizonne *et al.* 2003; de Gennes 2002). Regarding slip length, in case a gas layer between solid surface and liquid exists, it has been proposed that the slip length originating from the viscosity discontinuity is given by (Vinogradova 1995):

$$b = \left(\frac{\eta_w}{\eta_a} - 1\right) h_b \tag{17.23}$$

where η_w is the viscosity of water and η_a is the viscosity of air and h_b is the air height. Given the fact that water and air viscosities at 300 K are approximately $\eta_w = 851.5$ μPas and $\eta_a = 18.6$ μPas, respectively (Lide 2008). Apparently, the slip length b can be 45 times the height of air h_b because of the high ratio of viscosity coefficients.

The effect of periodic patterns has also been studied in the work by Sofos *et al.* (2009a, 2012), in which liquid argon flow between parallel krypton plates was studied for various lengths of periodic patterns with use of molecular dynamics simulations. The density diagrams close to the walls and particularly close to the walls with periodic patterns present an inhomogeneity. There are atoms that stay trapped for a long time span inside the periodic cavities. The velocity profiles varied towards the part of the channel located towards the periodic protrusions side. As protrusion length is reduced, velocity inside the protrusions is also reduced, resulting in a decrease in slippage.

In Figure 17.6 the effects of wall wettability, as expressed by its hydrophobic/hydrophilic character, and roughness on the effective slip length are summarized from the work of Sofos *et al.* (2012). Taking into account that local velocity calculation near the rough wall is somewhat imprecise since velocity profiles are difficult to fit by a polynomial curve, the effective slip length calculation may only reveal trends and not exact values. We observe that the effective slip length increases as the wall roughness factor p increases, that is, the protrusion and cavity length decreases. This conclusion was also found by Yang (2006). Moreover, as wall hydrophilicity increases, it appears that the effective slip length decreases. The effect of wall wettability on slip length is relatively small (Figure 17.6) and this is due to the small range of wall wettability values investigated in our simulations. It is noted that our results are in qualitative agreement with similar results reported by Nagayama and Cheng (2004) and Ziarani and Mohamad (2008). Finally, it was observed by Sofos *et al.* (2012) that atoms are trapped for some time inside the cavities and this trapping time seems to be related to the slip length value. In Figure 17.7, trapping time values *vs.* slip length are presented, for every p value. It is shown that as the trapping time increases, the slip length decreases.

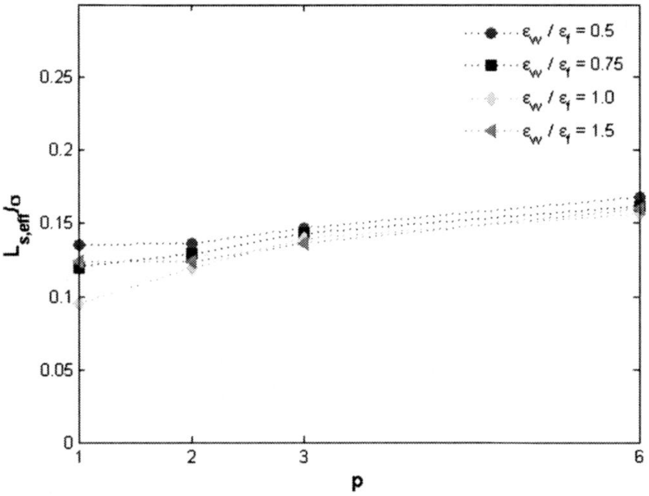

Figure 17.6 Effective slip length in various hydrophobic/hydrophilic rough wall nanochannels. For the interpretation of the parameter p see Figure 17.5 (Reproduced with permission from Sofos *et al.* 2012).

Jabarzadeh *et al.* (2000) using molecular dynamics simulations examined the behavior thin liquid films of alkanes in extreme conditions in the boundary lubrication regime. In their study, the walls are modeled as atomically rough sinusoidal surfaces and the effect of the roughness characteristics as described by the period and amplitude of the sinusoidal wall on slippage, was investigated as well as the effect of the length (size) of the fluid molecules. Their results show that the relative size of the fluid molecules and wall roughness determines the slip or no-slip regime. It was shown that as the period of roughness is increased, the degree of slip on the wall also increases. Also, they observed that with larger roughness amplitudes it is possible to decrease the slip and that for longer molecules slip is larger, indicating an interrelation between the structure of the fluid and the slip length.

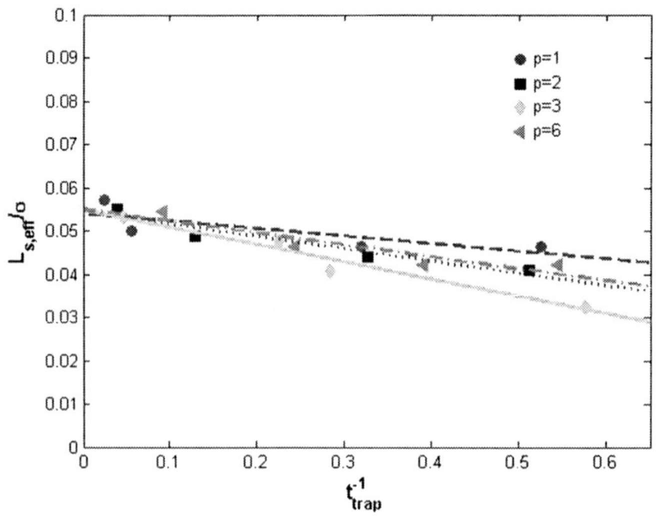

Figure 17.7 Effective slip length *versus* trapping time. For the interpretation of the parameter *p* see Figure 17.5. [Reproduced with permission from Sofos *et al.* (2012)].

17.5.3 Effect of nanostripes

The effect of periodic nanostrips embedded in the solid surface was also investigated. A schematic representation of a typical such surface is shown in Figure 17.8 where stripes of width a are present on the lower channel wall. The stripes are in general tilted relative to the direction of the flow velocity U, or they can be perpendicular or parallel to it.

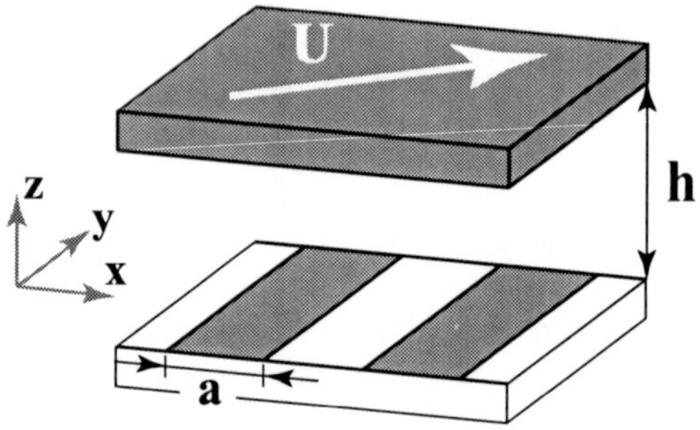

Figure 17.8 Schematic representation of the channel geometry for the study of the effect of nanostrips. Surface patterns consist of wetting regions (dark regions) and non-wetting regions (white regions). Steady shear flow is induced by the upper wall moving with a constant velocity U in the xy plane at an angle with respect to the x direction. The lower patterned wall is stationary (Reprinted with permission from [Priezjev 2011]. Copyright [2011], American Institute of Physics).

In the Yang (2006) study, solid walls with periodic nanostrips are used as rough walls. The simulation results suggest that both wettability and roughness of the interface are important and should be taken into consideration simultaneously in the analysis of flows in nanochannels. For flat hydrophobic surfaces they show that there is a layer of air or nanobubbles near the solid-fluid interface. The presence of nanobubbles reduces the momentum transfer between the fluid and the wall, leading to apparent slippage on the surface. For a given surface with a given wettability, fluid velocities increase with external force.

The work of Maynes *et al.* (2007) is of interest, since analytical calculations and an experimental study of laminar flow in a microchannel with superhydrophobic walls are presented. The pressure drop along the microchannel length was measured and the average f Re was calculated, where f Re is the mean product of the flow friction coefficient f and the flow Reynolds number. It appears that due to the superhydrophobic behavior, the fluid meniscus does not enter the nooks, resulting in a significant reduction of the contact angle between the flowing fluid and the solid walls. Both experimental and computational results indicate that a significant reduction in flow resistance is possible.

More recently Priezjev (2011) studied the velocity field and diffusion of fluid molecules near interfaces using simple fluids and surfaces patterned with anisotropic and random textures. The typical size of surface patterns is smaller than the channel dimensions. For flows over surfaces patterned with stripes of varying wettability, the heterogeneous surfaces induce wavy perturbations in velocity profiles and the slip velocity acquires a transverse component. In such cases the effective slip depends on the shear flow direction relative to the stripe orientation (see Figure 17.8). The interfacial diffusion coefficient of fluid molecules correlate well with the effective slip length as a function of the shear flow direction. On the other hand for random textured surfaces the results show that the effective slip length is determined by the total area of wetting regions following Eq.(17.24)

$$L_S(\phi) = \frac{b_\omega b_n}{\phi b_n + (1-\phi) b_\omega} \tag{17.24}$$

where b_w and b_n are, respectively, the slip lengths for wetting ($\phi = 1$) and non-wetting ($\phi = 0$) surfaces. The results for the slip length in various cases are presented in Figure 17.9. When the fraction of wetting regions is large, the diffusive motion of fluid atoms is strongly influenced by the surface potential and the slip length is nearly proportional to the in plane diffusion coefficient at equilibrium. On the limit of small wetting areas, the diffusion of fluid particles is less affected by the surface potential and slip length depends on the number of strongly attractive wall atoms.

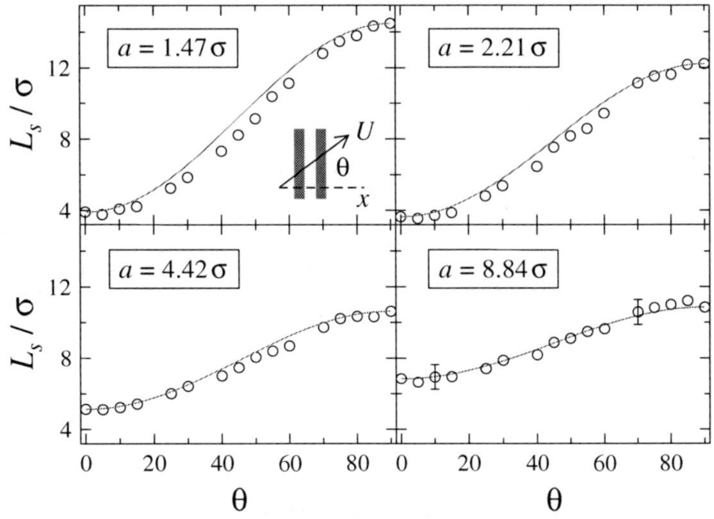

Figure 17.9 Effective slip length (expressed in units of σ) as a function of the angle θ of the flow and the stripe orientation for several stripe widths (Reproduced with permission, Preijzev, 2011).

17.6 CONCLUSIONS

The design and construction of micro- or nano-structured surfaces has attracted recently a great deal of interest in a wide range of applications. Characteristic examples include biotechnology, lab-on chip for the detection of various substances, desalination, just to mention a few. The role of slip in flows, especially at small scales, seems to be very important, since its existence is expected to lead to lower required pressure gradients to move the fluid and consequently lower energy consumption. However, the detailed nature of slip behavior is not easy to understand and measure for the time being. Thus simulation methods and especially Molecular Dynamics can provide insight in the behavior at the atomic level and contribute to a better understanding of experimental results and the design of devices.

Atomistic simulations show the existence of slip even in the case of atomically flat surfaces. However, the magnitude of the slip seems to be the result of complex interactions between several factors which include a) the type of interaction between the fluid and the solid wall as expressed by the hydrophobicity or hydrophilicity of the surface, b) the wettability of the surface (which seems to be affected by the geometry) c) the magnitude of the driving force or equivalently the shear rate d) the geometry of the surface as expressed by its atomic and thermal roughness as well as the patterns that are fabricated on the surface, such as periodic orthogonal protrusions, nanostrips and so on. The type of the fluid, that is, if it consists of simple molecules or long chain molecules (as it is the case for polymers or biological molecules) seems to play a role on the slip in conjunction with the structure of the wall surface. More recently it was found that the orientation of these structures relatively to the direction of the flow influences also the slip magnitude. Several simulation results indicate the interrelation between slip length and other properties such as the diffusion coefficient variation close to the wall and the volumetric flow rate that can help in the interpretation of experimental results with methods like the newly developed nano-PIV.

Experimental results as well as detailed simulation and theoretical work has led to the conclusion that a suitably modified surface can result in super-hydrophobic surfaces that reduce the drag force combining hydrophobicity and roughness. Although several interrelations have been documented, there is not yet a unified approach for the prediction of slip in each case. It seems that a promising idea is to create surfaces with hierarchical forms of roughness, that is have different patterns or atomic roughness at various scales. It is also of interest to simulate flows of complex liquids (as it is the case for chemical macromolecules or molecules employed in biological applications) as well as to implement more realistic potentials than the usually employed Lennard-Jones ones. MD simulations of flows necessitate in general long simulation runs applying it in more complex wall geometries and more sophisticated potential interaction leads to a significant increase of the computational burden. The increasing power of computers and the possibility to use large computer clusters in combination with the emerging use of GPUs (Graphical Processing Units) will enable researchers to study larger systems for larger real times via atomistic simulations employing more realistic interaction potentials. Thus MD can play a more active role in the design of nano and micro fluidic devices as well as in understanding thoroughly flow mechanisms at the nanoscale.

It is a clear that the field of nanotechnology and, more specifically that of the micro-fluidic and nano-fluidic systems, supported by suitable simulations and experimental testing, will play an important role in the design of innovative materials for advanced special purpose flow conduits for use in a wide spectrum of technological applications.

REFERENCES

Agrawal A., Park J., Ryu D. Y., Hammond P. T., Russell T. P. and McKinley G. H. (2005). Controlling the location and spatial extent of nanobubbles using hydrophobically nanopatterned surfaces. *Nano Lett.*, **5**, 1751–1756.

Allen M. P. and Tildesley T. J. (1987). Computer Simulation of Liquids. Clarendon Press, Oxford.

Asproulis N. and Drikakis D. (2010a). Boundary slip dependence on surface stiffness. *Phys. Rev. E*, **81**, 061503.

Asproulis N. and Drikakis D. (2010b). Surface roughness effects in micro and nanofluidic devices. *J. Comput. Theor. Nanosci.*, **7**(9), 1825–1830.

Asproulis N. and Drikakis D. (2011). Wall-mass effects on hydrodynamic boundary slip. *Phys. Rev. E*, **84**, 031504.

Barrat J. L. and Bocquet L. (1999). Large slip effect at a nonwetting fluid-solid interface. *Phys. Rev. Lett.*, **82**, 4671–4674.

Baudry J., Charlaix E., Tonck A. and Mazuyer D. (2001). Experimental evidence for a large slip effect at a nonwetting fluid-solid interface. *Langmuir*, **17**(17), 5232–5236.

Berendsen H. J. C. and van Gunsteren W. F. (1986). Practical algorithms for dynamic simulations. In: Molecular Dynamics Simulation of Statistical Mechanical Systems, G. Ciccoti and W. G. Hoover (eds), North Holland, Amsterdam, pp. 43–65.

Biben T. and Joly L. (2008). Wetting on nanorough surfaces. *Phys. Rev. Lett.*, **100**(18), 186103–186104.

Binder K., Horbach J., Kob W., Paul W. and Varnik F. (2005). Molecular dynamics simulations. *J. Phys.: Condens. Matter*, **16**, 429–453.

Bhushan B. (2007). Handbook of Nanotechnology. 2nd edn, Springer, Heidelberg, Germany.

Bhushan B., Wang Y. and Maali A. (2008). Coalescence and movement of nanobubbles studied with tapping mode AFM and tip-bubble interaction analysis. *J. Phys.: Condens. Matter*, **20**(48), art. no. 485004, 10.

Bocquet L. and Barrat J. L. (1993). Hydrodynamic boundary conditions and correlation functions of confined fluids. *Phys. Rev. Lett.*, **70**, 2726–2729.

Bocquet L. and Barrat J. L. (2007). Flow boundary conditions from nano- to micro-scales. *Soft Matter*, **3**, 685–693.

Bonaccurso E., Kappl M. and Butt H. J. (2002). Hydrodynamic force measurements: boundary slip of water on hydrophilic surfaces and electrokinetic effects. *Phys. Rev. Lett.*, **88**, 076103.

Borkent B. M., Dammer S. M., Schonherr H., Vancso G. J. and Lohse D. (2007). Superstability of surface nanobubble. *Phys. Rev. Lett.*, **98**, 204502.

Brooks B. R., Brooks C. L. III, Mackerell A. D. Jr, Nilsson L., Petrella R. J., Roux B., Won Y., Archontis G., Bartels C., Boresch S., Caflisch A., Caves L., Cui Q., Dinner A. R., Feig M., Fischer S., Gao J., Hodoscek M., Im W., Kuczera K., Lazaridis T., Ma J., Ovchinnikov V., Paci E., Pastor R. W., Post C. B., Pu J. Z., Schaefer M., Tidor B., Venable R. M., Woodcock H. L., Wu X., Yang W., York D. M. and Karplus M. J. (2009). CHARMM: the biomolecular simulation program. *Comput. Chem.*, **30**, 1545–1614.

Chao-Hua Xue, Shun-Tian Jia, Zhang Jing, Li-Qiang Tian, Hong-Zheng Chen and Wang Mang (2008). Preparation of superhydrophobic surfaces on cotton textiles. *Sci. Technol. Adv. Mater.*, **9**(3), 035008.

Churaev N. V., Sobolev V. D. and Somov A. N. (1984). Slippage of liquids over lyophobic solid surfaces. *J. Colloid Interface Sci.*, **97**(2), 574–581.

Cieplak M., Koplik J. and Banavar J. R. (1999). Applications of statistical mechanics in subcontinuum fluid dynamics. *Physica A*, **274**, 281–293.

Cieplak M., Koplik J. and Banavar J. R. (2001). Boundary conditions at a fluid-solid interface. *Phys. Rev. Lett.*, **86**, 803–806.

Cottin-Bizonne C., Barrat J. L., Bocquet L. and Charlaix E. (2003). Low-friction flows of liquid at nanopatterned interfaces. *Nature Mater.*, **2**, 237–240.

Cottin-Bizonne C., Barentin C., Charlaix E., Boequet L. and Barrat J. L. (2004). Dynamics of simple liquids at heterogeneous surfaces: molecular dynamics simulations and hydrodynamic description. *Eur. Phys. J. E*, **15**, 427–438.

Cottin-Bizonne C., Cross B., Steinberger A. and Charlaix E. (2005). Boundary slip on smooth hydrophobic surfaces: intrinsic effects and possible artifacts. *Phys. Rev. Lett.*, **94**, 056102.

Craig V. S. J., Neto C. and Williams D. R. M. (2001). Shear-dependent boundary slip in an aqueous Newtonian liquid. *Phys. Rev. Lett.*, **87**, 054504.

de Gennes P. G. (2002). On fluid/wall slippage. *Langmuir*, **18**(9), 3413–3414.

Daniello R. J., Waterhouse N. E. and Rothstein J. P. (2009). Drag reduction in turbulent flows over superhydrophobic surfaces. *Phys. Fluids*, **21**, 085103.

Evans D. J. (1986). Thermal conductivity of the Lennard–Jones fluid. *Phys. Rev. A*, **34**(2), 1449–1453.

Fan X. J., Phan-Thien N., Yong N. T. and Diao X. (2002). Molecular dynamics simulation of a liquid in a complex nano channel flow. *Phys. Fluids*, **14**, 1146–1153.

Feng L., Li S., Li Y., Li H., Zhang L., Zhai J., Song Y., Liu B., Liang L. and Zhu D. (2002). Superhydrophobic surfaces: from natural to artificial. *Adv. Mater.*, **14**, 24.

Galea T. M. and Attard P. (2004). Molecular dynamics study of the effect of atomic roughness on the slip length at the fluid-solid boundary during shear flow. *Langmuir*, **20**, 3477–3482.

Goldstein S. (1938). Modern Development in Fluid Dynamics. Clarendon, Oxford.

Goldstein S. (1969). Fluid mechanics in the first half of this century. *Annual Review of Fluid Mechanics*, **1**, 1–29. doi:10.1146/annurev.fl.01.010169.000245.

Hatzikiriakos G. S. (2011). Wall slip of molten polymers. *Prog. Polym. Sci.*, **37**, 624–643, doi:10.1016/j.progpolymsci.2011.09.04.

Hatzikiriakos G. S. and Dealy J. M. (1992). Role of slipand fracture in the oscillating flow of a HDPE in a capillary. *J. Rheol.*, **36**, 845–884.

Harting J., Kunert C. and Herrmann H. J. (2006). Lattice Boltzmann simulations of apparent slip in hydrophobic microchannels. *Europhys. Lett.*, **75**, 328–334.

Heinbuch U. and Fischer J. (1989). Liquid flow in pores: Slip, no-slip, or multilayer sticking. *Phys. Rev.*, **40**, 1144–1146.

Holmberg M., Kuhle A., Garnaes J., Morch K. A. and Boisen A. (2003). Nanobubble trouble on gold surfaces. *Langmuir*, **19**(25), 10510–10513.

Holt J. K., Park H. G., Wang Y., Stadermann M., Artyukhin A. B., Grigoropoulos C. P., Noy A. and Bakajin O. (2006). Fast mass transport through sub-2-nanometer carbon nanotubes. *Science*, **312**(5776), 1034–1037.

Honig C. D. F. and Ducker W. A. (2007). No-slip hydrodynamic boundary condition for hydrophilic particles. *Phys. Rev. Lett.*, **98**, 028305.

Horn H. W., Swope W. C., Pitera J. W., Madura J. D., Dick T. J., Hura G. L. and Head-Gordon T. (2004). Development of an improved four-site water model for biomolecular simulations: TIP4P-Ew. *J. Chem. Phys.*, **120**, 9665–9678.

Ishida N., Inoue T., Miyahara M. and Higashitani K. (2000). Nano bubbles on a hydrophobic surface in water observed by tapping-mode atomic force microscopy. *Langmuir*, **16**(16), 6377–6380.

Jabbarzadeh A., Atkinson J. D. and Tanner R. I. (2000). Effect of the wall roughness on slip and rheological properties of hexadecane in molecular dynamics simulation of Couette shear flow between two sinusoidal walls. *Phys. Rev.*, **61**, 690–699.

Jorgensen W. L., Chandrasekhar J., Madura J. D., Impey R. W. and Klein M. L. (1983). Comparison of simple potential functions for simulating liquid water. *J. Chem. Phys.*, **79**, 926–935.

Joseph P. and Tabeling P. (2005). Direct measurement of the apparent slip length. *Phys. Rev.*, **71**, 035303(R).

Karniadakis G. E., Beskok A. and Aluru N. (2005). Microflows and Nanoflows: Fundamentals and Simulation. Springer, New York.

Koplik J. and Banavar J. R. (1988). Molecular dynamics of Poiseuille flow and moving contact lines. *Phys. Rev. Lett.*, **60**, 1282–1285.

Koplik J., Banavar J. R. and Willemsen J. F. (1989). Molecular dynamics of fluid flow at solid-surfaces. *Phys. Fluids*, **1**, 781–794.

Lauga E. and Stone H. (2003). Effective slip in pressure-driven Stokes flow. *J. Fluid Mech.*, **489**, 55–77.

Lauga E., Brenner M. and Stone H. (2005). Ch. 19, Microfluidics: The No-slip Boundary Condition. Springer Handbook of Experimental Fluid Mechanics Cameron Tropea, Alexander L. Yarin, John F. Foss (eds.), Springer, New York.

Li D., Qin-Feng Di, Jing-Yuan Li, Yue-Hong Qian and Hai-Ping Fang (2007). Large slip length over a nanopatterned surface. *Chinese Phys. Lett.*, **24**(4), 1021.

Lide D. R. (2008). CRC Handbook of Chemistry and Physics. 88th edn, American Chemical Society, Washington.

Liu Q.-X., Jiang P. X. and Xiang H. (2006). Molecular dynamics study of the thermal conductivity of nanoscale argon films. *Mol. Simul.*, **32**(8), 645–649.

Maali A., Hurth C., Cohen-Bouhacina T., Couturier G. and Aime J. P. (2006). Improved acoustic excitation of atomic force microscope cantilevers in liquids. *Appl. Phys. Lett.*, **88**, 163504.
Maali A. and Bhushan B. (2008). Nanorheology and boundary slip in confined liquids using atomic force microscopy. *J. Phys.: Condens. Matter*, **20**, 315201.
Maali A., Cohen-Bouhacina T. and Kellay H. (2008). Measurement of the slip length of water flow on graphite surface. *Appl. Phys. Lett.*, **92**, 053101. Doi:10.1063/1.2840717.
Mackerell A. Jr (2004). Empirical force fields for biological macromolecules: overview and issues. *J. Comput. Chem.*, **25**, 1584–1604.
Maynes D., Jeffs K., Woolford B. and Webb B. W. (2007). Laminar flow in a microchannel with hydrophobic surface patterned microribs oriented parallel to the flow direction. *Phys. Fluids*, **19**(9), 093603–0936012.
Murad S., Ravi P. and Powles J. G. (1993). Anisotropic thermal conductivity of a fluid in a system of microscopic slit pores. *Phys. Rev. E*, **48**(5), 4110–4412.
Nagayama G. and Cheng P. (2004). Effects of interface wettability on microscale flow by molecular dynamics simulation. *Int. J. Heat Mass Transf.*, **47**, 501–513.
Neto C., Evans D. R., Bonaccurso E., Butt H.-J. and Craig V. S. J. (2005). Boundary slip in Newtonian liquids: a review of experimental studies. *Rep. Prog. Phys.*, **68**, 2859.
Ou J., Perot B. and Rothstein J. P. (2004). Laminar drag reduction in microchannels using ultrahydrophobic surfaces. *Phys. Fluids*, **16**, 4635–4643.
Ou J. and Rothstein J. P. (2005). Drag reduction and m-PIV measurements of the flow past ultrahydrophobic surfaces. *Phys. Fluids*, **17**, 103606.
Philip J. and Angew Z. (1972). Integral properties of flows satisfying mixed no-slip and no-shear conditions. *Math. Phys.*, **23**, 960.
Pit R., Hervet H. and Leger L. (2000). Direct experimental evidence of slip in hexadecane: solid interfaces. *Phys. Rev. Lett.*, **85**, 980–983.
Priezjev N. (2011). Molecular diffusion and slip boundary conditions at smooth surfaces with periodic and random nanoscale textures. *J. Chem. Phys.*, **135**, 204704.
Rappaport D. C. (1995). The Art of Molecular Dynamics Simulation. Cambridge University Press, Cambridge.
Reyes D. R., Iossifidis D., Auroux P. A. and Manz A. (2002). Micro total analysis systems. 1. Introduction, theory and technology. *Anal. Chem.*, **74**, 2623–2636.
Richardson S. (1973). On the no-slip boundary condition. *J. Fluid Mech.*, **59**(04), 707–719.
Sbragaglia M., Benzi R., Biferale L., Succi S. and Toschi F. (2006). Surface roughness-hydrophobicity coupling in microchannel and nanochannel flows. *Phys. Rev. Lett.*, **97**(20), 204503.
Schmatko T., Hervet H. and Leger L. (2005). Friction and slip at simple fluid-solid interfaces: the roles of the molecular shape and the solid-liquid interaction. *Phys. Rev. Lett.*, **94**(24), 244501.
Schmatko T., Hervet H. and Leger L. (2006). Effect of nanometric-scale roughness on slip at the wall of simple fluids. *Langmuir*, **22**(16), 6843–6850.
Schnell E. (1956). Slippage of water over nonwettable surfaces. *J. Appl. Phys.*, **27** (10), 1149–1152.
Shirtcliffe N. J., McHale G., Atherton S. and Newton M. I. (2010). An introduction to superhydrophobicity. *Adv. Colloid Interface Sci.*, **161**(1–2), 124–138.
Simonsen A. C., Hansen P. L. and Klosgen B. (2004). Nanobubbles give evidence of incomplete wetting at a hydrophobic interface. *J. Colloid Interface Sci.*, **273**(1), 291–299.
Sofos F. D., Karakasidis T. E. and Liakopoulos A. (2009a). Effects of wall roughness on flow in nanochannels. *Phys. Rev. E*, **79**, 026305.
Sofos F., Karakasidis T. E. and Liakopoulos A. (2009b). Transport properties of liquid argon in krypton nanochannels: anisotropy and non-homogeneity introduced by the solid walls. *Int. J. Heat Mass Transf.*, **52**, 735–743.
Sofos F., Karakasidis T. E. and Liakopoulos A. (2010). Effect of wall roughness on shear viscosity and diffusion in nanochannels. *Int. J. Heat Mass Transf.*, **53**, 3839–3846.
Sofos F., Karakasidis T. E. and Liakopoulos A. (2012). Surface wettability effects on flow in rough wall nanochannels. *Microfluidics and Nanofluidics*, **12**(1-4), 25–31.

Sokhan V. P., Nicholson D. and Quirke N. (2002). Fluid flow in nanopores: accurate boundary conditions for carbon nanotubes. *J. Chem. Phys.*, **117**, 8531–8539.

Soong C. Y., Yen T. H. and Tzeng P. Y. (2007). Molecular dynamics simulation of nanochannel flows with effects of wall lattice-fluid interactions. *Phys. Rev. E, Stat. Nonlinear Soft Matter Phys.*, **76**(3), 036303–14.

Sun M. and Ebner C. (1992). Molecular dynamics study of flow at a fluid-wall interface. *Phys. Rev. Lett.*, **69**, 3491–3494.

Taliadourou E., Georgiou G. C. and Alexandrou A. N. (2007). A two dimensional numerical study of the stick-slip extrusion instability. *J. Non-Newtonian Fluid Mech.*, **146**, 30–44.

Tang H. S. and Kaylon D. M. (2008a). Unsteady circular tube flow of compressible polymeric liquids subject to pressure-dependent wall slip. *J. Rheol.*, **52**, 507–525.

Tang H. S. and Kaylon D. M. (2008b). Time-dependent tube flow of compressible suspensions subject to pressure-dependent wall slip: ramifications on development of flow instabilities. *J. Rheol.*, **52**, 1069–1090.

Thompson P. A. and Robbins M. O. (1989). Simulations of contact line motion – slip and the dynamic contactangle. *Phys. Rev. Lett.*, **63**, 766–769.

Thompson P. A. and Robbins M. O. (1990). Shear flow near solids – epitaxial order and flow boundary conditions. *Phys. Rev. A*, **41**, 6830–6837.

Thompson P. A. and Troian S. M. (1997). A general boundary condition for liquid flow at solid surfaces. *Nature*, **389**, 360–362.

Travis K. P., Todd B. D. and Evans D. J. (1997). Departure from Navier-Stokes hydrodynamics in confined liquids. *Phys. Rev. E*, **55**, 4288–4295.

Travis K. P. and Gubbins K. E. (2000). Poiseuille flow of Lennard-Jones fluids in narrow slit pores. *J. Chem. Phys.*, **112**, 1984–1994.

Tretheway D. C. and Meinhart C. D. (2002). Apparent fluid slip at hydrophobic microchannel walls. *Phys. Fluids*, **14**, L9.

Tyrrell J. W. G. and Attard P. (2002). Atomic force microscope images of nanobubbles on a hydrophobic surface and corresponding force – separation data. *Langmuir*, **18**(1), 160–167.

Verlet L. (1967). Computer experiments on classical fluids. I. Thermodynamical properties of Lennard-Jones molecules. *Phys. Rev.*, **159**, 98–103.

Vinogradova O. I. (1995). Drainage of a thin liquid film confined between hydrophobic surfaces. *Langmuir*, **11**(6), 2213–2220.

Vinogradova O. I. and Yakubov G. E. (2003). Dynamic effects on force measurements. 2. Lubrication and the atomic force microscope. *Langmuir*, **19**(4), 1227–1234.

Voronov R. S., Papavassiliou D. V. and Lee L. L. (2006). Boundary slip and wetting properties of interfaces: correlation of the contact angle with the slip length. *J. Chem. Phys.*, **124**, 204701.

Wang Y., Bhushan B., Wang Y. and Maali A. (2009a). Atomic force microscopy measurement of boundary slip on hydrophilic, hydrophobic, and superhydrophobic surfaces. *J.Vac. Sci. Technol. A: Vac., Surf. Films*, **27**(4), 754–760.

Wang Y., Bhushan B. and Zhao X. (2009b). Nanoindents produced by nanobubbles on ultrathin polystyrene films in water. *Nanotechnology*, **20**(4), 045301.

Whitesides G. M. (2006). The origins and the future of microfluidics. *Nature*, **442**, 368–373.

Yang J. W., Duan J. M., Fornasiero D. and Ralston J. (2003). Very small bubble formation at the solid–water interface. *J. Phys. Chem. B*, **107**(25), 6139–6147.

Yang S. (2006). Effects of surface roughness and interface wettability on nanoscale flow in a nanochannel. *Microfluidics Nanofluidics*, **2**(6), 501–511.

Zhu Y. X. and Granick S. (2001). Rate-dependent slip of Newtonian liquid at smooth surfaces. *Phys. Rev. Lett.*, **87**, 096105.

Lightning Source UK Ltd.
Milton Keynes UK
UKOW010145281112

202812UK00004B/119/P

9 781780 401089